TUBE GUITAR AMPLIFIERS VOLUME 2
HOW TO REPAIR, MODIFY & BUILD GUITAR AMPS

IGOR S. POPOVICH

DISCLAIMER & COPYRIGHT NOTICE

The information contained in this book is to be taken in the context of general overview, not specific advice. You should not act on the information contained herein without seeking professional advice. Neither the author nor the publisher (or any other person involved in the publication, distribution or sale of this book) accepts any responsibility for the consequences that may arise from readers acting in accordance with the material given in the book. Professional advice about each particular case / instance should be sought.

Our choice of designs, parts, models and brands was based on their availability and educational value. We were not influenced or induced by anybody in our selection, and their use does not mean we actually endorse or recommend them. You should satisfy yourself that a particular device, component or method is suitable for your intended purpose.

Some circuit diagrams of commercial equipment discussed here were not published by the manufacturers, but were posted online by others, and we cannot confirm their accuracy or authenticity. They are used here for review and discussion purposes as "fair dealing", permitted by international copyright laws.

Designs marked with a copyright symbol are intellectual property of their copyright holders and should not be used without their permission. They are discussed here from educational perspective only.

INDEMNITY NOTICE

The consequences of any modifications to and deviations from the featured designs are at your own risk, and no responsibility will be accepted by us.

Tube amplifiers involve lethal voltages, high temperatures and other safety hazards. By purchasing and reading this book you agree to indemnify its author, publisher and retailer against any claims, of any nature, and for any reason. If unsure about any aspect of testing, repairing or constructing a tube amplifier, engage services a properly qualified and authorized expert.

© **Copyright Igor S. Popovich 2017**

All rights are reserved. No part of this publication may be used, reproduced or transmitted by any means, without the prior written permission from the publisher, except in the case of brief quotations in articles and reviews.

Published by Career Professionals

Second (revised) edition, 2022

Bulk purchases

This book may be purchased in larger quantities for educational, business or promotional use. Please e-mail us at sales@careerprofessionals.com.au

National Library of Australia Cataloguing-in-Publication Data:

Popovich, Igor S.

    TUBE GUITAR AMPLIFIERS, VOLUME 2: HOW TO REPAIR, MODIFY AND BUILD GUITAR AMPS

    ISBN: 978-0-9806223-6-2
    1. Electrical engineering 2. Electronics
    I Igor S. Popovich II Title III Index

621.3

CONTENTS OF VOLUME 2

1. OUTPUT AND INTERSTAGE TRANSFORMERS FOR TUBE GUITAR AMPS _____ Page 9
- OUTPUT TRANSFORMERS
- INTERSTAGE TRANSFORMERS
- DESIGN EXAMPLE: 5kΩ PUSH-PULL OUTPUT TRANSFORMER FOR GUITAR AMPS
- ANALYSIS: ORDINARY AND SIMUL-CLASS® MESA OUTPUT TRANSFORMERS
- TESTING AUDIO TRANSFORMERS
- LINE-MATCHING AUDIO TRANSFORMERS IN PUSH-PULL OUTPUT SERVICE
- DUAL PRIMARY POWER TRANSFORMERS AS OUTPUT TRANSFORMERS

2. LOUDSPEAKERS, OUTPUT ATTENUATORS & HEADPHONE CIRCUITS _____ Page 23
- ELECTRODYNAMIC (FIELD COIL) AND PERMANENT MAGNET SPEAKERS
- SPEAKER TESTS AND MEASUREMENTS
- HOW SPEAKERS IMPACT AMPLIFIERS' POWER LEVELS AND TONAL BALANCE
- ATTENUATORS
- OUTPUT ATTENUATOR AND HEADPHONE CIRCUIT CASE STUDY: BUGERA VINTAGE 5 AMP AND IBANEZ TUBESCREAMER TSA5TVR-S

3. TROUBLESHOOTING AND REPAIRING TUBE GUITAR AMPLIFIERS _____ Page 45
- TYPES OF FAULTS AND COMMON CAUSES OF FAILURES
- FAULT LOCATION AND SIGNAL TRACING METHODS
- POWER SUPPLY TROUBLESHOOTING
- CAPACITOR TROUBLES
- VARIOUS COMMON SYMPTOMS AND MOST LIKELY CAUSES
- ESSENTIAL TOOLS AND TEST GEAR
- AC (SIGNAL) TESTS USING A VOLTMETER (MULTIMETER)
- TESTING & MATCHING TUBES FOR GUITAR AMPS
- CHECKING AND MATCHING TUBES WITHOUT A TUBE TESTER

4. WIRING, SOLDERING & MODIFICATION PRACTICES _____ Page 67
- GENERAL GUIDELINES FOR MODIFICATION SUCCESS
- SAFETY RULES AND PRECAUTIONS
- WIRING OF THE AC CIRCUITS
- SOLDER & SOLDERING
- REDUCING HUM: ISSUES AND SOLUTIONS
- HOW TO POWER AN AMPLIFIER UP FOR THE FIRST TIME

5. POWER SUPPLY MODIFICATIONS AND IMPROVEMENTS _____ Page 83
- CLEANING UP THE POWER SUPPLY
- SIMPLE POWER SUPPLY MODS & IMPROVEMENTS
- MAKING TRANSFORMERLESS AMPS (WIDOMAKERS) SAFE & LEGAL: KAY K503A AND WINSTON
- HOW TO FIX DC VOLTAGES: JOYO "SWEET BABY"
- IMPROVING POWER SUPPLY FILTERING: PANAMA CONQUEROR
- THE OUTPUT TUBE OPERATED WAY ABOVE ITS MAXIMUM DISSIPATION LEVEL: EPIPHONE ELECTAR TUBE 10
- VARIABLE POWER CONTROL ADDED: BLACKHEART LITTLE GIANT

6. TONE TWEAKS _____ Page 111
- CHANGING PREAMP TUBES
- CHANGING POWER TUBES
- CHANGING RECTIFIERS
- ADDING LOUDNESS CONTROL USING A FREQUENCY-COMPENSATED ATTENUATOR

7. MODERN PUSH-PULL AMPS _____ Page 123
- ORANGE TINY TERROR
- JET CITY JCA20H (with Bitmo™ Sufra™ modifications)
- FENDER EXCELSIOR (PAWNSHOP SERIES)
- LORDEN TL-15R
- EPIPHONE ELECTAR CENTURY

CONTENTS OF VOLUME 2, continued

8. DIY PROJECTS: CONVERTING SOLID STATE GUITAR AMPS TO TUBES _____ Page 141
- EVALUATING POTENTIAL CABINET DONORS
- DIY PROJECT: THE FIFTH AVENUE
- DIY PROJECT: THE FOURTH FORCE
- DIY PROJECT: AGENT ORANGE
- DIY PROJECT: THE SCORPION STING
- DIY PROJECT: THE FRONTMAN

9. DIY PROJECTS: ULTRA-SMALL AMPS _____ Page 165
- COMMERCIAL BENCHMARKS: FLEA-POWERED AMPS
- DIY PROJECT: DIRTY THIRTY TUBED
- DIY PROJECT: EL COMANDANTE
- DIY PROJECT: TUBEFINDER

10. REBUILDING COMMERCIAL TUBE AMPS IN A HANDWIRED FASHION _____ Page 181
- EPIPHONE VALVE JNR. - LEGACY VALVE EDITION 5 - HARLEY BENTON GA5
- CRATE PALOMINO V8
- VOX AC4TV

11. DIY PROJECTS: QUIRKY & UNUSUAL DESIGNS _____ Page 199
- THE NEW YORKER: EVERYTHING OLD IS NEW AGAIN
- GOLDEN GECKO: A CHAMELEON AMP THAT CAN USE SIX DIFFERENT OUTPUT TUBES
- DOUBLE TROUBLE: BLENDED SE OUTPUT STAGES
- LOW VOLTAGE TUBE GUITAR AMPLIFIERS

12. CONVERTING VINTAGE TUBE GEAR INTO GUITAR AMPS _____ Page 219
- TUBE SIGNAL TRACERS AS LEARNING & CONVERSION PLATFORMS
- TUBE AUDIO GENERATORS
- TUBE PUBLIC ADDRESS (PA) AMPLIFIERS AS CONVERSION PLATFORMS
- AMATEUR TUBE TRANSCEIVERS
- SONY SSA-464 AMPLIFIER SPEAKER SYSTEM

INTRODUCTION TO VOLUME 2

Welcome to Volume 2 of "Tube Guitar Amplifiers." In Volume 1, we've laid the groundwork for more practical aspects of guitar amps that will be covered in this volume, namely amp troubleshooting & repair, modifications & improvements, and construction (building).

Out of necessity, Volume 1 was a bit more theoretical of the two. Understanding operational principles and design & analysis tools is necessary to fully master the hands-on stuff, so you will know not only what should be done and how, but also why or why not.

The first two chapters of this volume deal with the final stages in the amplification chain, namely output transformers, loudspeakers, attenuators, and headphone/line-out circuitry. Chapter 3 covers the basic troubleshooting and amplifier repair tools and methods.

Chapter 4 discusses practical wiring, soldering, and circuit modification issues seldom or never mentioned in other books. Although numerous circuit modifications and improvements have been dispersed throughout both volumes, chapters 5 and 6 cover two specific areas where mods are particularly effective and noticeable, namely in power supplies and tone changes through simple changes such as tube and component substitutions.

Chapter 7 reviews and analyzes a few modern push-pull amplifiers such as Orange Tiny Terror, Jet City JCA20H, Fender Excelsior, and Epiphone Electar Century. These case studies illustrate common issues and improvement options.

The DIY project section (last five chapters) will take you through a dozen or so unique amplifier designs, some single-ended, other of push-pull, using the different preamp and less common output tubes, such as 12FQ8, 12A6, 5687, 12AU7, 6SN7, 6AR6, 6BX7, 6EM7, ECL86 and PL508. Of course, there are also designs and DIY projects with much-loved favorites such as EL84, 6V6, 6L6, EL34, and 7027A.

Finally, we look at examples of vintage tube gear that already contain a tube amplifier inside and how to quickly and easily convert them into guitar amps.

The old maxim says, "When the student is ready, a teacher will appear." Well, I'm here, and you are ready!

Various types of frames used

MEASURED RESULTS:
- BW: 15Hz - 35 kHz (-3dB, at $10V_{RMS}$ into 8Ω)
- V_{MAX}: $11V_{RMS}$
- P_{MAX} = 15W

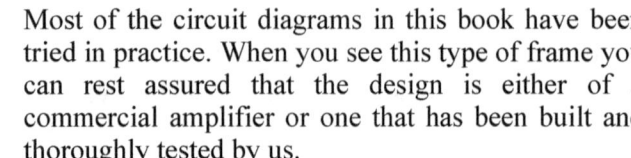

Most of the circuit diagrams in this book have been tried in practice. When you see this type of frame you can rest assured that the design is either of a commercial amplifier or one that has been built and thoroughly tested by us.

RULE-OF-THUMB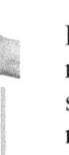
Load impedance for triodes:
$Z_{AOPT} \approx 3r_I - 4r_I$
r_I = internal resistance of a tube

For amplifier builders who don't want to bother with high-level maths, models and similar highbrow concepts, these Rules- of-Thumb are simple shortcuts. Easy to memorize, they approximate and summarize much more complex formulas, methods and concepts.

Although each detailed circuit diagram in this book could be a DIY project by itself, small projects are framed and marked with this soldering-iron symbol.
DIY PROJECT

TUBE PROFILE
Each of the tubes discussed or featured in the designs in this book will have its basic parameters and operational data summarized in a box such as this one.

Commercial designers and manufacturers wish to protect their practical knowledge and "insider secrets". Framed boxes of this kind will emphasize lesser-known practical tips and tricks. Although a magician's hat and a magic wand are used as symbols, there is nothing magical about these tips & tricks, all are underpinned by solid scientific and engineering principles.
TRADE TRICKS

MANUFACTURER'S SPECIFICATION  **SPECS**
A list of amp's features and technical parameters published by the amplifier's manufacturer or retailer

A WARNING OR A VERY IMPORTANT POINT!
Some issues, myths and warnings are so important that they warrant being emphasized in a frame of this kind.

IMPORTANT FORMULA
The calculator symbol indicates an important or often-used formula.

TRANSFORMERS
Both power & output transformers are critical for proper operation and the tone mojo of tube guitar amps. A few easy-to-measure parameters of commercial transformers are given in this type of a frame.

LOUDSPEAKER PROFILE
Just as with tubes, these frames outline the most important parameters of certain speaker drivers used in combo guitar amps and speaker cabinets.

DESIGN PROBLEM OR CALCULATION
Intuitive feeelings and practical experimentation are fine design and troubleshooting methods, but sometimes one cannot proceed without performing some basic calculations first. It will save you time, hassle and money!

INVESTIGATION or a CLOSER LOOK
The best defence against false claims, myths and misconceptions is to investigate issues deeper and closer, just any good detective would do, hence the magnifying glass!

Abbreviations used (in no particular order)

AC	Alternating current	MAX	Maximum	GND	Ground terminal
DC	Direct current	MIN	Minimum	COM	Common terminal
RMS	"Root-Mean-Square", effective value of an AC signal	BNC	Bayonet Neill–Concelman, video & test equipment connector for coaxial cable	ESR, ESL	Equivalent series resistance and inductance (capacitor)
THD	Total harmonic distortion	SQ	Special quality (tube)	AWG	American Wire Gauge
EM	Electromagnetic	PCB	Printed Circuit Board	SPL	Sound Pressure Level
BW	Bandwidth	PP	Push-pull (amplifier)	CF	Cathode follower
MF	Metallized film (capacitor)	OT	Output transformer	SPL	Sound pressure level
CW	Clockwise (rotation of a potentiometer)	CCW	Counterclockwise (rotation of a potentiometer)	DCR	Direct Current Resistance
PP	Peak-to-peak (AC signal)	PPP	Parallel push-pull	TUT	Tube under test
LOG	Logarithmic scale or taper (potentiometer)	NTC, PTC	Negative and positive temperature coefficient (of a resistor or other component)	QF	Quality factor (of a transformer)
LIN	Linear scale or taper (potentiometer)	SET, PSET	Single-ended triode, parallel SET	EMF, CEMF	Electro-magnetic force, counter EMF
TR, VR	Turns or voltage ratio (of a transformer)	E/S	Electrostatic (field, interference or shield)	RCA	Unbalanced audio connector
IR	Impedance ratio (of a transformer)	SLO BLO	Slow blowing fuse, delay fuse	NP	Non-Polarized (electrolytic capacitor)
RC	Resistive-Capacitive coupling between stages or a power supply filtering stage	IC	Internal connection (pin of a tube) or Integrated Circuit	MMF	Magnetomotive Force (of a speaker magnet or transformer)
LC	choke-capacitor power supply filtering stage	GOSS	Grain-oriented silicon steel (transformer lamination material)	FET, JFET, MOSFET	Field effect transistor, Junction FET, Metal Oxide FET
CG	Control grid (of a tube)	SG	Screen grid (of a tube)	SP	Suppressor grid (of a tube)
TPV	Turns-per-volt (of a transformer)	CT	Center tap (of a transformer)	DF	Damping factor
CRC	Capacitor-resistor-capacitor filter	CLC	Capacitor-inductor-capacitor filter	TX	Transfer curve (of a tube)
AF	Audio frequency	RF	Radio frequency	RFI	Radio frequency interference
TC	Tone control	TS	Tone stack	LCR	Inductance-capacitance-resistance meter or circuit
NOS	New Old Stock	NFB	Negative feedback	SB	Standby mode
IEC	International Electrotechnical Commission	DIL	Dual-In-Line (integrated circuit)	UL	Ultralinear (output stage)
IS	Interstage transformer	ZD	Zener diode	DIY	Do it yourself
SS	Solid state (semiconductor)	SW	Switch	TP	Test point
EQ	Equalization	VOL	Volume	HV	High voltage
SMD	Surface-Mounted Device	FX	Effects loop	LV	Low voltage

Currents, voltages and other markings on circuit diagrams

250V	DC voltage in the marked node (quiescent state, no signal)	A=55	Voltage amplification of the adjacent stage
1V ~	AC signal voltage in the adjacent node (RMS or effective value)		
→ 5mA	DC current through the adjacent branch (quiescent state, no signal)		
✗	WRONG - how not to do it!		
✓	RIGHT - how to do it!		

Symbols

PHONE (1/4') JACKS
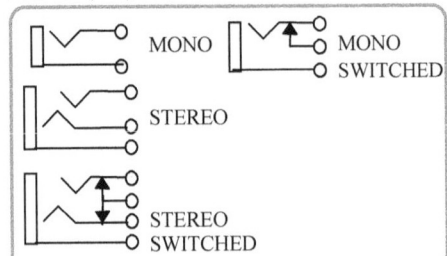

MAGNETIC COMPONENTS
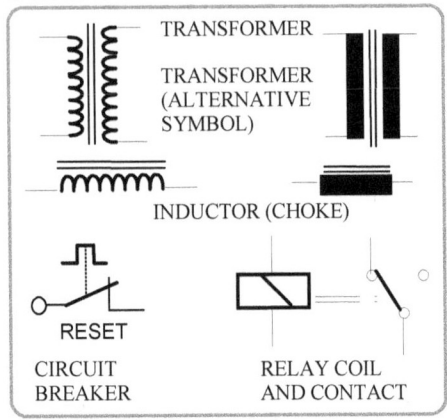

CAPACITORS
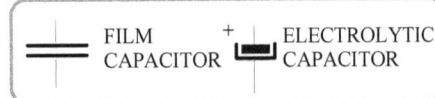

AC AND DC SOURCES
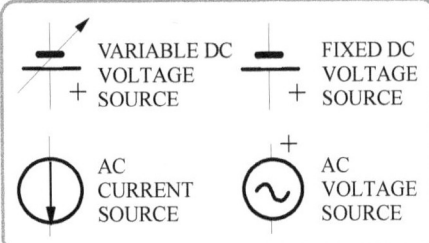

MISCELLANEOUS SYMBOLS
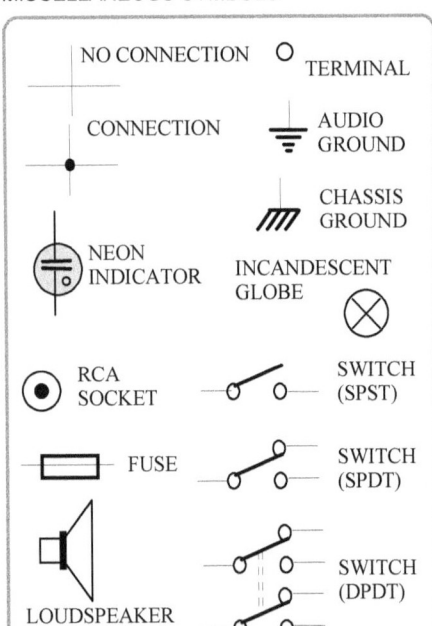

RESISTORS
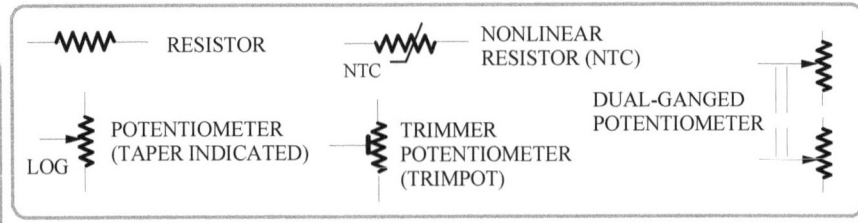

SEMICONDUCTORS
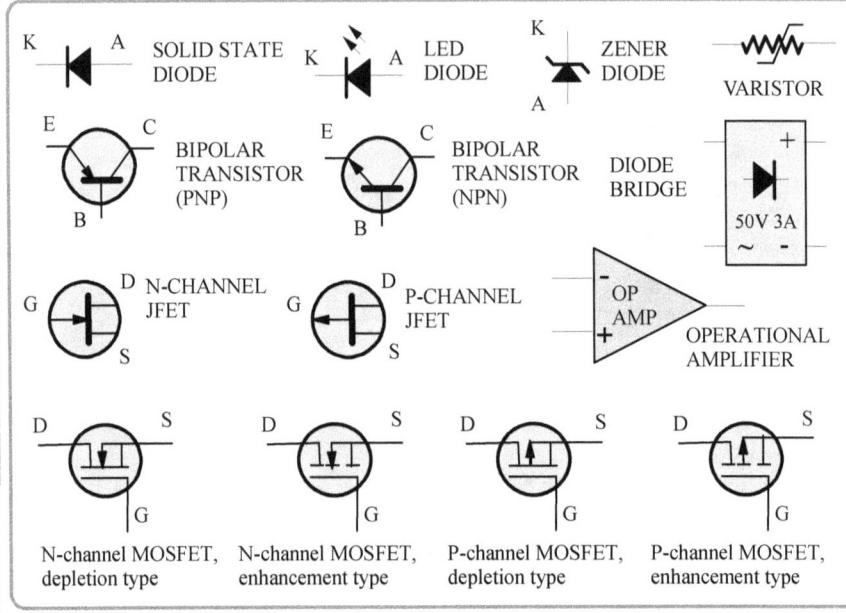

ELECTRON TUBES
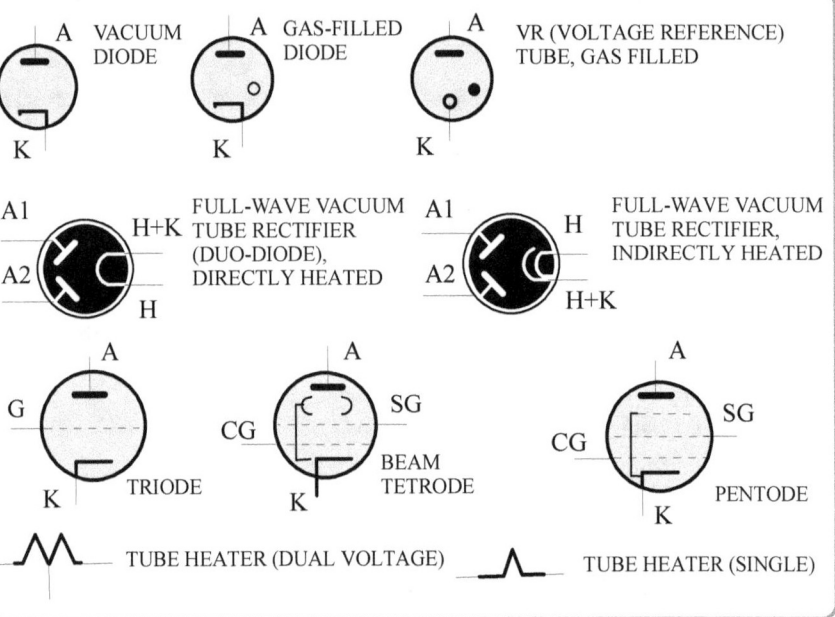

MATHEMATICAL & LOGICAL SYMBOLS

$\approx$ APPROXIMATE $\|$ PARALLEL CONNECTION $\underline{\wedge}$ EQUIVALENT

TEST INSTRUMENTS

OUTPUT AND INTERSTAGE TRANSFORMERS FOR TUBE GUITAR AMPS

- OUTPUT TRANSFORMERS
- INTERSTAGE TRANSFORMERS
- DESIGN EXAMPLE: 5kΩ PUSH-PULL OUTPUT TRANSFORMER FOR GUITAR AMPS
- ANALYSIS: ORDINARY AND SIMUL-CLASS® MESA OUTPUT TRANSFORMERS
- TESTING AUDIO TRANSFORMERS
- LINE-MATCHING AUDIO TRANSFORMERS IN PUSH-PULL OUTPUT SERVICE
- DUAL PRIMARY POWER TRANSFORMERS AS OUTPUT TRANSFORMERS

Since high impedance tubes must be matched to low impedance speakers, all tube guitar amps use output transformers for that purpose. Some have interstage and reverb-driver audio transformers as well. To get the best value for your money as a DIY amp builder or designer or the best tone as a guitar player, you don't need to become an expert in transformer design and winding. Still, you should understand at least the most important issues and aspects of their operation, design, and testing.

OUTPUT TRANSFORMERS

Tube amplifier power stages need output transformers to efficiently couple low impedance loudspeakers, typically in the 4 to 16 ohm range, to high impedance vacuum tubes in the output stage. Power triodes have a much lower internal impedance than pentodes or beam power tubes, but even they need output transformers. For example, the venerable 300B, named the queen of triodes by audiophiles, has an internal impedance of around 800Ω.

Assuming a push-pull output stage, most pentodes require a plate-to-plate impedance (total primary winding impedance) in the 3-10 kΩ region.

Assuming an 8Ω secondary that means their impedance ratios need to be in the 375-1,250 region (3,000/8 =375 and 10,000/8 = 1,250). Since the impedance ratio IR is a square of the voltage VR and turns ratio TR (VR=TR), the turns ratios between their primary and secondary windings are in the 19-35 range ($\sqrt{375}$ =19.4 and $\sqrt{1,250}$ =35.4).

Notice that impedance and turns ratios don't have any unit or "dimension", they are dimensionless ratios.

Salvaged vintage audio parts or overpriced modern replacements?

Watts Tube Audio (and many other small and not-so-small online parts retailers) sell a replacement output transformer for Epiphone Valve Junior (made by Heyboer) for US$75 and a replacement power transformer (also by Heyboer) for US$ 115.95. One can buy a used Epiphone Valve junior amp for US$116 and get all of its parts!

If you modify or built tube amps on a regular basis, this again illustrates the importance of buying cheap salvaged parts whenever an opportunity arises. You never know when you will need them. For example, if you see a pulled vintage output transformer for a single tube such as EL84, 6V6, or ECL86 for US$12.95, assuming it is in perfect working condition, buy it.

Although the opposite should be expected, many vintage transformers used a better magnetic material (laminations) than cheaply made-in-China currently produced ones. Also, modern transformers use plastic film for insulation and plastic bobbin to carry the windings, while the vintage ones use impregnated paper insulation and cardboard or Paxolin bobbins.

Sure, plastic film (called Mylar®, although strictly speaking, that is just DuPont's brand name for a particular type of plastic film) is a superior isolating material. Still, paper insulated transformers sound better, softer, and more "vintage." Newer does not always mean better!

CASE STUDY: Salvaged Telefunken SE output transformers

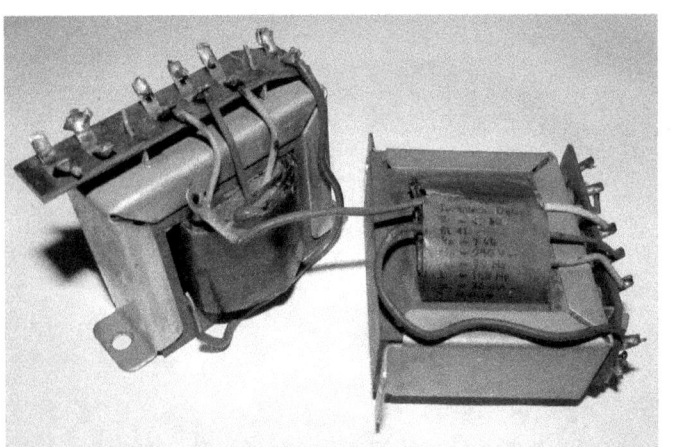

TELEFUNKEN OUTPUT TRANSFORMER
- EI54 laminations
- a=18 mm, stack thickness S=20mm
- Center leg cross section A=aS = 3.6cm^2
- Power rating: P=A^2= 3.6^2= 13W
- Output tube: EL41 (I$_A$=36mA)
- Impedance: 7kΩ/3.6Ω
- Primary DCR: R$_P$=690Ω
- Primary inductance @1kHz L$_P$= 11.2H
- Leakage inductance @1kHz L$_L$=121mH
- Quality factor: QF = 11.2/0.121 = 93

There are still thousands of vintage tube radios, mono amps, and stereo consoles all over Europe, if not in homes, then in sheds and attics. A few entrepreneurial recyclers (mainly from Germany) specialize in salvaging power and output transformers from such gear and selling them on eBay. They also sell tubes, sockets, and other parts. This particular pair, however, was one of three identical pairs for sale by an Australian seller, only AU$30 (US$22) for two, so a very good value for a budget-conscious amp constructor.

The pedantic German tube gear makers of the 1950s and 60s had a habit of printing all sorts of helpful information on their transformers, something USA and UK transformer makers did not do. Here they managed to publish pretty much the whole application note on the tiny transformer!

Apart from the transformer parameters such as 7kΩ primary and 3.6Ω load, 11.9H primary inductance and f$_{QU}$ of 79Hz, it says that the transformer was made for EL41 output tube with 36mA idle DC current and 250V anode voltage. You now have all the info you need to decide if these oldies are suitable for your project.

Their primary DCR is usually relatively high (690Ω in this case), meaning they were wound with many primary windings of a small diameter wire to fit so many turns into their smallish winding windows.

The power rating of the lamination stack is around 13 Watts, and the leakage inductance is quite high, 120mH, resulting in a low quality factor of less than 100. That would be shockingly bad for a hi-fi transformer but less important for a guitar amp application. It simply means the transformer will start attenuating very higher frequencies (above say 12 or 15 kHz), compared to higher fidelity transformers that will pass frequencies up to 25-45 kHz. Those extra high frequencies usually need to be tamed down anyway; otherwise, your guitar amp may sound shrill and unpleasant.

Leakage inductance is a parameter that indicates how much magnetic flux between the primary & secondary winding is not coupled through their shared magnetic core but instead leaked into the surrounding air. Ideally, L_L would be zero, meaning there is no leakage, and the windings share 100% of the flux.

As the audio signal's frequency and thus the magnetic flux it creates in the audio transformer increases, L_L also increases; that is why it should be measured using the highest test frequency.

NSC022848-T MEASUREMENTS

- EI75 laminations, S=26mm
- Primary inductance : L_P= 10.25H @ 120Hz, 10.16H @ 1kHz
- Leakage inductance: L_L = 26.2 mH
- 1/2 Primary DC resistance blue-red (CT): R_P=86.6Ω
- 1/2 Primary DC resistance brown-red (CT): R_P=67.1Ω
- Secondary DC resistance: R_S=0.2Ω

CASE STUDY: NSC022848-T output transformers

NSC022848-T transformers are made in Taiwan and sold by the New Sensor Corporation in the USA for US$37.64 each (in 2016). They've been listed as replacement output transformers for Tremolux, Pro Reverb, Vibrolux®, Bandmaster, and Vibroverb models. Rated at 35 Watts, the primary impedance is 4kΩ PP and the output impedance is 4Ω, so 1,000 impedance ratio or 31.6 voltage or turns ratio.

We got a pair for $20 from an eBay seller who listed them as 50 Watt transformers (!). Let's see what the power rating of their core is. The laminations are EI75, meaning their center leg is 23mm wide. The stack thickness is 26mm wide, so the gross cross-sectional area of the center leg is A = a*S = 2.3*2.6 = 5.98cm². Assuming a stacking factor of 0.96, the net or effective area is A_{EFF} = 0.96*A = 5.74 cm², so the power rating of the core is P = A_{EFF}^2 = 33VA.

Notice a huge difference in DC resistance between the halves of the primary, 87 versus 67 Ω! This means the DC voltage drop on one half (87 Ω) will be higher, and the anode voltage of that tube will be lower. There will also be a dynamic unbalance at higher currents and with AC signals present.

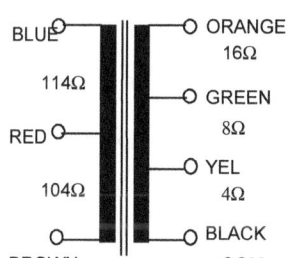

BASLER OUTPUT TRANSFORMER

BLUE — ORANGE 16Ω
114Ω
RED — GREEN 8Ω
104Ω — YEL 4Ω
BROWN — BLACK COM

- EI76 laminations, a=25.3 mm, S=25mm
- A=aS = 6.3cm²
- Power rating: P=A²= 39W
- Primary inductance L_P= 6.8H@120Hz, 8.7H@1kHz
- L_L=6.9mH
- QF = 8.7/0.007 = 1,243

CASE STUDY: Basler output transformer

Basler output transformers are made in Mexico and sold on eBay. This model is a push-pull transformer with 4,500 Ω primary impedance and three secondary taps, 4, 8 and 16 Ω.

Rated at 40 Watts, it would suit octal power tubes such as 6L6, 7027A, and EL34.

The primary inductance is very low, under 7 H at 120Hz, but since its leakage inductance is extremely low, under 7 mH, the resulting quality factor is very high for a guitar amp transformer, above 1,200!

CASE STUDY: VOX AC15 output transformer

VOX AC15 combo amps, both vintage and reissue, are expensive. So, instead of buying a whole amp, we only purchased this genuine reissue output transformer, firstly to measure its parameters for this case study and then to use it in one of our push-pull amps.

The transformer is quite large for its 15 Watt output - its core is rated for 100 Watts! Compare that with tiny output transformers in other modern amps, for instance, the minuscule 13 Watt-rated transformer in Fender Vaporizer.

We haven't tested the output transformer for AC30, but this baby could easily cope in a 30-50W amp judging by its size. However, its primary impedance of around 8.5 kΩ would be too high for paralleled EL84 output stage as in AC30 or lower impedance tubes such as EL34 or 6L6.

The two primary halves are well balanced, with only a 2Ω difference in DCR (DC resistance). The primary inductance is relatively high, around 24 Henry, meaning this transformer's frequency range at full power will extend into very low frequencies for a full and powerful bass.

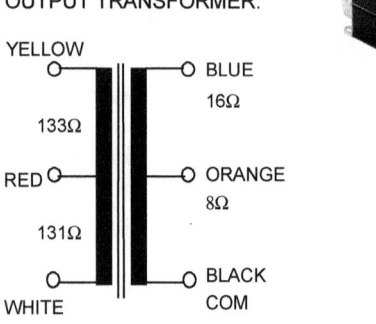

VOX AC15 OUTPUT TRANSFORMER:

- YELLOW — BLUE 16Ω
- 133Ω
- RED — ORANGE 8Ω
- 131Ω
- WHITE — BLACK COM

- EI76, a=25.3 mm, stack thickness S=40mm
- A=a*S = 10.1cm^2, P=A^2 = 10.1^2 = 102W
- Impedance ratio (blue-black): 533
- Impedance ratio (orange-black): 1,048
- Primary impedance with 8Ω speaker: 8.4kΩ
- 1/2 Primary DCR (red-yellow): R_P=133Ω
- 1/2 Primary DCR (red-white): R_P=131Ω
- Primary inductance @1kHz L_P= 24.1H
- Leakage inductance @1kHz L_L=101mH
- Quality factor: QF = 24.1/0.101 = 239
- Primary-to-secondary capacitance: C_{PS}=1.2nF
- Primary-to-case capacitance: C_{PC}=189pF

Its 100 mH leakage inductance is on a high side (anything above 25-30mH), meaning its upper-frequency limit will be curtailed (limited).

At 1.2 nF, the primary-to secondary parasitic capacitance is also relatively high. Both are consequences of the large lamination size (EI76) and the thick lamination stack (4cm). The smaller the transformer, the lower these leakage inductances and parasitic capacitances usually are.

So, you can now see the frustrating compromise needed in the design and construction of audio transformers. To get good bass (superior low-frequency reproduction), the transformer should be as large as possible. However, for an extended high-frequency range (sparkling and detailed treble) the transformer should be as small as possible.

Push-pull winding arrangements

The first test you should perform on any push-pull transformer you have or are considering buying is to measure the DC resistance of the two halves of the primary winding (anode #1 to CT and CT to anode #2). If there is a significant difference (say 180Ω compared to 130Ω), as illustrated below, the transformer is of "el cheapo" winding design and will always have a certain degree of unbalance.

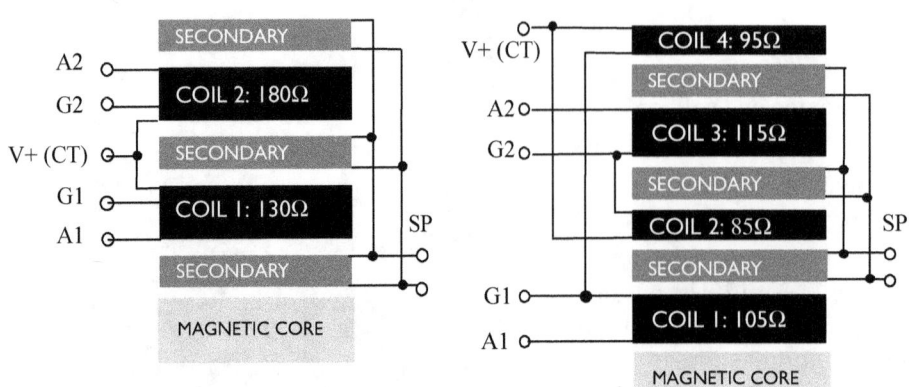

A bifilar winding ensures complete balance in push-pull or interstage transformers, both in DC and signal (AC) terms.

Alternatively, connecting sections 1 and 4 together, as well as sections 2 and 3, ensures a much better DC and AC balance compared to the simple winding that budget transformers use!

105+95=200Ω

115+85=200Ω

INTERSTAGE TRANSFORMERS

Input and interstage transformers were seldom used in vintage guitar amps and are unheard of in modern commercial designs. However, they are highly regarded in hi-fi tube amps, where transformer coupling between the stages is claimed to sound better than the capacitive coupling.

CASE STUDY: Thodardson 20A19 interstage transformer

Vintage designs by Triad and other manufacturers of yesteryears, such as the Thordarson and STC models illustrated below, are unsuitable for hi-fi use but can be successfully incorporated in guitar amp designs. These are regularly available on eBay. Since tube hi-fi amp builders don't want them and guitar amp builders usually copy commercial designs with other types of phase inverters, they are cheap, usually selling for under $20.

The Thordarson transformer has a turns ratio (same as the voltage ratio) of 1:3, meaning each secondary voltage is three times higher than the primary. The transformer is a voltage amplifying stage, providing voltage gain of 3 times or 20*log3 = 9.5dB!

Remember, although the voltage & turns ratio is 1:3, the impedance ratio is a square of the voltage ratio, or 1:9, so a 10kΩ primary impedance reflects onto the secondary side as 90kΩ!

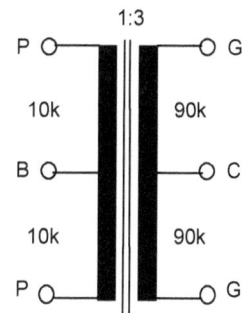

FAR LEFT: The printing on the box and the look of the Thordarson 20A19 interstage transformer
LEFT: The winding diagram with impedances marked

CASE STUDY: A-52-C by Chicago Standard Transformer Company

A-52-C, made in 1965 by Chicago Standard Transformer Company (also known just as STC) is a single plate to push-pull grids interstage transformer. The maximum allowed primary DC current is 10mA and the "overall turns ratio" (primary-to-secondary) is 1:2, which means (since there are two secondaries) that the voltage ratio between the primary and each secondary is 1:1. So, unlike the Thordarson transformer, A-52-C provides no voltage gain.

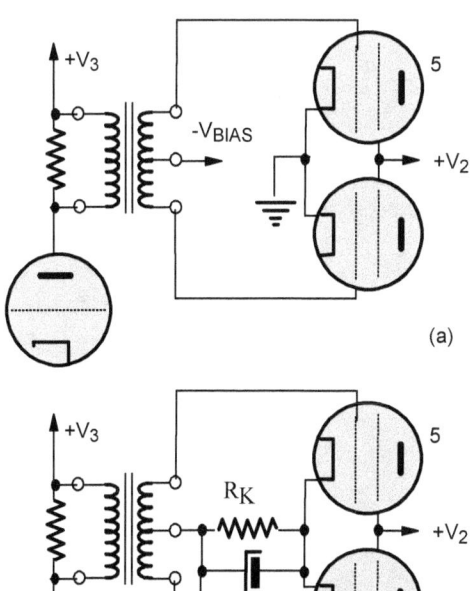
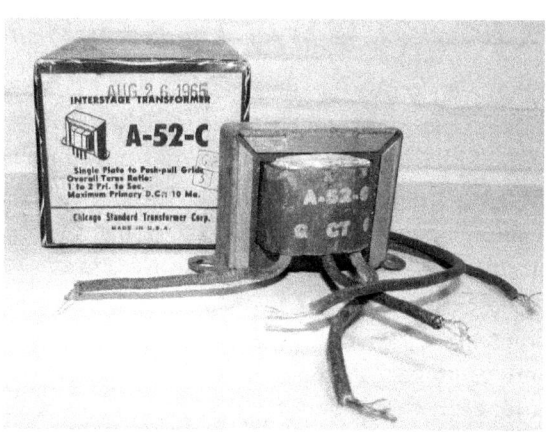

LEFT: Interstage transformers as phase splitters lend themselves equally well to fixed-biased (a) and to cathode biased (b) output stages!

A-52-C INTERSTAGE TRANSFORMER

- EI41, a=13 mm, S=13mm
- A=a*S = 1.7cm^2, P=A^2= 2.9W
- Primary DCR=485Ω
- Secondary DCR=874Ω + 1,026Ω

DESIGN EXAMPLE: 5kΩ PUSH-PULL OUTPUT TRANSFORMER FOR GUITAR AMPS

Designing, winding, and assembling power and audio transformers is a specialist topic and beyond the means or even interests of a typical guitar DIY constructor. However, it is not black magic; it can be understood (providing you have a solid knowledge of electrical and magnetic circuits) and, with practice, mastered.

Let's design a simple push-pull output transformer for a pentode output stage with 6L6 tubes to show you what is involved. We'll select 5kΩ anode-to-anode (or "plate-to-plate" as our American friends would say) primary impedance, to suit an 8-ohm speaker (load), with 100mA maximum primary current in each half.

Choosing the laminations and the winding arrangement

Obtaining transformer laminations, bobbins (plastic or paper), winding wire, and insulation is surprisingly tricky in many parts of the world, even in large cities. It seems that nobody outside China makes transformers these days!

However, there is a solution. Get some good quality small mains transformers, take them apart and rewind them as output transformers. Most use non-GOSS steel (Grain Oriented Silicon Steel), but some do, and those will have a wider frequency range, higher primary inductance, and better sound.

These 30VA (30V/1A) Taiwan-made mains transformers are great for conversion into guitar output transformers. You get the frame, the magnetic laminations, the plastic bobbin, and often you can even reuse its winding wire!

ABOVE: These and similar small (30VA-rated) power transformers with low voltage secondaries are cheap and widely available. Some use better quality magnetic laminations and sound great when rewound as push-pull output transformers.

The bobbin is divided vertically in half, but this divider can be cut and filed off. I say vertically, although it seems horizontal on the photo because we look at the bobbin with the winding window in the horizontal plane, just as in the EI66 profile drawing on the right.

Since the frequency range of a guitar output transformer needs to be only 80 - 8,000 Hz, the cheapest output transformers do not have sectionalized windings at all, as is mandatory in hi-fi transformers. The primary winding is wound first, followed by the secondary. A better option for push-pull transformers would be to split the primary into two halves and wind the secondary between them. This conceptual winding diagram is illustrated below right.

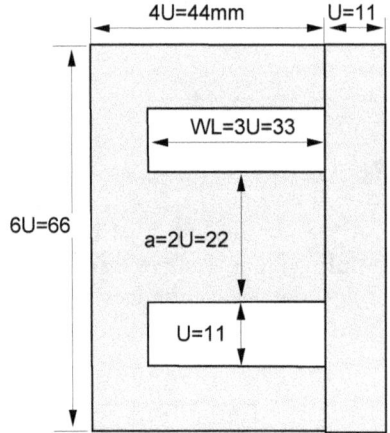

ABOVE: EI66 laminations have 33mm wide windows (winding length WL)

BELOW: Each design project starts with an envisaged or initial (in principle) winding arrangement

Impedance and turns ratios and wire diameters

Since we know the desired primary (5kΩ) and nominal secondary impedances (8Ω), we can immediately calculate the impedance ratio as IR=5,000/8 = 625, so the voltage or turns ratio is TR = $\sqrt{(IR)}$ = 25!

We'll design this transformer "backward", first figuring out the secondary winding and then designing the primary. We want the amplifier to produce a maximum of 40 watts into 8Ω load, so the maximum secondary current will be $I_2 = P_2/Z_2$ = 40/8 = 5A

From the wire tables (available online) we can see that 1mm diameter wire can take 2 A continuously, so two 1mm diameter wires in parallel will be OK for 4A continuously, or 5 A momentarily. The amp will never work continuously at its maximum power level anyway.

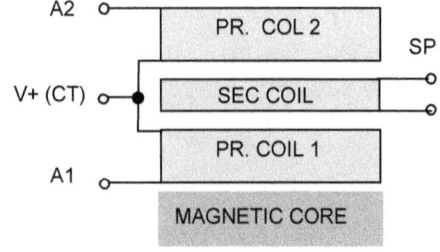

It is easier to wind 1 mm wire than 1.5mm wire which would be needed for 5A. The larger the wire, the stiffer it is and the harder to tension it manually.

Secondary TPL, primary turns, primary current and wire diameter

We have to figure out how many turns we can fit in one layer and how many layers we need. The window length is 33mm, and subtracting 2 mm for the thickness of bobbin walls (plastic or paper frame that holds the winding) on each side leaves us with the maximum coil length of CL = 33-4=29mm. We will work with 28mm!

How many turns of 1mm wire can we fit into one layer? $TPL_{MAX}= CL/d_S = 28/1 = 28$ is the maximum secondary turns-per-layer that we can fit, but that would be impossible in real life, no matter how precisely we wind it. So, let's work with $N_2 = 24$ turns, which will nicely fill the whole width of the bobbin. We will have two layers connected in parallel to double the current capacity. Impedance does not change when layers with the same number of turns are connected in parallel!

Since the TR=25, for 24 turns on the secondary, we need $N_1 = TR*N_2 = 25*24 = 600$ turns of the primary. Now, if the secondary current is I_2=5A, the maximum primary current I_1 will be $I_1=I_2/TR = 5/25 = 0.2A$.

From the wire tables, we see that d=0.3mm wire is suitable for 181mA continuous service, which seems fine for up to 200 mA occasionally. Again, a guitar amp never works at its maximum level for long.

How many turns of 0.3mm wire can we fit into one layer? $TPL_{MAX}= CL/d_S = 28/0.3 = 93$! Since we need 600 turns in total, that is 300 turns for each half of the primary winding, and 300/93 = 3.23, meaning we will choose four layers for each primary half (two sections).

TPL= 600 turns / 8 layers = 75. The ratio 75 turns versus 93 turns (theoretical maximum) is the horizontal fill factor, 75/93 = 81%, which is great for novice winders. Anything over 90% is considered a tight fit and is not for fainthearted!

The winding diagram

Instead of winding both layers of the secondary one after the other (together), which would result in 2/1 sectionalizing, you can wind one section (4 layers) of the primary, then one half (one layer) of the secondary, followed by the second half of the primary (4 layers) and finally the second half (second layer) of the secondary on top, as per the winding diagram. It is always good to have a winding with a large diameter wire on top, to give the coil mechanical strength and rigidity.

Only four layers of insulation are used between the windings; there is no insulation between layers. This technique is called "bulk winding" and is much faster than laying down an insulation layer after every layer!

Also, with less insulation, more turns or wire layers can be used with bulk winding, which will result in higher primary inductance and better & cleaner bass response.

Notice the winding direction, which is always the same, from the left of the bobbin to the right. The two adjacent primary taps are connected together (CT or Center Tap), and that is where the high voltage Va will be brought in.

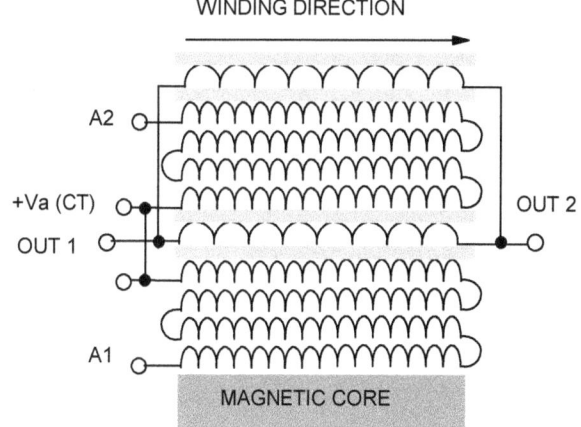

EACH SECONDARY SECTION: 1*24T, d_2=1.0 mm
EACH PRIMARY SECTION: 4*75T, d_1=0.3 mm

ABOVE: The winding diagram is the end result of the design process. It specifies all details required for the winding of the transformer's coil.

Notice how the currents in two primary sections will be in the opposite direction, meaning that their fluxes through the common magnetic core will be equal and cancel each other out. Hence, the resulting DC flux is zero. Of course, that assumes that the two power tubes are 100% identical or "matched," which is never the case at all power levels and operating points; there is always some DC unbalanced current that magnetizes the core and increases B (flux density).

Checking for vertical fit into the window

Our window (the rectangular opening in laminations) is only 11mm tall, and we lose around 2 mm on the bobbin thickness, leaving us with a maximum coil height of 9 mm. We need to check if all those layers can fit it in vertically.

PRIMARY HEIGHT: PH= 8 layers * 0.3mm = 2.4mm

SECONDARY HEIGHT: SH = 2 layers * 1 mm = 2mm

Insulation height IH = 4* 0.2 mm Mylar insulation = 0.8mm

Total coil height CH = PH+SH+IH = 5.2mm

If we allow for a 15% bulging factor (the winding usually bulges out in the middle), the thickest part of the winding will be 1.15*5.3 = 6.1mm tall.

The vertical fill ratio is 6.1/9 = 68%, which is great, with plenty of clearance left.

ANALYSIS: ORDINARY AND SIMUL-CLASS® MESA OUTPUT TRANSFORMERS

To get a feel of how commercial amp manufacturers (or their specialist suppliers, since most guitar amp makers don't have enough knowledge of transformer craft!) design and wind their output transformers, let's look at the output transformer from Mesa 290 amplifier, marked 562004R-1 EIA606-540.

Without having the actual MESA transformer in our hands, we can only go by some information from the Internet fora. The transformer was dissected, and the results were published online by one pedantic explorer. We will base our analysis on that unverified source.

There are no two pairs of taps on the primary winding of that particular output transformer as with some other Simul-Class® transformers. Both triode and pentode plates are connected to the same points, A1 and A2.

It seems that some output transformers in Mesa's Simul-Class® amplifiers have separate primary taps, so triodes and pentodes operate with different anode impedances, while others don't, meaning their pentode pair works with the same load (primary impedance) as the triode pair. 2:3 sectionalizing is used; the two 4 ohm secondary sections are paralleled, while the 8 Ω section is tacked on in series.

The measured primary DCR was 86Ω, while its inductance was only 9H, with 16mH leakage inductance with 4Ω tap shorted to ground. Thus, the quality factor was 9/0.016= 562, which is pretty good, certainly above average for a guitar amp transformer.

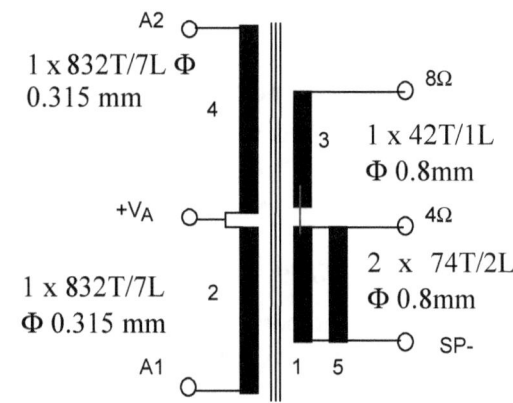

ABOVE: The measured parameters of 562004R-1 EIA606-540 transformer from Mesa 290 amplifier. The numbers signify winding order.

108 pairs of EI96 laminations were used, resulting in 4cm stack thickness. We know from tables that the width of their center leg is 3.2cm. Using our tried-and-tested Rule-Of-Thumb estimation, the center leg cross section is $A=a*S = 3.2 \times 4 = 12.8$ cm^2, which means that transformer's magnetic core power rating is $P=A^2= 12.8^2= 164$ VA.

Step 1: Calculating turns and impedance ratios

$N_1 = 2* 832T = 1,664T$, $N_2 = 74$ T (for 4 Ω), so TR=1,664/74 = 22.49 and IR=TR^2 = 505

Thus, a 4 Ω speaker would be reflected back onto the primary side as Zp (4 Ω) = 505 x 4 = 2,022 Ω plate-to-plate. For 8 Ω output: N_2 = 74T + 42T = 116T, TR=1,664/116 = 14.34, IR=TR^2 = 205.8 and Zp (8 Ω) = 205.8 x 8 = 1,646 Ω plate-to-plate.

We didn't get 2 kΩ, so it seems that in this case, the 8Ω tap is not exactly 8Ω but simply an approximation, caused by the transformer maker's inability to fit the whole section #3 (42 turns) into one layer!

Step 2: Checking the horizontal and vertical fit of the windings

EI96 laminations were used (3.75", 9.53 cm), with stack thickness S = 1.6" (4.06 cm)

PRIMARY: Window length WL=48, Coil Length CL=44mm, so TPL$_{max}$=CL/ϕ_P = 44/0.355 = 124 Turns per layer. There are 7 layers * 124 T = 868 Turns. That is the estimated maximum that could fit. Since there are 832 turns, each layer has 832/7 = 119 turns or 119/124 = 96% horizontal fill.

Anything over 90% would be too high for amateur winders winding transformers by hand, but professional transformer manufacturers could achieve them if they use programmable winding machines.

I have used d=0.355 mm instead of 0.3 mm as claimed, since 0.3mm wire is only good up to about 200 mA, and 0.355 mm wire can take 250 mA continuous!

The 4 Ω secondary has 74 turns in 2 layers, 37 turns per layer so the maximum diameter we can fit is ϕ_S = 44/37 = 1.18 mm, so let's keep the 0.8 mm wire as claimed.

Let's check the vertical fit: P_H= 14*0.355 = 5 mm, S_H = 5*0.8 = 4 mm, W_H = P_H+S_H= 9 mm

There is no insulation between layers; bulk winding will be used). A 16 mm window, less 2mm for the bobbin's thickness, leaves us with 16-2-9 = 5 mm for the insulation between sections and for the bulging factor (the winding is always thicker in the middle, creating a "bulge), which is plenty.

Designing a triode/pentode output transformer based on Mesa Simul-Class® benchmark

Another enthusiast measured a few parameters of the primary of the actual Simul-Class® transformer in his Mesa amp and got the following results, which seem to confirm the previous data:

IR = Z_P/Z_S = 3,300/4 = 825   TR = $\sqrt{(IR)}$ = 28.7

N_1 = 28.7 * 74T = 2,125 Turns

AP denotes anode of the pentode, AT is the anode of a triode.

How would one design and wind a similar homemade transformer based on that information?

First, we need the value of plate currents. For 6L6 pentodes the currents are between 116 mA (zero-signal or idle state) and 210 mA (full power) per tube, while for EL34 (triode connected) in Class A the current is in the order of 71 - 74 mA per tube. So, there is up to 290 mA flowing in the shared part of the primary and up to 210mA in the end sections.

This trafo has 2,125 primary turns, compared to 1,664 in the previous design, so it needs many more primary windings, meaning so many turns of the 0.355mm diameter wire may not fit, so let's start with ϕ_P =0.315 mm primary wire.

The window length is 48mm, but allowing 4 mm for bobbin wall's thickness and clearance we will work with 44mm:

TPL_{max}= CL/ϕ_P = 44/0.315 = 139 T, so we need 2,125 T / 139 T = 15.29 layers. 16 layers would demand very precise winding, if we could fit (vertically) 18 layers, that would make the winding job much easier and would result in TPL = 2,125/18 = 118 or the horizontal fill factor of 118/139 = 85%

So, each half of the primary will have 9 layers or 1,062 turns.

Sections #3 and #5 will have 7 layers each: After 868 turns we need our triode taps, so 832/118 = 7.05 layers, close enough to 7 full layers! So, sections #2 and #6 will have 2 layers each.

Checking vertical fit:

P_H= 18*0.315 = 5.7 mm, S_H= 5*0.8 = 4 mm

Total winding height without insulation: W_H = P_H+S_H= 9.7 mm

With a 16mm high window, less 2mm for the bobbin wall thickness, so we are left with 16-2-9.7 = 4.3 mm for the insulation between sections and allowing for the bulging factor, which is just manageable.

A better winding order is illustrated on the right. We start with one primary section, AP1-AT1, followed by the first secondary section, then the opposite inner primary section (+V_A-AT2), the second secondary section, then the outer primary section AT2-AP2, and so on. That way the DC resistances of the primary halves will be almost identical, since section 1 has the lowest DCR (it is the closest to the core), its MLT (mean length of turn) is the lowest, and section 7 has the highest DCR since it is the furthest from the bobbin and the core.

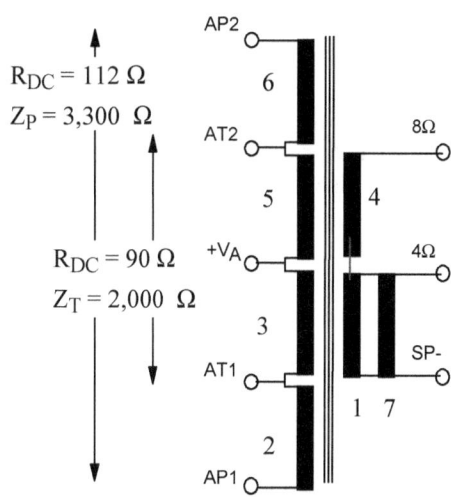

ABOVE: The measured parameters of Mesa Simul-Class® output transformer

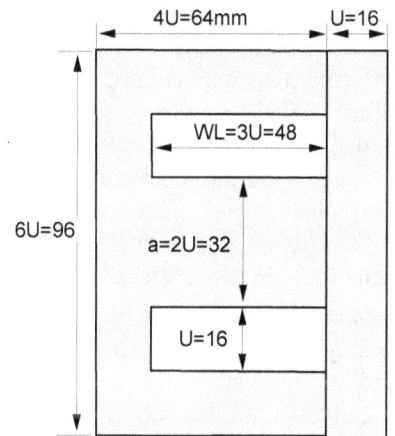

ABOVE: EI96 laminations

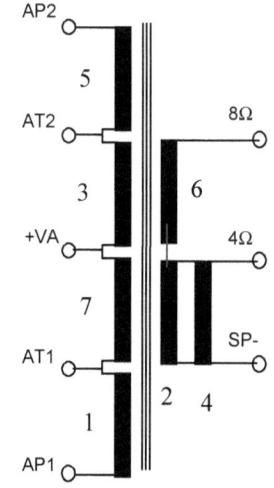

ABOVE: A better winding order for this transformer

FURTHER READING

For deeper and more detailed coverage of power and audio transformer design and winding, please consult my book "Audiophile Vacuum Tube Amplifiers", Volume 2.

Three chapters deal with this subject, "Fundamentals of magnetic circuits and transformers", "Mains transformers and filtering chokes", and "Audio transformers".

TESTING AUDIO TRANSFORMERS

Measuring the primary and leakage inductance

In the first approximation, the easiest and fastest way to measure both the primary inductance and the leakage inductance is with a digital LCR meter. The inductance changes with a DC current flowing through the primary (as in single-ended output transformers). Thus, a more precise way to measure these parameters would be by using an LCR bridge, especially ones with the capability to adjust the DC current through the coil. For estimation purposes, the results without any DC flowing are usually close enough.

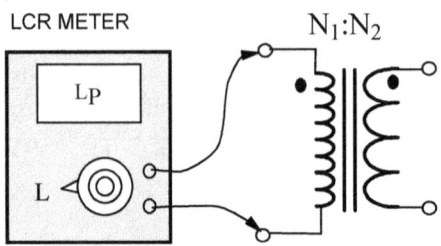

ABOVE: The primary inductance is measured with open secondary winding(s).

The primary inductance should be measured at the lower frequency, such as 120 Hz since its importance is only to the bass frequencies in the audio range. This measurement is performed without any load on the secondary (open secondary).

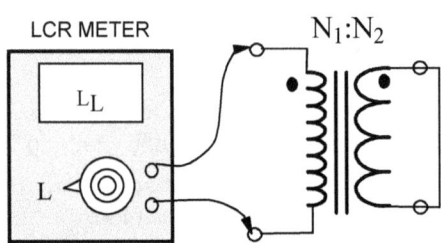

ABOVE: The leakage inductance is measured with shorted secondary winding(s).

The leakage inductance should be measured at as high a frequency as possible, which is only 1kHz in most LCR meters, but some models have a 10 kHz measurement option as well. Secondary terminals must be shorted for this measurement.

Measuring inter-winding capacitances

Transformer capacitances are important for hi-fi output and interstage transformers because these need to have an extended upper-frequency limit, way above 20 kHz. The parasitic capacitances between primary and secondary windings shunt high frequencies to the ground and reduce transformers' upper-frequency limit. Since guitar output amps don't have to go that high (8 kHz is considered a minimum), these capacitances are less critical.

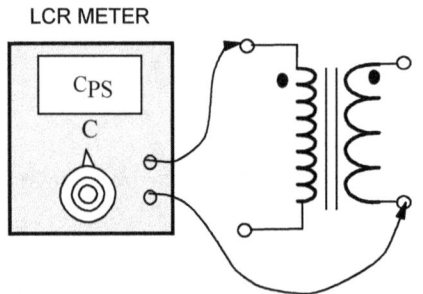

Capacitances should also be measured at the highest possible test frequency of your LCR meter, usually 1kHz or 10kHz. Using crocodile clips and performing these tests without you touching the test leads or the transformer is recommended. The capacitance between your body and the LCR meter and/or the transformer could affect the results.

The capacitance of the primary winding is measured between its ends, as is the capacitance of the secondary. Of more interest is the capacitance between the primary and secondary, and since there are at least four terminals, there are four possible measurements. The results are very close in most cases, so you don't have to perform all four tests.

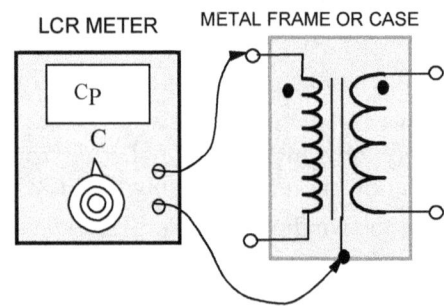

ABOVE: The primary-to-secondary and primary-to-ground (metal case or core) capacitance measurements

The capacitance between the primary/secondary winding and the magnetic core and/or the metal frame or case, considered "ground" since they are usually earthed, is measured similarly. The two ends of a winding may have different capacitance to the ground, so you may need to test both.

How to determine transformer's primary impedance

When set on the "R" or resistance range, digital LCR meters measure AC impedance, not DC resistance. So, if you terminate a transformer with its nominal load R_L (usually 4, 8, or 16Ω) using a dummy (resistive) load, the "R" reading will be its reflected impedance onto the primary side.

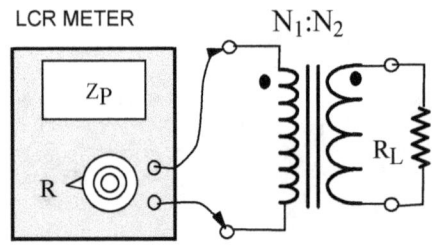

ABOVE: The primary impedance is measured with secondary loaded by its nominal load

To see how much the two impedances differ, perform this test on both 120 and 1kHz frequencies of a typical LCR meter. Yes, they will always be different due to transformers' complex behavior and frequency-dependent nature of their parameters!

OUTPUT AND INTERSTAGE TRANSFORMERS FOR TUBE GUITAR AMPS

LINE-MATCHING AUDIO TRANSFORMERS IN PUSH-PULL OUTPUT SERVICE

What is a 100V or 70V line system and what do line-matching transformers do?

In PA (public address) systems where long cable runs are typical, connecting low impedance speakers (4-16Ω) would mean high circulating currents and increased power losses. Instead, a 70V or 100V constant voltage system is used. This high voltage-low current distribution line feeds multiple speakers, each with its own line-matching or step-down transformer. It is a similar concept as in tube amplifiers, where the output transformer converts high voltage low current primary signals into low voltage high current secondary signals through the loudspeaker.

Since different rooms require different power levels, the step-down transformers usually have multiple primary taps, typically marked in power levels. In the example below, speaker SP1 is connected to the 25W tap, SP2 to the 100W tap and SP3 to the 50W tap.

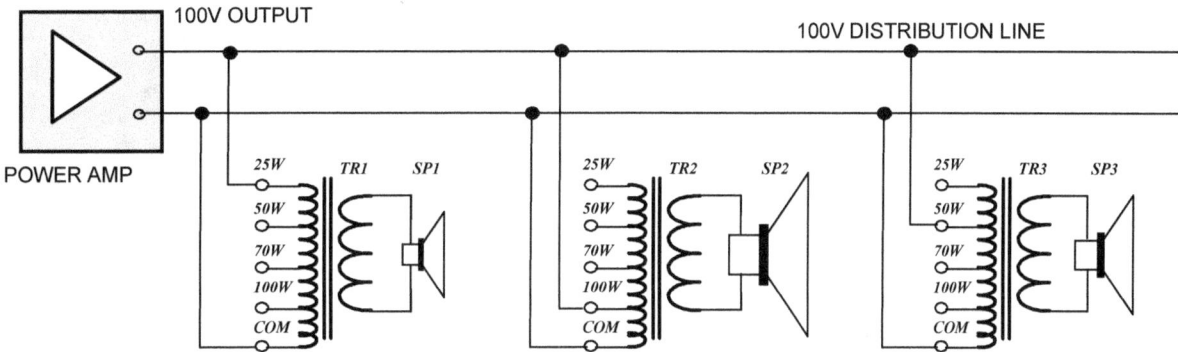

A 100V distribution system with three speakers of various power ratings, all using identical step-down transformers with multiple primary taps

Making sense of line transformers' specifications

These Atlas Soundolier line step-down transformers have primary power levels marked, as all such transformers do. Unusually though, these also have corresponding resistances (impedances) marked.

How can we tell if it's a 70V or a 100V line transformer? Power is $P=V^2/R$ so $V^2=P*R$ Using any tap as an example, for instance the 10W tap: $V^2=10 \text{ Watts}*500Ω =5,000$ so $V= \sqrt{5,000} = 70.7V$, meaning it is a 70V transformer. Secondly, notice that, for some reason, the power level for the 45Ω tap is not specified. However, now we know that $P=V^2/R = 70.7^2/45 = 111W$!

Line matching transformers as push-pull output transformers

Output transformers for tube guitar amps can get expensive, so someone had an idea of using cheaper alternatives. Since line transformers usually use GOSS (grain-oriented silicon steel) laminations of higher quality than those made of ordinary (non-oriented) silicon steel, they come to mind first.

Australian retailer and distributor Altronics sells a range of 100V transformers, most likely made in Taiwan, using GOSS laminations. Let's take as an example a higher power transformer, model M-1130 and its specs: 5-40W, 100V Line, EI core transformer, frequency response: 30Hz - 20kHz ±3dB, secondary taps: 2, 4, 6 & 8 Ω, primary (power) taps: 5W, 10W, 20W, 40W.

Since $P=V^2/Z$, the impedance is $Z=V^2/P$, for 5W tap Z=2kΩ, for 10W tap Z=1kΩ, for 20W tap Z=500Ω, for 40W tap Z=250Ω.

For push-pull use, we need symmetrical transformers. The impedance between the center tap and each end must be 1/4 of the total anode-to-anode (end-to-end) impedance. In this case, the A-A impedance is between 0 (COM) and the end tap, which is the 5W tap. This impedance is 2kΩ, so now we look for a tap with a quarter of the impedance or 500Ω. Yes, we have it; it is the 20W tap, which will be our CT.

ABOVE: M-1130 terminals when used as a push-pull output transformer

There are also toroidal step-down line transformers, but they are of higher power ratings. As you may have noticed from the discussion above, the higher the power rating, the lower the impedance, so most are unsuitable for output use! If a line transformer has a large core, it is of a high power rating and low impedance and is thus of no use. If it is of the right impedance, it is inevitably on a smaller core and, hence, a lower power rating. Therefore, this option is limited to low-power push-pull amplifiers, less than 20 Watts, for tubes such as 6V6, EL84, and 6L6.

CASE STUDY: Two low power line transformers tested

M1115 is 100V line transformer with an 8Ω secondary and primary taps marked 0-1.25-2.5-5-10-15W. M1120 is the next model up, apparently rated 5 Watts higher, at 20 Watts, and although it uses a different size magnetic core and stack thickness, its frequency response is also specified as 30Hz - 20kHz (±3dB), which seems unlikely and makes you question the acuracy and truthfulness of these specs.

M1120 has 4, 8, and 16Ω taps. Both transformers reflect an 8Ω load as 9kΩ anode-to-anode load onto the primary side, more specifically, 8.7k load for M1115 and 8.9k load for M1120. In both cases, the 5 Watt lug is the CT (Center Tap) while "0" and "1.25W" lugs are the anode connections.

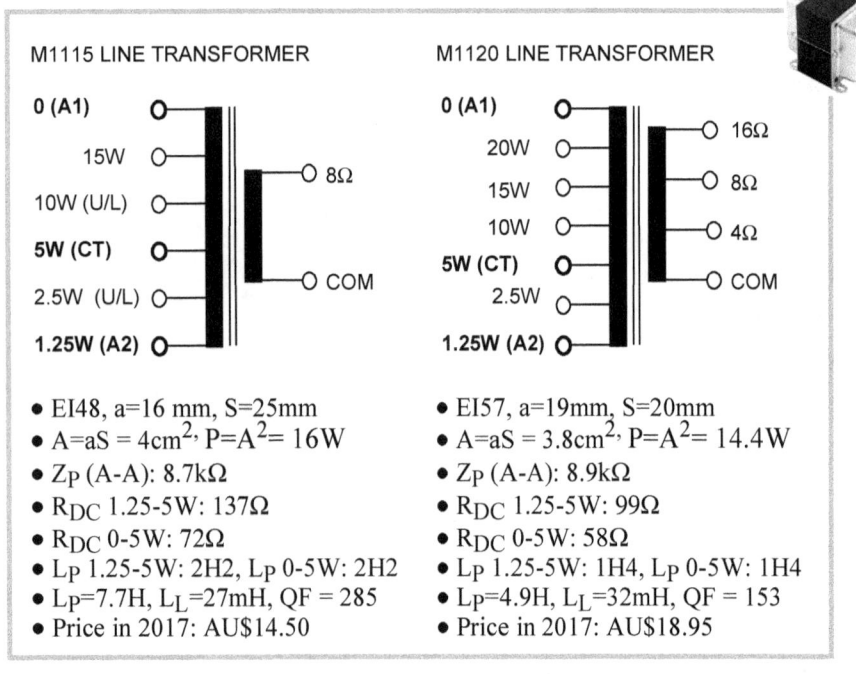

Notice that while the AC impedances of the two halves are equal, their DC resistances are very different. For instance, in the M1115 case, one side has a DCR of 72 Ω, the other almost double, 137 Ω!

Assuming the same anode current flowing through both, say 40mA for 6BQ5 (in idle state), the voltage drops on the two primary halves will be 2.9V and 5.5V, which is not an issue, especially not in guitar amps.

The power rating of the smaller transformer is 16 watts, in accordance with its 15 Watt specified power output, but the 20W-rated bigger transformer has a core that can only support 14.4 Watts.

To make things worse, the bigger M1120 has a much lower primary inductance, 4.6 Henry, compared to 7.7H for M1115. Since the leakage inductances are in the same ballpark (27 and 32 mH), M1120's quality factor is much lower, 153, while that of M115 is 285.

So, unless you need various secondary impedances, M1115 is a better value. As luck would have it, the remaining primary taps are also suitable for ultra-linear connections. For instance, on M1115, the 2.5W tap would be the U/L tap for one, and the 10W tap would be the U/L tap for the other tube.

A closer visual inspection indicates a different lamination color and texture for the two transformers. Lugs and bobbins are also different, so it seems that two different suppliers made them. M1120 has "VRK" stamped on the metal frame, M1115 doesn't. VRK Spectrum Co., Ltd. is a Korean company making transformers in Thailand and capacitors in their Korean and Thai plants.

CASE STUDY: ATLAS SOUND T-18 line transformer

The enclosed data sheet dates these vintage transformers back to 1972. EI66 laminations with a 24 mm stack indicate a power rating of 28 VA. The primary inductance is low, 2.1H @ 120Hz and only 1.45 H at 1 kHz, which is sufficient for a guitar amp output transformer application.

Since 500Ω is a quarter of 2,000Ω, the 500Ω would be the center tap (CT) in the push-pull stage, and the two anodes would be connected to the 0 (COM) tap and the 2,000Ω tap. Thus, these transformers can be used as PP output transformers.

With an 8Ω load connected to the 8Ω tap, the primary impedance is around 2.3kΩ. That can be doubled to 4.6 kΩ by connecting the 8Ω load to the 4Ω tap or halved to 1.15 kΩ by connecting the 8Ω load to the 16Ω tap.

Such a low plate-to-plate impedance usually is not used but is needed when tubes are working with low anode voltages (24-60 V_{DC}), as in our low voltage push-pull guitar amp project, for which these transformers are perfect.

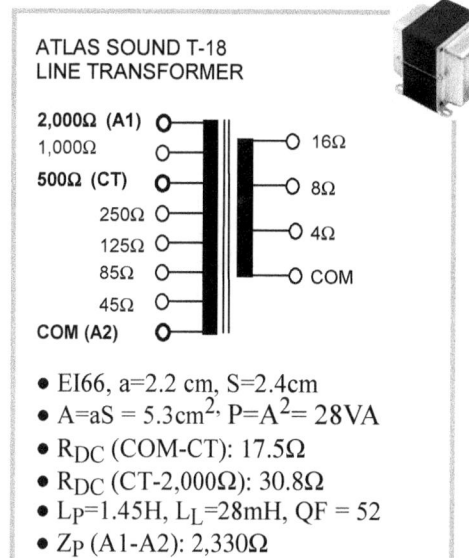

ATLAS SOUND T-18 LINE TRANSFORMER

- EI66, a=2.2 cm, S=2.4cm
- A=aS = 5.3cm², P=A² = 28VA
- R_{DC} (COM-CT): 17.5Ω
- R_{DC} (CT-2,000Ω): 30.8Ω
- L_P=1.45H, L_L=28mH, QF = 52
- Z_P (A1-A2): 2,330Ω

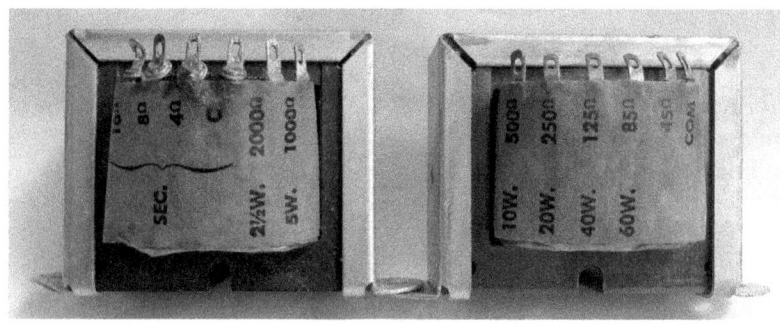

CASE STUDY: ELA-T50 line transformer

This 50 Watt 100V line transfromer sold on eBay seems to be made by Omnitronic in Germany. The primary plate-to-plate impedance is only 1k6, but when an 8-ohm load is connected to a 4-ohm tap, it increases to 4k2. The primary inductance is low, under 2H, so the bass performance will suffer.

The leakage inductance is also low, so the Q-factor is decent. The price is low, so it's worth trying it out with higher-powered output tubes such as 7591, 6L6, or paralleled EL84s for 30 Watts out.

DUAL PRIMARY POWER TRANSFORMERS AS OUTPUT TRANSFORMERS

RN20 is an R-core power transformer made in China. Rated at 30VA, it has two identical primary windings for 115V mains voltage or 230V. Each of the two identical secondary windings is rated at 9V @ 1.66A.

Our idea was to use the two identical primaries as push-pull output primaries and to parallel the two identical secondaries. That would give us a voltage ratio of VR=230V/9V=25.6, so the impedance ratio would be IR=VR² = 25.6² = 655. An 8Ω speaker impedance would reflect onto the primary side as 655*8 = 5.24kΩ, perfect for two 6L6 or EL34 tubes or four 6V6 or EL84 in a parallel push-pull output stage.

The 16Ω speaker would be a 655*16 = 10.5kΩ load on the power tubes, perfect for 6V6 or EL84 push-pull stages. Two secondaries in parallel would be able to supply 2x1.66A*9V = 29.9 Watts of power, as per the 30VA specs.

If you study the construction of R-core transformers, they always have two identical windings on two longer sides, so these can be connected in series or parallel.

ELA-T50 LINE TRANSFORMER

- EI66, a=2.2 cm, S=3.0cm
- A=a*S = 6.6cm², P=A² = 43.5VA
- L_P=1.7H, L_L=6.7mH, QF = 253
- Z_P (A-A) = 1.6kΩ

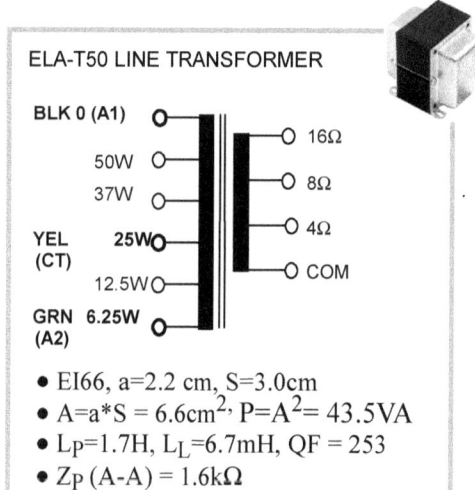

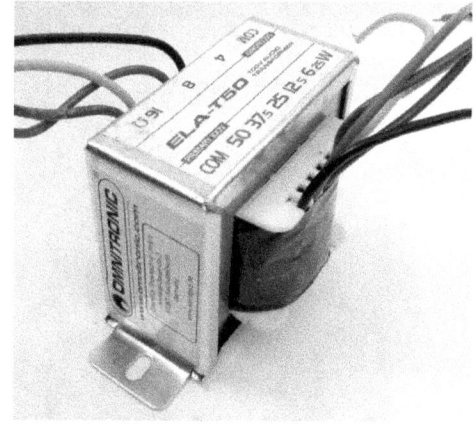

The transformer weighs only 500g and measures 79 × 68 × 40 mm. Its low profile (40mm height) means it can fit even under the slimmest of chassis.

Dual primary EI transformers can also be used for this purpose, but R-core transformers are superior due to their lower magnetic leakage and the fact that windings are perfectly balanced. Notice that the two primaries have identical DC resistances, which would only be possible if bifilar winding was used on an EI power transformer, which is unheard of.

The R-50A transformer (pictured on the right) is rated at 65 VA and features two 12V/2.7A secondaries, so VR=230/12= 19.2 and IR=368. An 8Ω speaker reflects as 3kΩ load onto the primary and a 16Ω speaker as a 6kΩ anode-to-anode load.

The measured parameters of these transformers, particularly their very high primary inductance and above-average quality factors, surpass most brand name output transformers for guitar amps (always of inferior EI variety), selling for $120-200, so an incredible value here!

DUAL PRIMARY 2x9V SECONDARY MAINS TRANSFORMER AS PP OUTPUT TRANSFORMER:

- Primary resistance WHT-BLK: 49.1Ω
- Primary resistance RED-BLU: 49.1Ω
- Primary inductance RED-BLK: L_P= 29H
- Leakage inductance L_L=36mH
- Quality factor: QF = L_P/L_L = 806

LOUDSPEAKERS, OUTPUT ATTENUATORS & HEADPHONE CIRCUITS

- ELECTRODYNAMIC (FIELD COIL) AND PERMANENT MAGNET SPEAKERS
- SPEAKER TESTS AND MEASUREMENTS
- HOW SPEAKERS IMPACT AMPLIFIERS' POWER LEVELS AND TONAL BALANCE
- ATTENUATORS
- OUTPUT ATTENUATOR AND HEADPHONE CIRCUIT CASE STUDY: BUGERA VINTAGE 5 AMP AND IBANEZ TUBESCREAMER TSA5TVR-S

Loudspeakers and output or power attenuators are the two final links in the audio chain. Of course, not all amps incorporate output attenuators; in fact, most don't. Also, many amps don't even have an inbuilt speaker, they are "heads" only (in contrast to "combo" amps), but they still must be connected to a speaker cabinet for sound to be produced.

Apart from the operating principles of permanent magnet and electrodynamic loudspeakers and attenuators, this section looks at some crucial issues such as amplifier-speaker interface and interaction.

ELECTRODYNAMIC (FIELD COIL) AND PERMANENT MAGNET SPEAKERS

How they work

A permanent magnet loudspeaker is an electromechanical transducer; it transforms the electrical energy of the signal current flowing through its voice coil ("spool") into a mechanical sound wave. A permanent magnet produces a steady uniform magnetic field where the voice coil is placed, surrounded by a tiny air gap.

With no audio signal, the coil is at rest. An AC signal current flowing in the coil (from the output of a guitar amp) produces its pulsating magnetic field, which interacts with the permanent magnet field and results in a mechanical force that is transferred onto the speaker's cone. The cone's movement follows the audio signal and creates sound waves of the same frequency as the audio signal.

The first permanent magnet speakers used Alnico (Aluminium-Nickel-Cobalt mixture), an alloy whose chemical composition is 8–12% Al, 15–26% Ni, 5–24% Co, up to 6% Cu (copper), up to 1% Titanium, and the rest is iron. In the 1970s, "ceramic" or "ferrite" magnets were developed, which replaced Alnico almost completely.

Electrodynamic (field coil) speakers

Alnico and ceramic magnet permanent speakers are relatively recent inventions. In the first half of the 20th century, speakers did not have permanent magnets. Instead, a magnetic yoke with a center insert was used, and the flow of DC current provided the magnetization through the "field coil." The field coil would have a few thousand turns N_F of a suitably sized lacquer-insulated copper wire, wrapped around a spool or cylindrical bobbin, which would then snugly fit over the centerpiece (pole).

The magnetomotive force (MMF) of the field coil or any electromagnet is $MMF = 1.257 * N_F * I_F$ [AmpereTurns]. It is also called "total field". The constant 1.27 is actually $4\pi/10$, N_F is the number of turns in field coil and I_F is the DC field or "excitation" current.

Magnetomotive force should not be confused with magnetizing force H, but the two are related by $H = MMF/LMP$ where LMP is the "Length of the Magnetic Path", or how long the average magnetic line of force is.

Finally, the magnetic flux density B is the number of the lines of magnetic fields through one unit of cross-sectional area: $B = \Phi/A = \mu * H = \mu N_F * I_F / LMP$.

Greek letter μ is the symbol for the magnetic permeability of the ferromagnetic material used for the centerpiece. When signal current flows through the voice coil VC placed in the uniform magnetic field, the force acting on the voice coil is $F = B * I_S * \ell_S * \sin\theta$.

Since the magnetic lines of force are perpendicular to voice coil windings, the angle between them is $\theta = 90°$, and $\sin\theta = 1$, so $F = B * I_S * \ell_S$, where B is the magnetic induction or the flux density, I_S is the signal current and ℓ_S is the total length of the coil wire (length of one turn multiplied by the number of turns N_V).

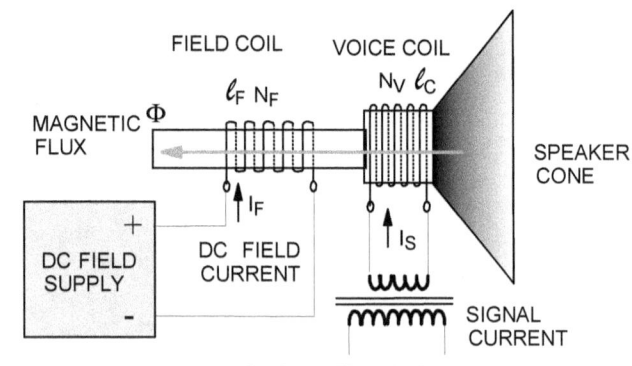

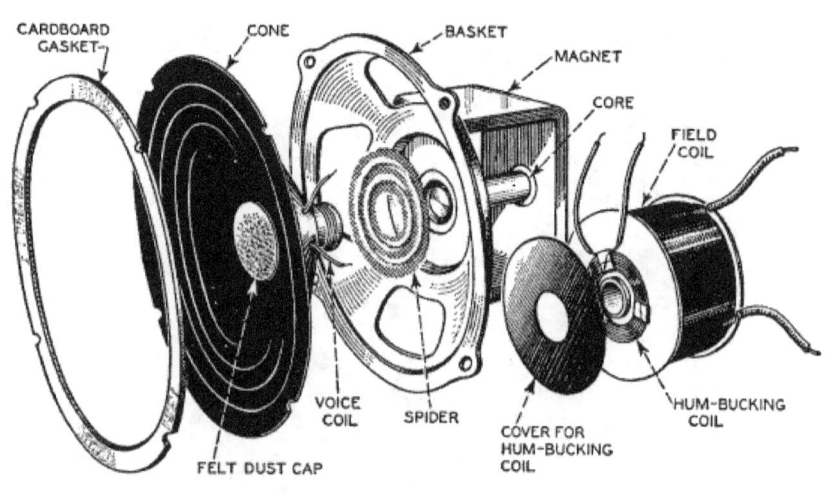

ABOVE RIGHT: The operating principle of electrodynamic speakers. For clarity, only the center slug is shown without the surrounding soft iron magnetic yoke needed to close the magnetic flux path. See a detailed cross-section on the next page.

RIGHT: A breakup view of a vintage electrodynamic speaker with major parts named & marked, source and copyright: "Radio and Television Receiver Circuitry and Operation", Alfred Ghirardi, 1955

LOUDSPEAKERS, OUTPUT ATTENUATORS & HEADPHONE CIRCUITS

Since ℓ_S is constant for each voice coil, the force is directly proportional to B! If B is reduced, the force on the voice coil is also reduced, which results in smaller movement (excursions) of the cone. Smaller movement means reduced air pressure created by the cone and lower loudness.

With permanent magnet speakers, B depends on the type and size of the magnet ring and is thus fixed, and so is their efficiency and "loudness".

Once the manufacturer chooses the magnetic material, its magnetic permeability is fixed, as its size and shape, and thus LMP is also fixed, so $B=k*N_F*I_F$ (where constant $k=\mu/LMP$). This means that with electrodynamic speakers, we can vary B by varying either the number of turns N_F or the excitation current I_F and thus vary their "loudness".

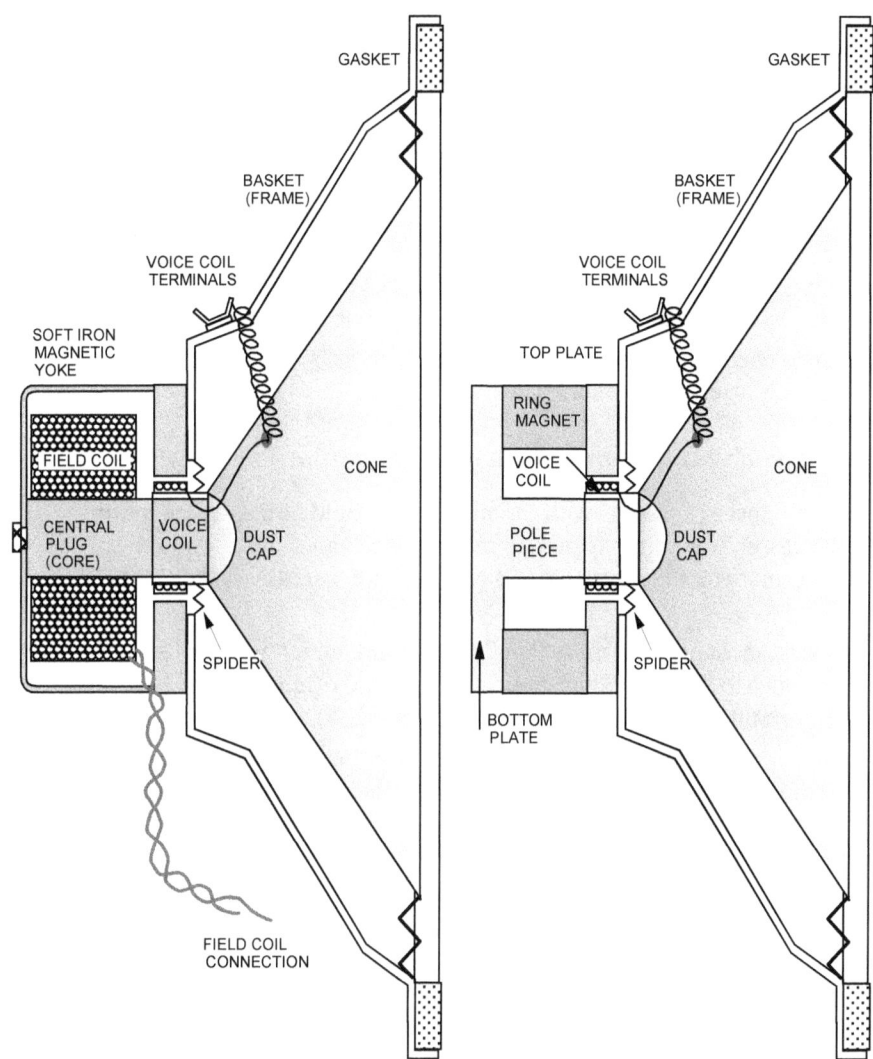

ABOVE: A cross-sectional view of a typical permanent magnet (RIGHT) and electrodynamic loudspeaker (LEFT).

Abb. 553. Lautstärkeregler für dynamische Lautsprecher.

ABOVE: An illustration from a vintage German book showing a commercial pre WWII rheostat used to vary the field current and thus loudness of an electrodynamic speaker.

Once we have the actual speaker, we cannot vary the number of turns, that is a given, so the only remaining practical way of changing the loudness of a field coil speaker is by altering its field current. The field coil of a pre-WWII Telefunken speaker is marked "Tel. Bv. 665 a" and "Feldspule 100V/67 mA". *Feld* is German for field and *spule* means coil, whose resistance can be easily calculated as $R=V/I=100/0.067= 1,493\Omega$.

A slightly newer (postwar) Radio-Graetz speaker does not have its voltage or current specified, only $1,600\Omega$ resistance. However, we do gain a further insight into its coil, whose number of windings and wire diameter are given, 14,700 turns of 0.16mm lacquered (L) copper (Cu) wire.

As for the maximum current allowed, various copper wire tables often give you slightly different results for 0.16mm wire, but they generally fall around 330mA mark.

Finally, as another example, the field coil of another German (Sachsenwerk) speaker is marked 75mA/6W. Since the power dissipated in the coil is $P=I^2*R$ we can calculate its resistance as $R=P/I^2 = 6/(0.075^2) = 6/0.005625 = 1,067\Omega$.

Sadly, most teachers aren't very good at exciting young minds and illustrating the boring theory with real life examples. So, if you ever wondered as a child why in the world you needed to learn Ohm's and Kirchoff's Laws in physics class, now you know: so that twenty or thirty years later, you can confidently calculate other, non-specified parameters of loudspeakers and amps!

Commercial benchmark: FluxTone electrodynamic speakers

USA company FluxTone manufactures a range of electrodynamic speakers and electronic controllers named *Variable Magnet Technology*. A 25 dB range of adjustment is claimed. The controller can be positioned on top of an amp or even inside it, a much user-friendlier solution.FluxTone even patented (in the USA) this general principle of changing SPL of a speaker by adjusting the field coil excitation using a variable power supply with simple one-knob control. How they could patent something that was widely known and practiced before WWII is mind-boggling. Whatever happened to the concept of "prior art"?

However, as always, there is the minor issue of price. For instance, their Model 2 ("Original Jensen F-12-N") in 2016 sold for a cool US$1,125! Yes, that's just for one bare speaker driver and the VMT controller box.

Is there a cheaper, DIY option? There certainly is, and that is to buy a vintage electrodynamic speaker and build your own field current controller.

Buying vintage electromagnetic speakers online

There are three problems with buying vintage field coil speakers online. For some reason, they are expensive, much more than similar vintage speakers with permanent magnets. On eBay, they are usually listed at $100-200 each.

Secondly, most are now more than 70 years old, which is a very long time for paper cones to survive. Thus, their old cones are often weak and brittle, even if they are intact with no rips or holes.

Finally, transporting bare speaker drivers by themselves is inherently risky since there is no speaker cabinet to protect them. If the seller does not pack them well, the damage is almost assured, which is what happened to the 12" speaker illustrated. Most sellers on eBay either have no idea of how to pack fragile items or simply don't care. Anyway, we got reimbursed by eBay and kept the output transformer.

We then bought a smaller 8" speaker made in 1937 by Motorola, which was well packed by the seller and arrived in perfect condition. Compared with the pedantic German speaker makers, no specs were printed on it at all, but we measured the speaker's field coil resistance of 700 ohms. Let's test it and get a better feel for the behavior of such speakers.

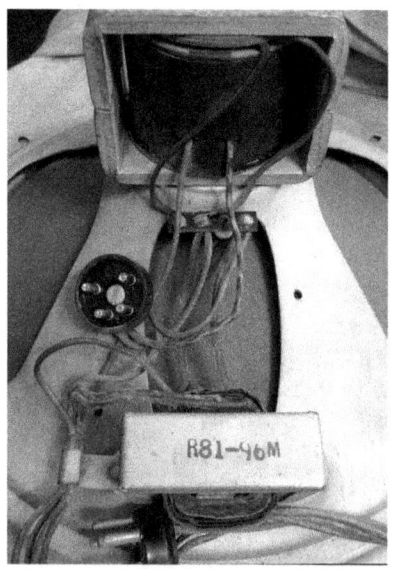

ABOVE: The 12" electrodynamic speaker (with integral output transformer) whose paper cone got destroyed in transport.

EXPERIMENT: Varying the SPL of a an electromagnetic speaker

You can either use a variable voltage DC power supply of suitable voltage and current rating or a fixed DC power supply and a rheostat for this test. If you don't know the required voltage and current for the field coil of your speaker, it is safer to use a variable DC power supply. Start with zero volts and gradually ramp the voltage up while monitoring the current. Many benchtop power supplies have meters for both voltage and current.

The output signal voltage of the amplifier must remain constant. Since you are not changing the frequency of the signal, that should be the case.

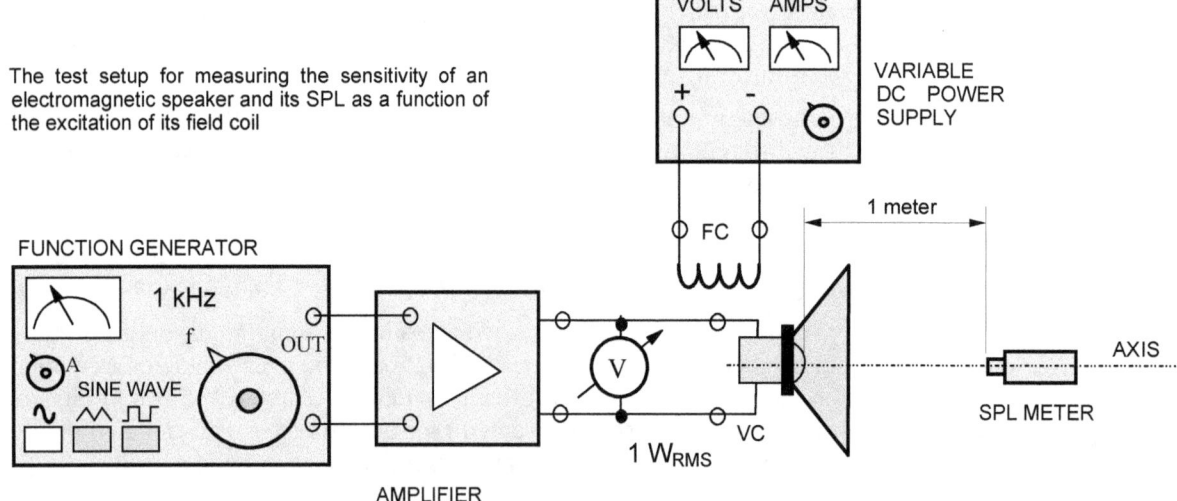

The test setup for measuring the sensitivity of an electromagnetic speaker and its SPL as a function of the excitation of its field coil

The required signal voltage at the amp's output will depend on the impedance of the speaker. The first step is to use a digital LCR meter set on the "R" range (which is not DC resistance but impedance) and on 1kHz test frequency, which is standard on all such LCR meters. Then, adjust the amp's volume control and thus the input voltage to the speaker to get precisely 1 Watt input power.

Say you measured an impedance of R=9.24 Ω at 1 kHz. Since $P=V^2/R$, the required voltage is $V=\sqrt{(P*R)} = \sqrt{(1*9.4)}$ = 3.07 V_{RMS}. This is the effective or RMS value of the sine signal, not its peak value! The sound pressure meter should be placed in line with the speaker's center (on its "axis").

We used a Legacy combo amp (a rebadged Epiphone Valve Junior) for this test since it had an 8" speaker. First, we measured the SPL of the existing speaker. Since we weren't interested in the absolute sensitivity values and our test bench had less than 1 meter of space, we set the SPL meter 0.6m away. We chose a much lower input voltage, 0.6 V_{RMS}, because 1 Watt of input power produced a deafening and annoying sound level. The relative comparisons still apply, of course. With such a setup, the original speaker for the Legacy combo amp produced 98 dB of sound pressure, which was very loud indeed.

The final question is about the required voltage and current capabilities of the DC voltage source. Most of these vintage speakers were used in pre-WWII radio consoles and didn't have a separate DC power supply. The field coil would be used as a filtering choke, resulting in substantial savings. Chokes and power supplies were and still are relatively expensive.

If you know the radio the speaker was salvaged from, you can try to find its circuit diagram and ascertain the total current through the field coil from it. However, even that is not necessary. It can be safely assumed that a small single-ended triode or pentode output stage was used, pulling between 40 and 80 mA. The current draw of preamp stages and the tuner section is less than 5mA in total and can be neglected.

So, in our case, we had a DCR (DC resistance of the field coil) of 700Ω, and assuming 70mA through the coil, that would require an excitation voltage of $V_{DC}=I_{DC}*R_{DC}= 0.07*700= 49V$!

So, a DC power supply of the standard 1A current capacity would be fine, but most only go up to 20, 25 or 30V, you need one that can supply up to at least 50V_{DC}!

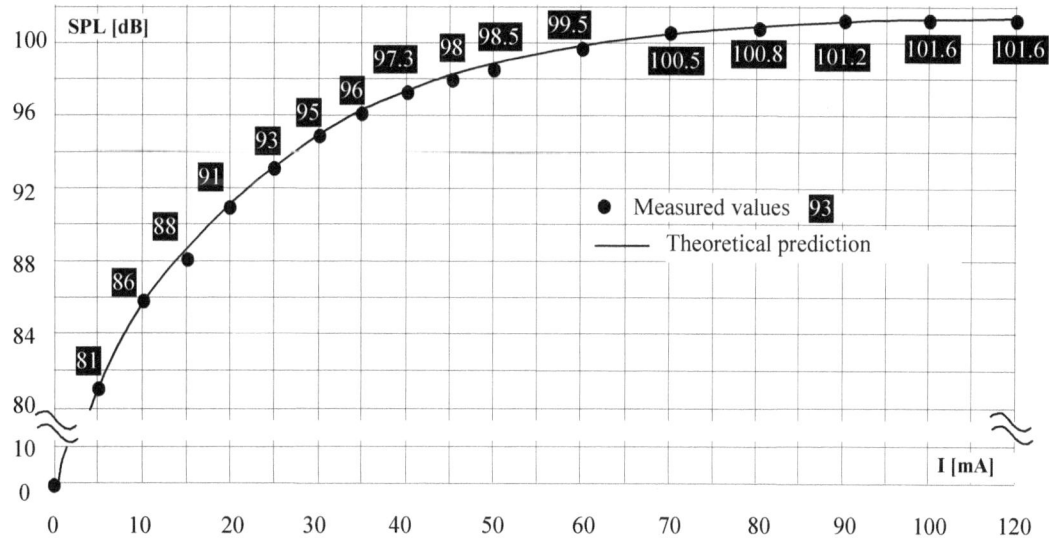

ABOVE: Test results of the 8" Motorola electrodynamic speaker, 0.6V_{RMS} signal, SPL measured at 0.6m distance

The graph above shows the results of our experiment. The horizontal scale (field coil current) is linear, and the vertical scale (SPL in decibels) also seems linear, but that is not the case since dB is a logarithmic unit. The formula says that the magnetic field strength and thus the sound pressure (loudness) increases linearly with the field current. Still, such a linear curve on a graph with a vertical log scale has a logarithmic shape.

Even at a very low excitation level of, say, 5mA (the first point measured), the SPL had already reached 81 dB! Doubling the current to 10mA increased the SPL by 5dB to 86 dB, and so on. Test points (marked as dots) will never perfectly align with the theoretical prediction (the smooth curve); there are always certain errors and imprecisions, but they should be reasonably close, and the overall trend should be clear.

With 45 mA of field current, the Motorola speaker reached the SPL level of the Legacy's speaker (98 dB). The maximum SPL was around 101.6 dB, reached with about 95mA of field current. Further increase in current did not increase output loudness.

Of course, you need to be careful not to exceed the nominal excitation current too much and for too long since such a high current may burn the field coil out. However, it only took a few seconds to measure each test point, so even at 120 mA, there was no overheating or smoke from the field coil.

The first important conclusion is that by replacing a permanent magnet speaker in your amp with an electrodynamic speaker of higher sensitivity, you can raise the loudness of your amp, in this case from 98 dB to 101.6 dB, an increase of 3.6dB. Since power in dB is P=10log(P1/P2), we get P1/P2=10(P/10) = 10(3.6/10) = 100.36 = 2.3

To get the same loudness from the original Legacy's speaker, you'd need to increase the power of that amp (in Watts) 2.3 times. Since that is a 5 Watt EL84 amp, with the replacement Motorola speaker at its maximum excitation, that amp would be as loud as a 5*2.3= 11.5 Watt amp!

Thus, the first possible benefit of replacing the original fixed magnet speaker in your amp with a field coil speaker is higher output levels, not in terms of Watts but in terms of SPL or "loudness," which is what ultimately matters. The second indication is that, should you decide to add variable excitation to such a speaker, a dynamic control range of around -20 dB is achievable. A 20 dB reduction in loudness is significant. Again, since power in dB is P=10log(P1/P2), we get P1/P2=10(P/10) = 10(20/10) = 10^2 = 100 times!

This means you could effectively attenuate your 5 Watt amp down to a 0.05W level or a 100W amp to a 1 Watt level (not with this tiny 8" speaker, of course, but with one or more suitable rated electrodynamic drivers in series or parallel connection).

A simple and cost-effective variable DC power supply for field coils

For under AU$10 on eBay (in 2017), this assembled 80 x 40mm PCB module seemed a good value. One could not buy the parts for that money in Australia, let alone a printed circuit board and labor. The design is simple, two 3-terminal regulators, LM317 (3) providing the positive and LM337 (4) supplying the negative voltage. Each output voltage can be separately controlled by its own 5k trimmer potentiometer, (1) and (2).

However, for our application, we need both to very in unison and also via potentiometer mounted on the front or at the back panel of a guitar amp. So, we unsoldered the two trimmer pots and soldered three wires taken out to a dual-ganged 5k pot.

LM317 and LM337 can provide only up to 37 V_{DC} on their outputs, but with the effective doubling of that voltage due to dual polarity outputs, this design can provide up to a maximum of +/-37V or 74 V, which should be enough for many if not most field coils. The two regulators are heatsinked and can easily provide up to 1.5 A of load current.

BELOW and RIGHT: The PCB-mounted symmetrical regulated DC power supply module sold online can be used as the field control module for your electrodynamic speaker.
Although 2x18V AC supply is shown (for 2x15V_{DC} out), for higher DC output voltages higher AC values are needed, such as 2x24V_{AC} or 2x30V_{AC}.

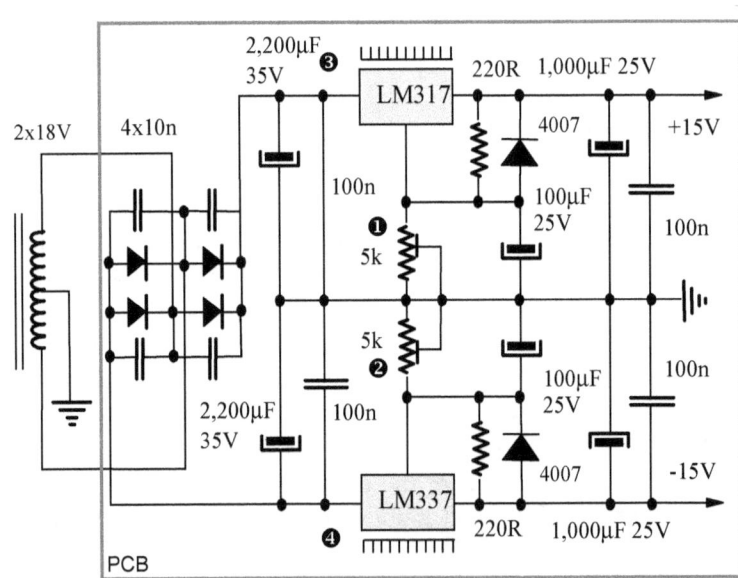

LOUDSPEAKERS, OUTPUT ATTENUATORS & HEADPHONE CIRCUITS

SPEAKER TESTS AND MEASUREMENTS

Quick speaker checks

There are four quick and simple tests to check loudspeaker drivers' mechanical/structural and electrical health. The first is the visual inspection. There should be no holes, tears, or rips in the paper cone. If there are, do not buy such a speaker; re-coning can be done by experts (if a speaker is rare and valuable and hard-to-replace) but is costly.

The second test involves gently pushing the speaker cone by hand. The cone should move freely and return to its original position. If any sudden resistance can be felt or if any "scraping" sound can be heard, that means the voice coil is rubbing against its surroundings. The "spider" or speaker's suspension has been damaged and misaligned. Such speakers may work to some extent but will sound distorted and will eventually fail.

Once a speaker passes these two mechanical tests, it is ready for electrical checks. If you have a multimeter, set it on a low resistance range and measure the resistance across the speaker's terminals. Since this is DC resistance, it will be lower than the speaker's nominal impedance. For instance, an 8ohm speaker may show a DCR of 5.7 ohms. That is fine and confirms that the speaker's voice call has continuity, i.e., that it hasn't burned out.

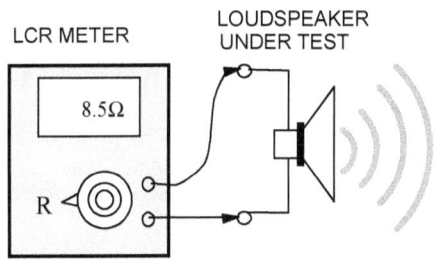

With an LCR meter set on "R" or impedance range, a pulsating tone should be heard in the speaker. The higher the speaker's sensitivity or SPL level (dB/W), the louder that tone will be.

A 1.5V or 9V battery can be used for another quick test. Take two leads terminated with crocodile clips and connect the battery across the speaker's terminals. With battery's + pole on speaker's + terminal, speaker's cone should quickly move forward (outward) and stay there! With the battery's + pole on the speaker's - terminal, the speaker cone should promptly move back (inward) and stay there!

Finally, the best and most conclusive test is by connecting a digital LCR (on "R" or resistance range) meter across speaker terminals. The LCR meter will send current pulses through the voice coil and show the voice coil impedance on its LCD at the selected frequency (60Hz or 1kHz). These pulses should be audible; the speaker should go beep - beep-beep ... about once a second.

A simple way to plot the loudspeaker impedance curve

You are measuring the difference between the readings of two AC voltmeters: the output voltage of a function generator (sinewave) and the AC voltage drop across the loudspeaker's impedance at various frequencies.

Had the speaker's impedance been constant with changing signal frequency, the ratio of the two measured voltages would remain constant. However, since the speaker's impedance is highly frequency-dependent, the ratio of the measured voltages will also change with frequency.

Since the 220Ω series resistor has a much higher DC resistance than the impedance of the measured loudspeaker, this simple circuit approximates a CCS (Constant Current Source). That same current flows through both the series resistor and the speaker.

Better function generators keep the amplitude of their output sine voltage constant regardless of its frequency and load impedance. In that case, you only need one voltmeter, V_2, since you don't have to read and record V_1 at each point; simply measure it at the first test frequency, and it will remain the same throughout this test.

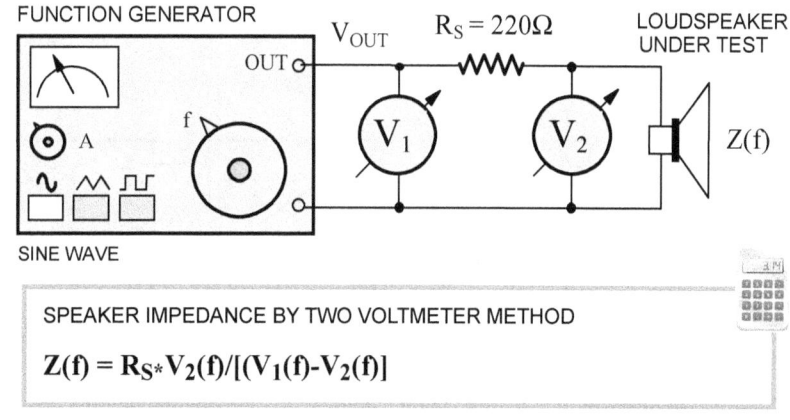

SPEAKER IMPEDANCE BY TWO VOLTMETER METHOD

$Z(f) = R_S * V_2(f) / [(V_1(f) - V_2(f)]$

ABOVE: Test setup for manual (point-by-point) plotting of a loudspeaker impedance versus frequency curve

After all this trouble, dozens of measurement points, typing those figures into a spreadsheet, and then drawing the impedance graphs, we can make three conclusions about the Celestion Super 8 speaker impedance curve (next page). The impedance is pretty constant throughout the bass and midrange regions (up to 1kHz), 4.8 to 5.2Ω, except the resonant peak at 125 Hz, where the impedance rises to around 20Ω. After 1 kHz, the impedance increases linearly, reaching 15.4Ω at 14 kHz and 18.2 Ω at 18kHz.

EXPERIMENT: *Loudspeaker impedance curve for Celestion Super 8*

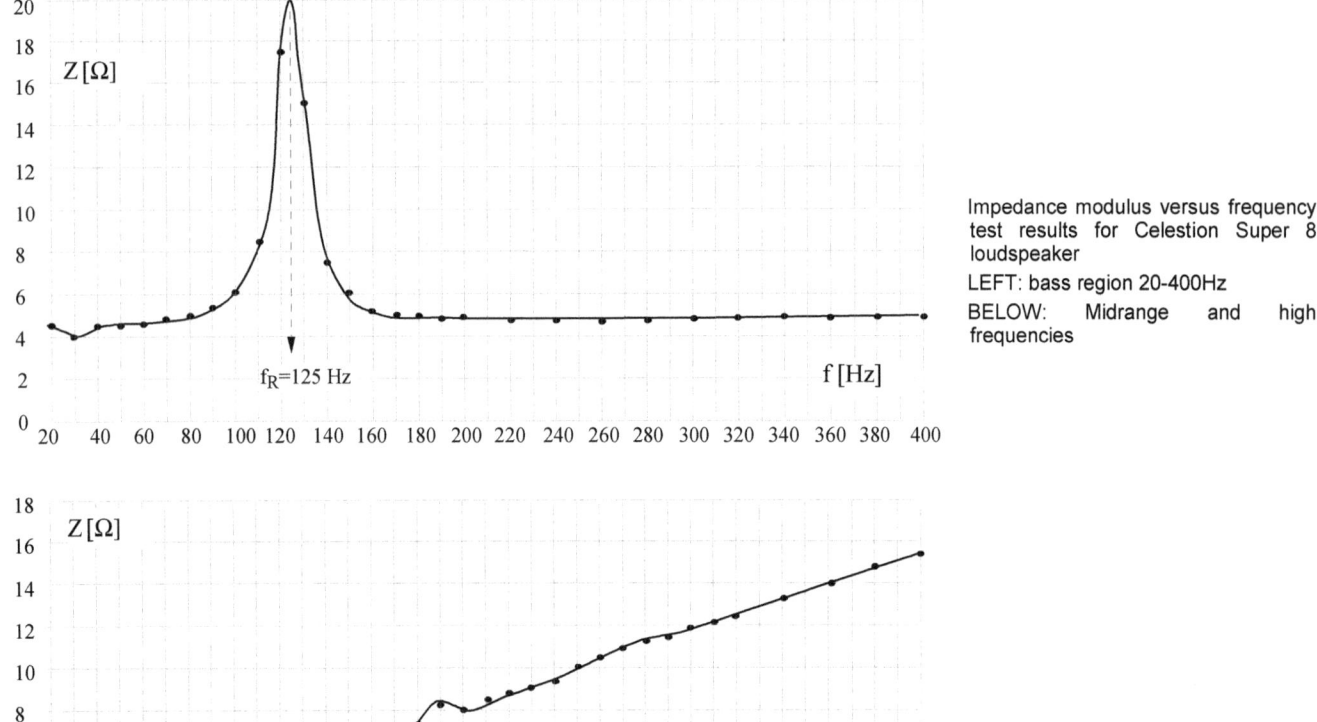

Impedance modulus versus frequency test results for Celestion Super 8 loudspeaker
LEFT: bass region 20-400Hz
BELOW: Midrange and high frequencies

HOW SPEAKERS IMPACT AMPLIFIERS' POWER LEVELS AND TONAL BALANCE

The signal chain inside a typical tube guitar amplifier

Let's consider a simple combo guitar amp without tone controls, with only one preamp stage amplifying 100 times (a pentode), driving the SE output stage that amplifies eight times (including the step-down output transformer). Since the output voltage is 8 Volts, the power on an 8-ohm load is $P = V^2/R = 8*8/8 = 8$ Watts. The current output stage must push through the speaker is $I = P/V = 8/8 = 1$ A!

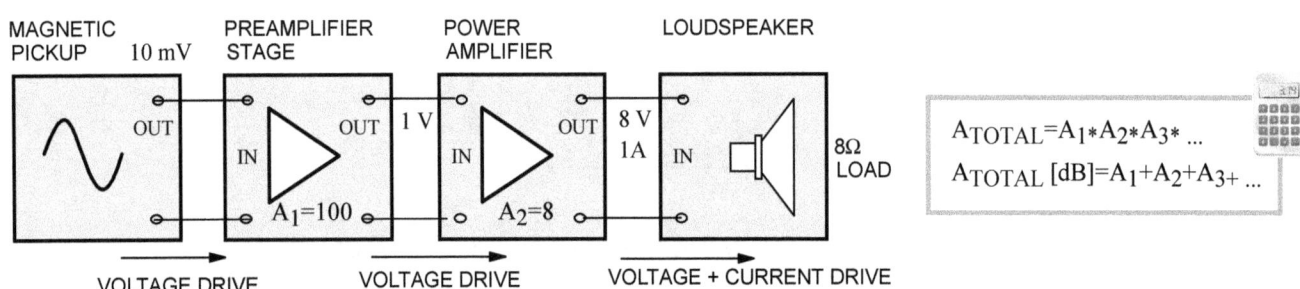

All amplification devices upstream of the power amplifier provide only voltage amplification, while the power amplifier provides both the voltage and current amplification.

There are two ways to calculate amplification factors (gains) in a signal chain. With gains expressed as dimensionless numbers, as is the case above, the overall gain is $A_{TOTAL} = A_1*A_2 = 100*8 = 800$, or in dB, $P = 20\log(V_{OUT}/V_{IN}) = 20*\log(8/0.001) = 20*\log(800) = 20*2.9 = 58$ dB

When individual gains are in dB, the overall gain is simply the sum of individual stage gains: $A_{TOTAL} [dB] = A_1 + A_2$!

Since $A_1 = 20\log(100) = 20*2 = 40$dB and $A_2 = 20\log(8) = 20*0.9 = 18$dB, we have $A_{TOTAL} [dB] = A_1 + A_2 = 40 + 18 = 58$ dB. Either way, the result must be the same, of course.

LOUDSPEAKERS, OUTPUT ATTENUATORS & HEADPHONE CIRCUITS

The importance of speakers' sensitivity

Consider two guitar amps, both heads. One is rated at 100 Watts (uses two pairs of 6L6 power tubes in a push-pull output stage) and is hooked up to a speaker box of the low sensitivity of 83 dB/Watt.

The other amp is a 10 Watt single-ended boutique head with a single 6L6 power tube, connected to a 2x2 cabinet of 93 dB/W sensitivity. Assuming the same guitar is used, which amp-speaker combination will sound louder in the same room?

If we express the ratio of power outputs of the two amps as a ratio, $P_2/P_1=100/10=10$ or in dB terms $10\log P_2/P_1 = 10\log10 = 10$ dB!

To make the maths easy, I deliberately chose nice & easy numbers (but very realistic). Ten times higher power of the push-pull amp is +10 dB above the output level of the low-powered single-ended amp.

However, its speakers have 10 dB/W lower sensitivity, so to get the same SPL (sound pressure level) or "loudness" as the low-powered amp with high-efficiency speakers, the more powerful amp must have 10 dB higher amplification, which it does.

Therefore, both setups will sound equally loud. What we gained in amplifying power we lost by using very inefficient speakers.

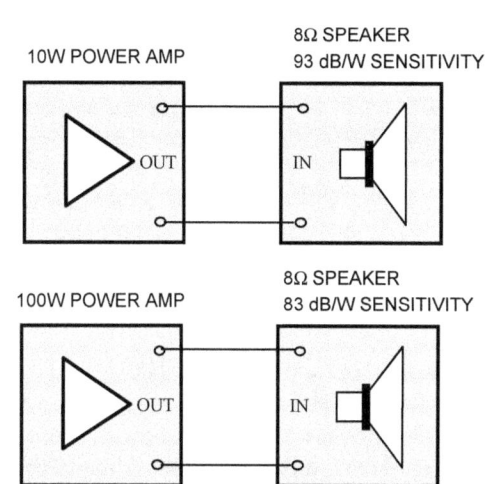

ABOVE: The two audio systems illustrated produce the same SPL!

The importance of amplifiers' low output impedance

The following discussion is somewhat simplified, assuming that the amplifier's output impedance is constant throughout the whole frequency range and that in all other respects, the amplifier is ideal, that it can provide any current demanded by the load. Neither is absolutely true for all loads and frequencies. Nevertheless, for this purpose, it's close enough to reality.

This scenario illustrates issues arising from the high output impedance of an amplifier (especially single-ended amplifiers without negative feedback) and its interaction with varying load impedance (the voltage divider effect). At frequencies where the speaker's impedance at its minimum (say 2Ω), $V_L/V_0 = R_L/(R_L+Z_{OUT}) = 2/(2+3) = 0.4$ or 40%, meaning only 40% of the actual voltage will drive the speaker, the rest will be lost on amplifier's internal impedance.

At resonant frequencies where speaker's impedance is at its peak (say 22Ω), $V_L/V_0 = R_L/(R_L+Z_{OUT}) = 22/(22+3) = 22/25 = 0.88$ or 88%, so a much better situation of only 12% loss. The output voltage will vary $20\log(0.88/0.40) = 20\log2.2 = 6.85$ dB!

Assuming an ideal amplifier with $Z_{OUT}=0$, the output voltage of a single-ended guitar amplifier capable of producing 8 Watts on an 8Ω load would be $V_L = V_0 = \sqrt{(PR_L)} = \sqrt{(8*8)} = 8$ V

A real amplifier, driving a speaker whose impedance dropped to 2Ω would have its output voltage divided between the output impedance of the amplifier and the load, so the voltage on the load at would be $V_L = V_0 R_L/(R_L+Z_{OUT}) = 8*2/(2+3) = 8*2/5 = 3.2V$

The power fed to the load would be only $P=V_L^2/R_L = 3.2^2/2 = 5.12W$, not its nominal 8W!

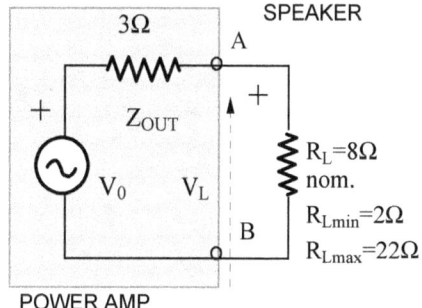

ABOVE: The output of an amplifier behaves as a frequency-dependent voltage divider

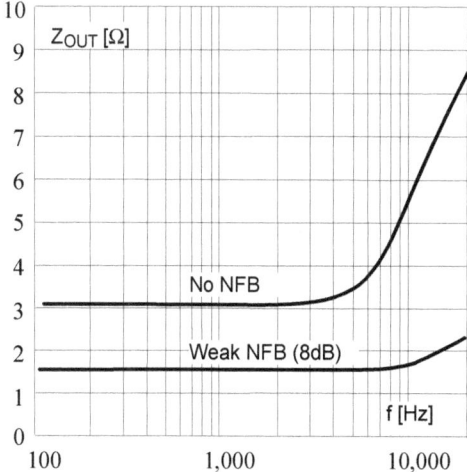

ABOVE: How the output impedance of a typical single-ended triode amplifier (8Ω output) varies with frequency with and without negative feedback.

Output impedance versus frequency

Damping factor (DF) helps us predict how well a particular amp would control the speaker cone. The higher the damping factor, the better an amp would control a speaker cone, the faster it will drive and the faster it will dampen its response and bring it to a stop. It is defined as a ratio of load impedance and amplifier's internal or output impedance: $DF = R_L/Z_{OUT}$.

Due to the output transformer's leakage inductance that increases with frequency, the output impedance of a tube amplifier rises with frequency, especially in the absence of negative feedback. Even with relatively mild feedback, the same amplifier exhibits a much lower rise in the output impedance than the rise when NFB is removed.

Finally, the damping factor is of most importance at low or bass frequencies. An amplifier must supply lots of current to the woofer coil. The damping of the woofer's cone is of paramount importance for the bass's speed, definition, and prominence. Thus, it is of much more importance in bass amps, and it usually isn't a concern in guitar amps at all.

12" speaker shoot-out

Let's study Celestion G12T100 12" guitar speaker's behavior across audio frequencies and compare it to the Japanese Dai-chi IS12000 instrument speaker from Laney TF300 amp. Both are 12" speakers selling around the US$129 mark (but often discounted down to $70-90 range), so our comparison will be meaningful.

Celestion's efficiency in the 100Hz-1kHz range is around 97dB, as specified (2). Below that it drops uniformly down to 70dB at 20Hz (1). There is a wide fluctuation in the critical 1-2 kHz midrange region (3) and a few peaks and dips in the 2-5kHz range (4). After that, the efficiency drops abruptly, followed by a not-so-pretty curve in the 5-20 kHz treble range. The peak around 14 kHz may be audible, resulting in some brightness (5).

Celestion G12T100 12" Guitar Speaker

- Power rating 100 W_{RMS}
- Nominal impedance: 4, 8 or 16Ω
- DC resistance: 3.4Ω, 7.4Ω or 14.83Ω
- Sensitivity: 97dB/W
- Magnet: 1kg Ceramic
- Frequency range 80Hz-5kHz
- Resonant frequency: 86Hz
- Weight: 3.2kg

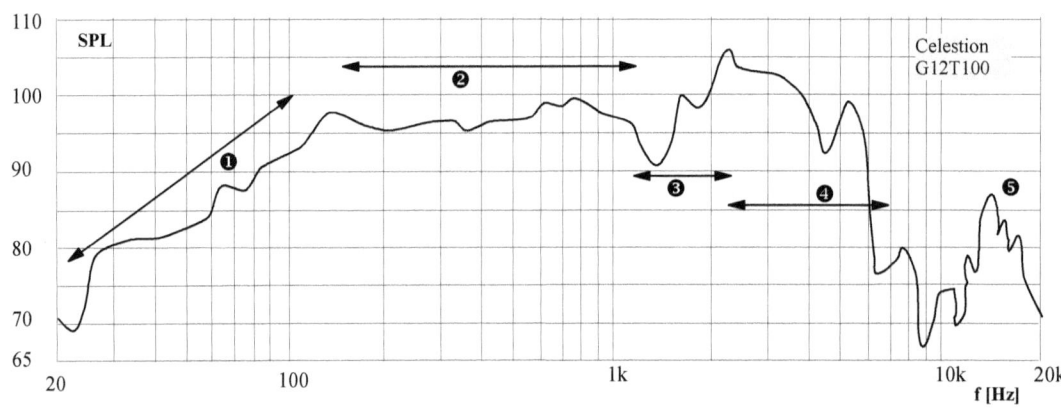

LEFT: SPL vs. frequency characteristics of Celestion G12T100 speaker

Dai-ichi's curve is smoother, but the efficiency is 2-5dB lower. The lower bass (6) and upper bass-lower midrange behavior (7) is similar to Celestion's, but there is a uniform drop in efficiency from 1 to 3 kHz (8), followed by a mild peak around 5 kHz. This suck-out in the upper midrange will result in a weaker upper midrange, so the speaker is bass-heavy.

Dai-ichi declares maximum power of 250/500 Watts, which is most likely peak power instantaneous (PMPO), but does not specify its rated RMS power (100 Watts in Celestion's case). The Dai-ichi magnet is heavier, 1.4 kg versus 1kg for Celestion, so perhaps it could be good for 120-130W_{RMS}.

Dai-ichi IS12000 12" Instrument Speaker

- Power rating 250W
- Nominal impedance: 8Ω
- DC resistance: 5.7Ω
- Sensitivity: 95dB/W
- Magnet: 1.42 kg Ceramic
- Frequency range 32Hz-5kHz
- Resonant frequency: 45Hz

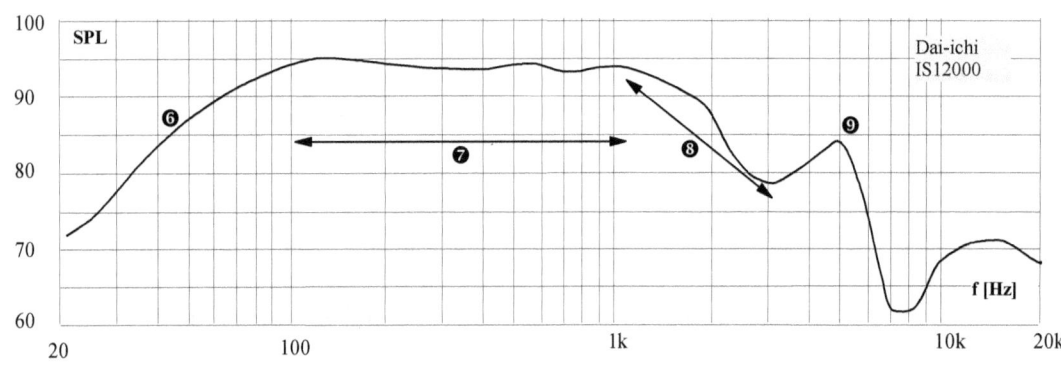

LEFT: SPL vs. frequency characteristics of Dai-ichi IS12000 speaker

LOUDSPEAKERS, OUTPUT ATTENUATORS & HEADPHONE CIRCUITS

ATTENUATORS

Attenuators are devices that attenuate or weaken the electric signal coming out of a guitar amplifier, passing only part of the signal to the loudspeaker. This attenuation can be in steps, using a multi-position switch and fixed passive components such as resistors, capacitors, and inductors, or in a continuous fashion, using a high-power potentiometer (rheostat).

The only reason attenuators are used is guitar players' desire for output stage distortion (which only happens at very loud levels) at manageable (meaning safe and legal levels that will not damage hearing or get neighbors to call police) loudness levels, such as at practice and during recording sessions.

However, no matter how well designed or intricate an attenuator may be, they always change the sound of the amp-speaker combination since they change the amp-speaker interaction, frequency response, and tonal balance. In electronics, just as in life in general, there is no such thing as "free lunch"; there is always a price to be paid!

L-pad attenuators

Rated at 50 Watts, this Dayton Audio speaker L-pad attenuator was designed to be used with 8 ohm speakers and has a 1" long shaft and 2" square body. The faceplate, volume control knob and all mounting hardware are included. It can be found at Parts Express (part number 260-255, $10.90). There is also a 100W version priced at $12.95.

At its maximum (fully CW), the wiper of the 8Ω section is at the input terminal (3) and the wiper of the 30Ω section is at the output terminal (2). The 8Ω resistor is out of circuit and the 30Ω resistor is connected across the load.

As the control knob is moved counterclockwise towards the MIN position, the larger and larger upper section of the 8Ω resistor is connected in series between the IN and OUT terminals, thus acting as a voltage divider and reducing the voltage across the load.

At the same time a smaller and smaller section of the 30Ω resistor is connected between the OUT and COM terminals (in parallel with the speaker). Due to that reducing resistance more and more of load current is diverted away from the load and shunted to COM.

At the minimum position (fully CCW), the 30Ω resistor is out of circuit and the OUT terminal is shorted to GND completely bypassing the load. The 8Ω resistor is now connected between the IN and COM terminals, so the amplifier "sees" this resistor as dummy load and the speaker terminals are short circuited so there is no sound.

For an 4Ω load a dual (stereo) 8Ωm L-pad can be used, simply connect both sections in parallel.

For a 16Ω load the two sections of a dual 8Ω L-pad can be connected in series as per diagram below.

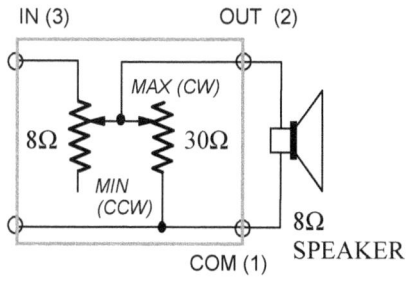

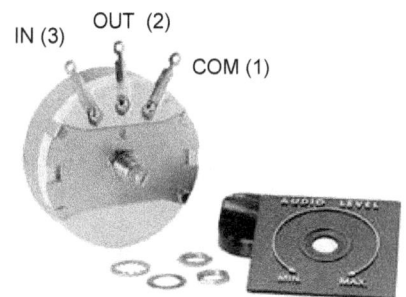

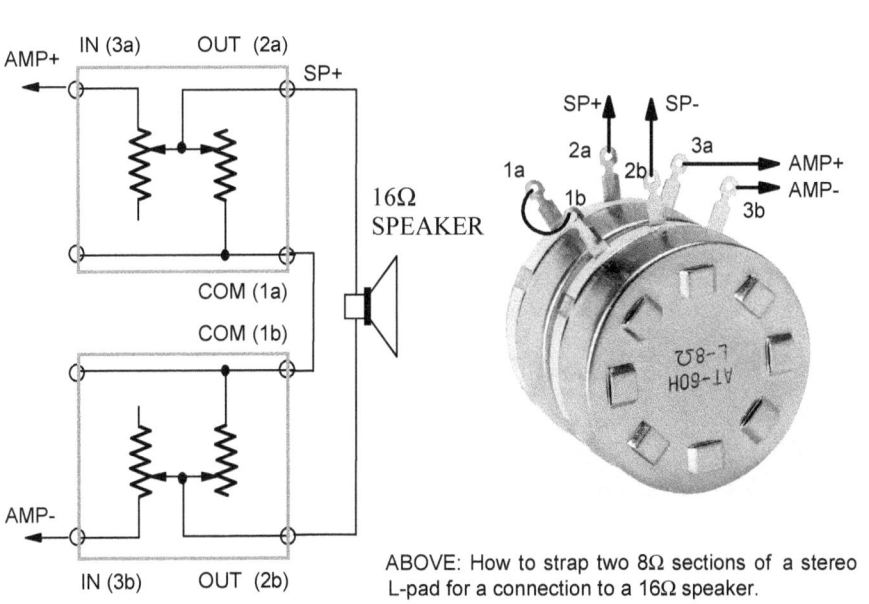

ABOVE: How to strap two 8Ω sections of a stereo L-pad for a connection to a 16Ω speaker.

Weber MASS Lite attenuator

Ted Weber started "Weber Speakers" in 1995. His company, now known as "Ted Weber's Famous Loudspeakers," apart from speakers, amplifier parts, speaker cabinets, and tools, make and sell a whole range of attenuators: MASS, MASS Lite, MiniMASS, MicroMASS, MASS 200.

MASS Lite comes in 50 Watt and 100 Watt versions, with three optional features: balanced line out, headphones output, and footswitch bypass. Our unit had the first two options installed.

The basic 100 Watt Mass Lite sold for US$192 in 2016. The unit pictured is the older version; the current model has different cosmetics, but we believe it is practically identical inside.

The "Lows-Mids" rheostat acts as the primary volume control, while the "Mids-Highs" knob provides treble control. Its action, however, is always dependent on the "Lows-Mids" control.

The higher the "Lows-Mids" attenuation, the less effect "Mids-Highs" or treble compensation will have on the sound.

With two main rheostats set at maximum attenuation, fully CW in "10" position, MASS Lite can be used as a dummy load for amplifier testing.

Since some amps don't have the headphones or the line outputs, this versatile device also adds those two features to its primary attenuation purpose.

As with any attenuator, the heart of the unit is a couple of 50W rheostats (rated at 100W on a 100W version, of course). Rheostat (1) controls "Lows-Mids" frequencies, and rheostat (2) "Mids-Highs." There is also a third rheostat (3) at the back, labeled "HP Volume," which controls the headphone output volume.

In the center is a voice coil assembly (VCA) of an actual dynamic loudspeaker (4). There is no metal basket or paper cone, but the VCA does behave like a "real" speaker.

LEFT: The internal view of Weber MASS Lite attenuator (50W version)

LOUDSPEAKERS, OUTPUT ATTENUATORS & HEADPHONE CIRCUITS

Weber is one of the rare attenuator makers that includes VCAs in their designs. Most guitar players would agree that simple resistive attenuators significantly change the sound of an amp, and that seems to be due to many factors, the most important being the absence of the speaker-amp interaction. With resistive loads, there is no induced back EMF (Electromotive Force) that a dynamic speaker feeds back into the output transformer's secondary. How successful or otherwise has Weber been in mitigating this problem? We will leave it to guitar players and their personal opinions. Sound, just like beauty, is in the eyes (ears) of the beholder.

Tucked in between the "Line out" and "Speaker" jacks is a small audio transformer (5) that converts the single-ended (unbalanced) signal to a balanced output. This is an optional feature. The transformer's primary is fed from the attenuator's input via a resistive voltage divider, two film resistors joined in a messy fashion and simply hanging in the air (6).

Notice two NP (Non-Polarized) electrolytic capacitors (7), the kind used in speaker crossovers. This means there is some kind of filtering and some frequency-dependent manipulation of the signal, which will enable a degree of control over the tonal characteristics of the output signal. One of the caps is in parallel with two 135W 15 Watts power resistors (8).

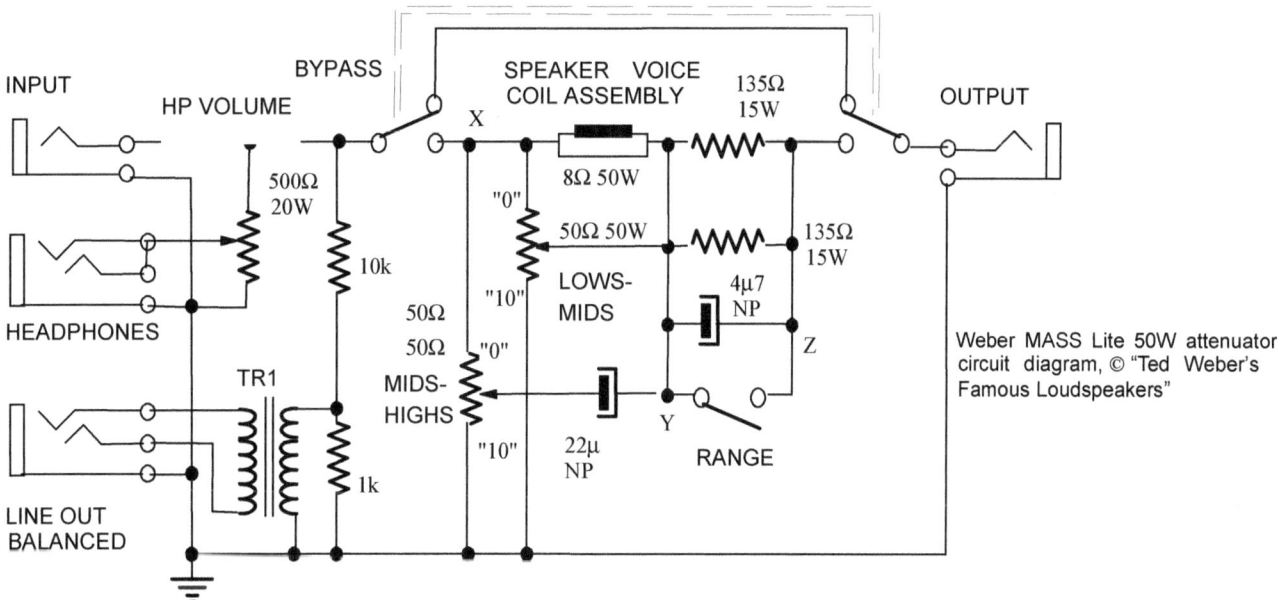

Weber MASS Lite 50W attenuator circuit diagram, © "Ted Weber's Famous Loudspeakers"

The top switch (9) is the bypass switch, which, as the name suggests, bypasses the whole attenuator when engaged, feeding its input signal straight out. The headphone and line out stay active even in bypass mode. The HP and line-out circuits are independent of the attenuator.

The line out signal on the primary side of the transformer is $1/(1+10)*100 = 9.1\%$ of the input signal from the amp's output. Since power is $P=V^2/R, V=\sqrt{PR}$. For instance, with 5 Watts coming in and an 8Ω amp output, the RMS voltage is $V=\sqrt{(5*8)} = 6.3V$ and the line out is $6.3*0.091 = 0.57V$.

Notice three main points or nodes as they are called in electrical network analysis. "X" is the input into the attenuator, "Z" is the output. When closed, the Range switch bypasses the two paralleled 135W resistors (62.5W total) and thus shorts points "Y" and "Z" together, increasing the output level. With the Range switch open (as shown), the 62.5W resistance is in series with the load (speaker), and the volume is somewhat reduced.

The reactance of the 4m7 parallel capacitor will decrease as the frequency rises, so its bypassing effect will be more pronounced. With rising frequency, the series 62.5W resistance will be gradually taken out of the circuit. HF attenuation will be somewhat reduced, thus providing a mild HF boost or compensation. This cap was most likely included to prevent the attenuator from sounding "dull." Without it, the "Lows-Mids" attenuator would be a simple resistive affair.

The *"Mids-Highs"* rheostat's wiper feeds the outgoing signal through a capacitor, which forms a simple low-pass (CR) filter with the load's resistance. Assuming a closed *Range* switch and an 8Ω load, the lower -3dB (half power) frequency of that filter would be $f_L=1/(2\pi RC) = 904$ Hz, meaning that "Mids-Highs" refers to frequencies above approx. 1kHz.

Finally, a handy feature of Weber attenuators is that there is no need for separate units for each speaker or amplifier output impedance, say 4, 8 or 16 ohms. These units can be used with any of those load impedances.

Dr Z Air Brake attenuator

The Air Brake is a small and relatively simple attenuator with two basic operating modes. The first five positions of the Attenuator switch one resistance in series with the signal and the load, and one resistance in parallel with the load as a simple resistive voltage divider.

The series resistances are tapped points along one large wire-wound resistor R1. The taps are simply metal buckles that touch the exposed wire turns on the resistor. The value of this series resistance varies according to the position of the switch, 0Ω in positions "0"(which is really a bypass setting) and "1", 4.1Ω in position "2", 10.6Ω in position "3" and 17.4Ω in position "4". The three resistor clamps on R1 marked with * are movable, the end connections are fixed. That way, you can adjust the attenuation in various positions to suit your needs.

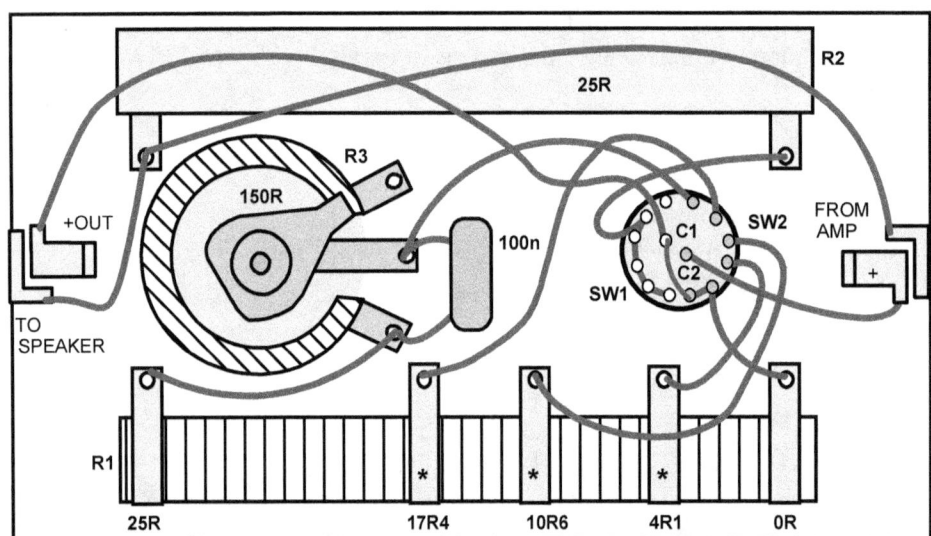

LEFT: The internal components and wiring of Dr. Z Air brake attenuator

The 6 terminals of SW1 and the common terminal C1 are depicted using white circles, the 6 terminals of SW2 and its common terminal C2 are gray circles.

The "Bedroom Level" rheostat is shown in MAX position, fully CCW when viewed from the front, as in the photo of its control panel below. MAX means maximum attenuation, MIN means minimum attenuation, not the output level!

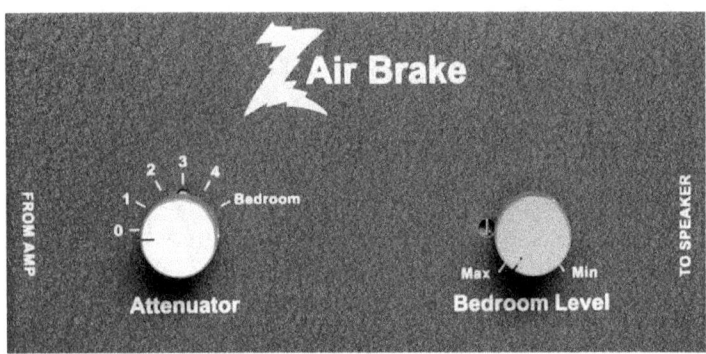

ABOVE: The front panel of Dr. Z Air Brake attenuator
BELOW: Its circuit diagram

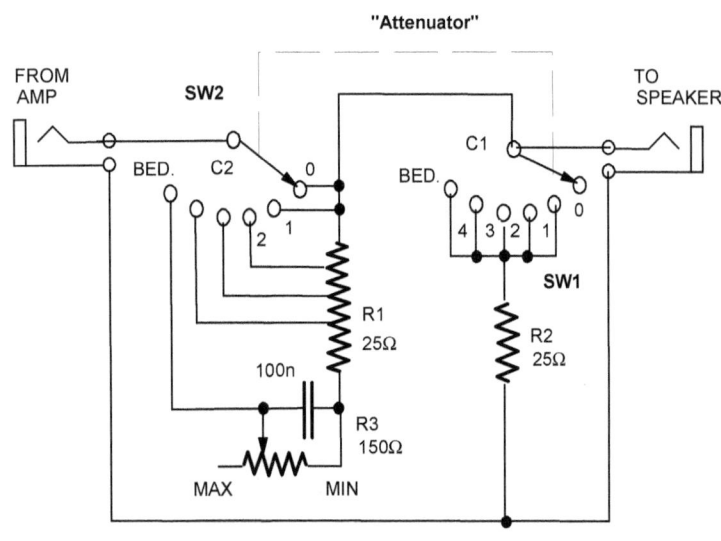

In the last position, named *Bedroom*, the total R1 resistance is in the circuit, plus the 150R rheostat whose active section (in the circuit, in series with R1) is bypassed by 100n frequency compensation cap. This enables continuous output level adjustment since the previous four attenuating positions were all in predetermined jumps.

Except in position "0", resistor R2 (25Ω) is always in parallel with the load.

Let's draw the circuit for each of the five settings of the *Attenuator* switch. For simplicity's sake, we will assume a fixed 8R load (speaker) and a purely resistive one. This isn't the case in real life (a loudspeaker is a complex impedance whose modulus and phase vary with frequency). Still, this exercise is to practice our quick estimation skills and gain insights into the mind works of other designers.

In all cases, the 8R speaker in parallel with that 25R resistor makes a fixed 6R parallel branch, which greatly simplifies calculations.

There is no voltage divider in position "1"; a small portion of the current that would otherwise pass through the 8R load is now diverted through the 25R shunt, but that attenuation is minimal.

The following three settings are simple voltage dividers, with attenuation ranging from 0.6 (the output voltage is 60% of the input signal), to 0.26 or 26%, or, in decibels, from -4.4dB to -11.8dB. Remember, attenuation in dB is depicted by the negative sign, meaning the output is lower than the input. The formula is $A = 20*\log(V_{OUT}/V_{IN})$!

In the *Bedroom* case, the minimum attenuation means the 150R rheostat is out of the circuit, as is its bypass cap. So, we have 25R and 6R producing 0.19 or -29.6dB attenuation. In the MAX position, the full 150R of resistance is added in series with the 25R fixed resistor (175R total), so only 3.3% of the voltage signal actually reaches the load.

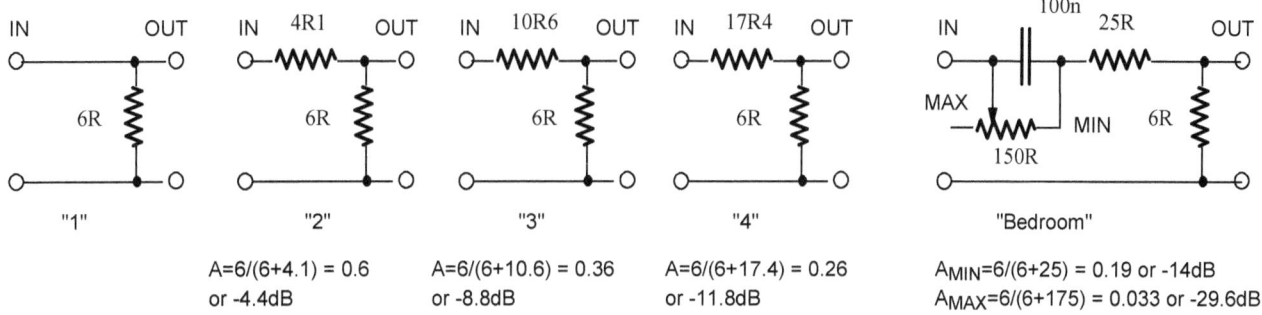

ABOVE: The five attenuation scenarios with 8R speaker (25R||8R=6R)

In the last (*Bedroom*) scenario, we conveniently ignored the 100n cap, assuming its reactance is much higher than the 0 to 150 ohms resistance it's bypassing. Now let's see for which frequencies that assumption is correct and for which frequencies such reactance must be taken into account; in other words, let's look at how a capacitor provides frequency compensation.

Normally if its reactance is ten times larger than the resistance the cap is bypassing (one order of magnitude), we can consider the cap an open circuit and ignore its presence. With a fixed bypassed resistance, we can calculate the frequency at which this happens.

For instance, with the maximum attenuation (150R in parallel with the cap), we are looking for f where $X_C = 1/(2\pi fC) = 10R$ or $10R*2\pi fCR = 1$! By expressing f we get $f = 1/(20\pi CR) = 1,061$ Hz, meaning the capacitor can be ignored for all frequencies lower than around 1kHz. Above that frequency, the reactance drops as the frequency rises, which reduces the total resistance in the series branch and thus reduces attenuation, so HF (high frequencies, in this case, anything over 1kHz) are attenuated less than LF (low frequencies).

Let's take one position of the rheostat's slider along the way between the two ends, say we have 25Ω in parallel with the capacitor. Since 125/25=5, the new R is 5 times lower, the frequency of interest will now be 5X higher: $f = 1/(20\pi CR) = 1,061$ Hz*5 = 5,305Hz!

We could have intuitively arrived at the same conclusion, albeit without any feeling for the numbers. Since CR circuit is a high pass filter (meaning it attenuates LF more), as soon as you see a cap in series with the signal you know high frequencies will be passed through less attenuated (stronger) than low frequencies!

This simple frequency compensation is implemented due to the human ear's sensitivity, which is the highest for MF (midrange frequencies), which include speech, and the sensitivity drops significantly towards the bass (LF) and treble (HF) frequency regions. In other words, our ears are logarithmic, not linear transducers, filtering out LF & HF much more than the MF band of frequencies.

Marshall Power Brake

Weighing at around 13 lb., the now discontinued Marshall Power Brake was a fan-cooled passive attenuator capable of handling 100 Watts of audio power. Its 12-step attenuator (SW2) provides 3dB jumps in attenuation, 0dB to -30dB. In the 12th position, Air Brake works as a passive load box with no output.

The only other control is a DPDT switch (SW1), selecting either 8W or 16W impedance. The rectifier bridge, capacitive filter and Zener diode voltage regulator supply DC power to the internal fan. The power comes from the amplifier's rectified output signal.

The two crucial parts of the attenuator are the 8W2 100W load resistor and the tapped autotransformer TR1.

Quite a few users reported the 160μ 100V elco blowing up, so it should be replaced with one of a higher DC voltage rating, such as 160V, 200V or 250V.

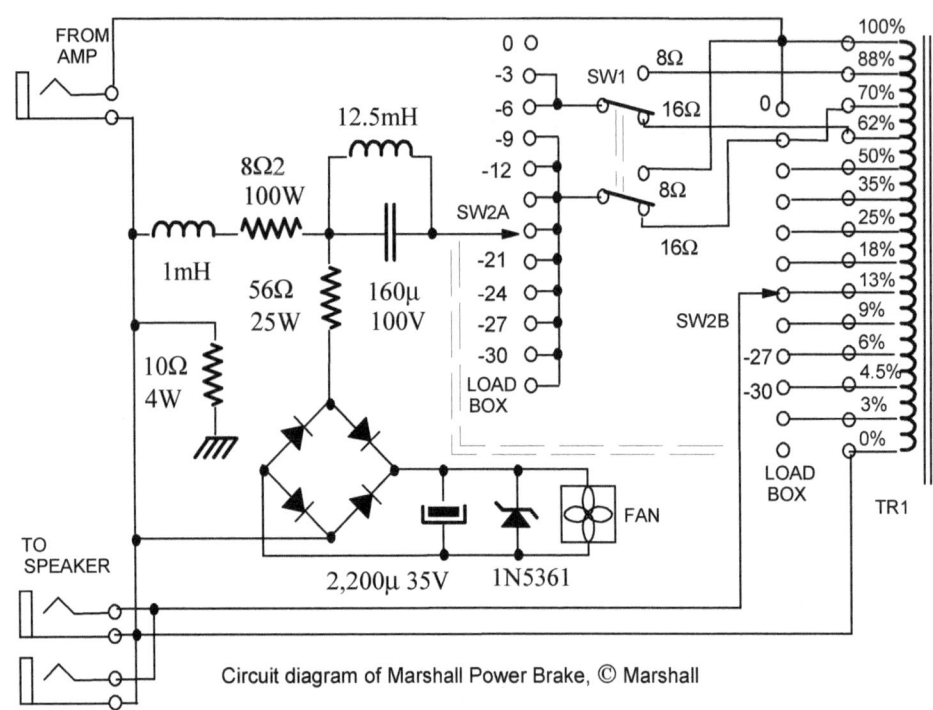

The 0, -3, ..., -30 are attenuation figures in dB, and the % figures next to autotransformer taps are the percentages of its total number of turns or the full voltage between the 0% and 100% ends, which are connected to the Power Brake's output and input, respectively.

The analysis of its operation is quite complex and not within the scope of this book.

Circuit diagram of Marshall Power Brake, © Marshall

ATTENUATOR AND HEADPHONE CIRCUIT CASE STUDY: BUGERA VINTAGE 5 AMP AND IBANEZ TUBESCREAMER TSA5TVR-S

For some reason, this amp sat on the shelf of a local music store for six years. Spotting a potential bargain, we asked one of the salesmen (they are all guitar players) about it, and he murmured something along the lines of the amp not sounding right and "doing nothing well," which to us, seasoned negotiators and price choppers, was a clear invitation for a low-ball offer, so we got it for a song.

There have been some negative online comments about Bugera and Behringer (both brands are owned by the Music Group, based in Manila, The Philippines, and chaired by Uli Behringer), so we wanted to put at least one of their amps through our MMS (The Merciless Mill of Scrutiny).

This little guy looked cute; the creamy top-half front panel made a nice contrast with the bland blackness. Why most guitar amp makers insist on all black amps is beyond comprehension.

The knock or percussion test

By the way, "the knock check" is the electronic equivalent of a doctor's tap or "percussion," as it's officially called. This is when they tap on your back, chest, or abdomen and make qualitative judgments if it sounds like air, solid, or liquid.

The test is inconclusive, of course, only the first indication that something could be amiss, just like doctors feeling your internal organs and tissues through palpation, which reveals their size, position, and consistency.

Perform the knock test on various combo amps' cabinets, and you will notice a wide variety of sound "signatures," which may be an early indication of the sound quality you could expect out of them.

BUGERA VINTAGE 5
- 1x 12AX7, 1xEL84 tube
- Solid state rectification
- PCB construction
- Output power: 5 Watts
- Controls: Tone, gain, volume, reverb
- Built-in digital reverb
- Output attenuator (5W, 1W, 0.1W)
- 8" speaker
- Dimensions: 36x22.5x40 cm
- Weight: 10.3 kg

LOUDSPEAKERS, OUTPUT ATTENUATORS & HEADPHONE CIRCUITS

The chassis is well made and with plenty of room for add-ons. The components are distributed across three circuit boards, one with the power supply and audio chain (1), the attenuator components are on their own board (2). In contrast, the "output board" (3) carries headphone and speaker jack and associated components.

The digital reverb board is a plug-in module on this older version (4). It is incorporated into the mainboard in newer versions. The bad news is that it is impossible to repair should it fail, but the good news is that the rest of the amp will continue to function without it.

The terminals of the mains switch (5) were insulated with electrical tape, an amateurish and, in many countries, illegal practice. Such tape dries out and falls off within a few years; heat-shrink should have been used instead.

Notice a myriad of fuses (6), in addition to the mains fuse on the chassis (not shown).

Immediately noticeable are also three clusters of four diodes (7), so even without a circuit diagram one could reasonably assume that one is a HV (High Voltage) rectifier for the anode voltage, one LV rectifier for tube heater voltage and the third one another LV rectifier for solid state and digital circuitry.

There is also a 3-terminal voltage regulator (8) on a small heatsink (LM7815), most likely for the LV supply for digital circuitry since its 15V would not be the correct voltage for tube heaters.

Positioning large heat-producing power resistors a millimeter or two away from heat-sensitive electrolytic caps is one of the most common mistakes modern designers of tube amps make; sure enough, we have not one but two examples here (9).

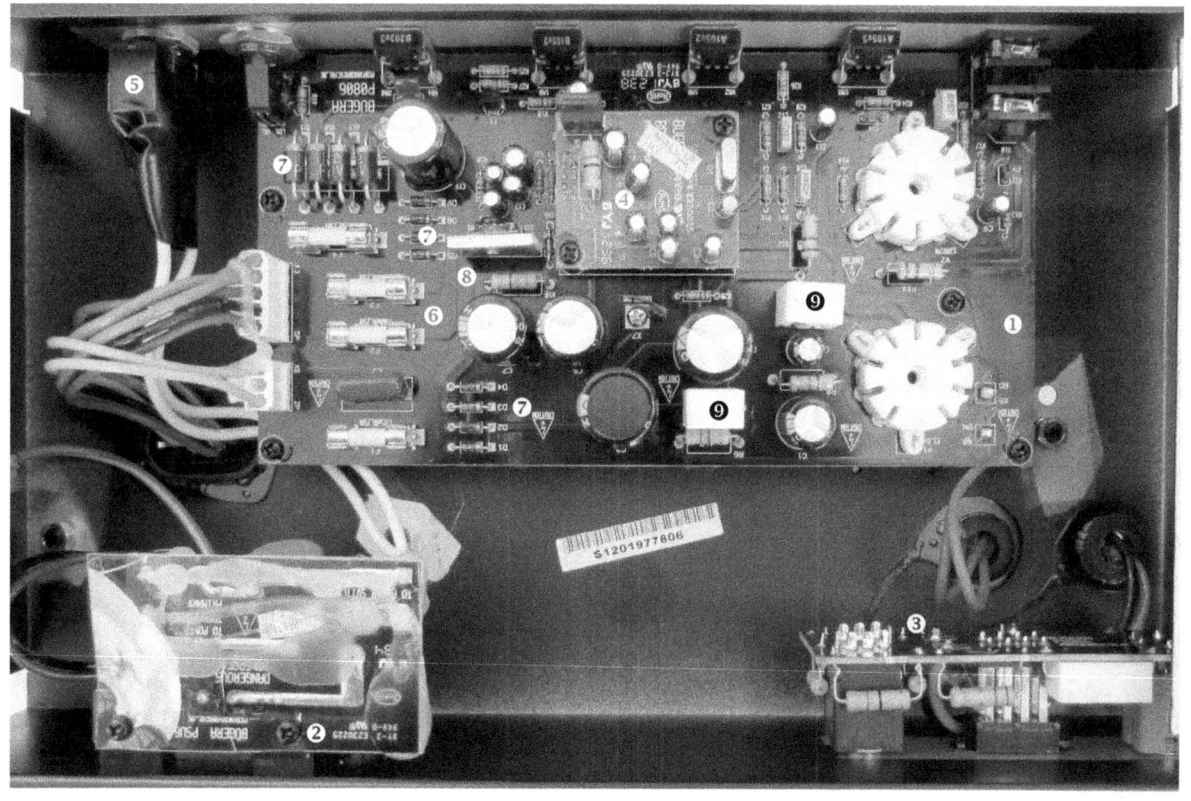

The circuit diagram

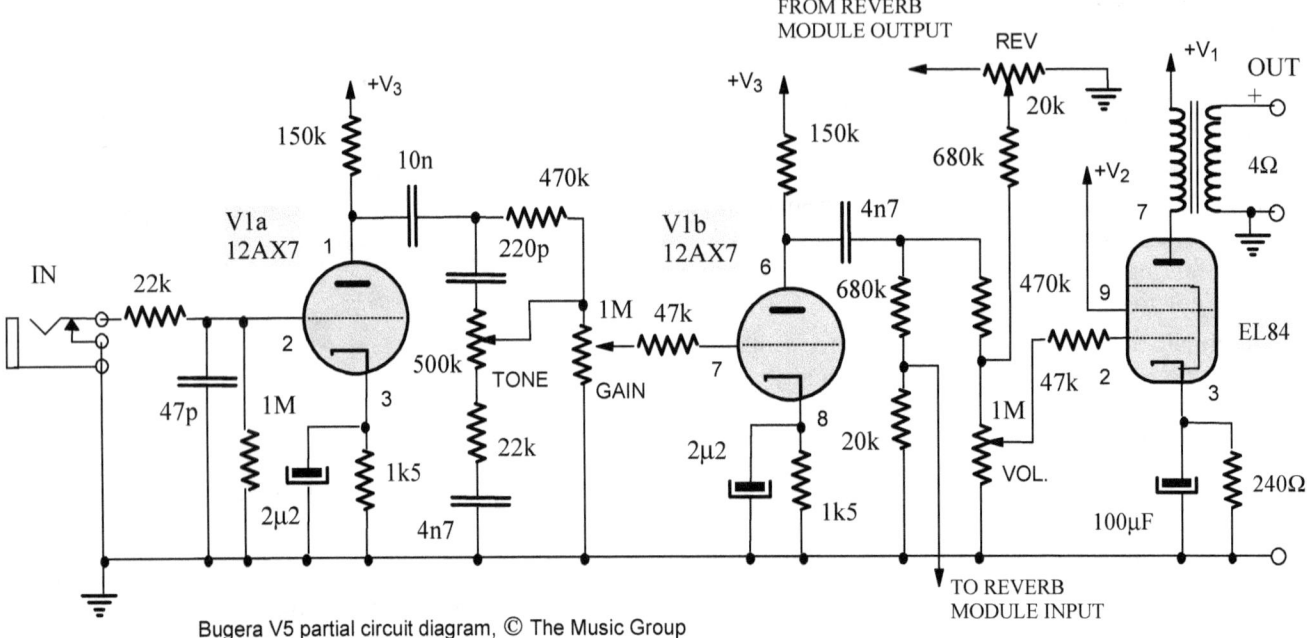

Bugera V5 partial circuit diagram, © The Music Group

The 22k input series resistor and 47pF cap to GND form a 1st order low pass filter with the -3dB upper-frequency $f_U = 1/(2\pi RC) = 154$ kHz. This is an RF filter that attenuates radio frequency interference that may be picked up by the guitar, cable, or the amp's input circuit.

Both input stages are identical, the lower -3dB frequency of the cathode circuit being $f_L = 1/(2\pi R_K C_K) = 48$ Hz, the partially bypassed cathode resistor and its capacitor act as a high pass filter, attenuating the bass signals.

The voltage divider at the output of the second stage takes $20/(680+20) = 20/700 =$ of the signal and feeds the input to the digital reverb module. The output of the reverb module or the "wet" signal comes back through an NPN transistor stage (not shown) and feeds the 20k *Reverb* pot, whose output (wiper) feeds a series 680k resistor and then enters the main signal chain at the top of the *Volume* potentiometer.

The output transformer

Compared to similar 5 Watt amps, the Bugera's output transformer, rated at 16 Watts, is on a larger side, although by our standards, still undersized.

A very high primary DC resistance of around 400 ohms indicates either lots of primary winding turns or a very thin wire used (or both). However, the relatively low inductance of 5.5 Henry means low-grade EI laminations were used (non-grain-oriented, low permeability) or a low number of primary turns. The leakage inductance is very high for such a small transformer (almost 30mH), indicating a lack of any sectionalizing of the windings, resulting in a low Q-factor.

BUGERA V5 OUTPUT TRANSFORMER:
- Brand: NRE, EI48 laminations
- a=16 mm, S=25mm, a=aS = 4cm^2
- Power rating: P=A^2= 16W
- Primary DCR: R$_P$=403Ω
- L$_P$= 5.5H @120Hz, 5.3H @1kHz
- Leakage inductance @1kHz L$_L$=26.6mH
- Quality factor: QF = 5.3/0.027 = 196

Sonic impressions

Despite the salesman's disparaging remarks, the amp sounded fine after we changed its defective EL84 power tube. After replacing the Bugera imprinted Chinese preamp tube with a vintage RCA 12AX7, it sounded even better.

Of course, just as with 99% of its brethren, this small amp's 8" internal speaker is its major bottleneck. Connect it to a 12" or 15" speaker or to a 2x2 cabinet and hear it really sing. The digital reverb sounded very good indeed, in my opinion much better than the usually noisy and temperamental spring reverb units. Another reminder to take salesmen's and other guitar players' comments with a great dose of skepticism.

Finally, a warning to those of you who may now rush to get one of these babies. The new version uses SMD technology for the whole amp, so it is impossible to repair or modify! If you want this amp for modification purposes, ensure you get the old version as illustrated here and not the later "Infinium" version.

LOUDSPEAKERS, OUTPUT ATTENUATORS & HEADPHONE CIRCUITS

The inbuilt output attenuator

We've included this amp as a case study in this chapter because of its relatively complex headphones output and its 3-position output power attenuator, so let's analyze them in some detail.

The circuit diagram shows the attenuator in the "5 Watt" position. There are four resistors in play, and in this position, due to the switching arrangement, they are all out of the circuit. Let's redraw the attenuator circuit in the other two positions. This is an opportunity to practice your circuit analysis skills, and secondly, let's see if the output power levels in those cases are really 1 Watt and 0.1 Watt as claimed by Bugera.

To be able to draw a circuit diagram for other positions, we need to understand how various contacts and switches work and how their operation is depicted with graphic symbols. Here we have two types of symbols, one for the three-position sliding switch and the other one for the input phone sockets (also known as 1/4" jacks). We have covered the operation of these jacks elsewhere in the book, so we will just summarize it here.

There are two sockets, X1 and X2. Both are stereo jacks (two tips, indicated by the V parts). X1, the speaker jack, is simpler; the two switchable contacts (small black arrow tips), each activated by one V tip, are both connected together to the two V tips. So, either a mono or a stereo speaker jack can be plugged in, and in both cases, the speaker(s) will be connected between points "H" (for "HOT" and "GND" for ground.

Jack X2 has two switchable contacts, but these are now of the "changeover" type. Notice the contact activated by tip "T." Its middle terminal (small circle "o") is indicated as connected to the bottom black arrow. That is the case when headphones are *not* plugged in.

Once the headphone jack is plugged in, one V tip (the upper one) will hit tip "T" and push that contact up, thus disconnecting the middle flap from the bottom arrow and connecting it to the top arrow to whose terminal the 47R resistor is soldered.

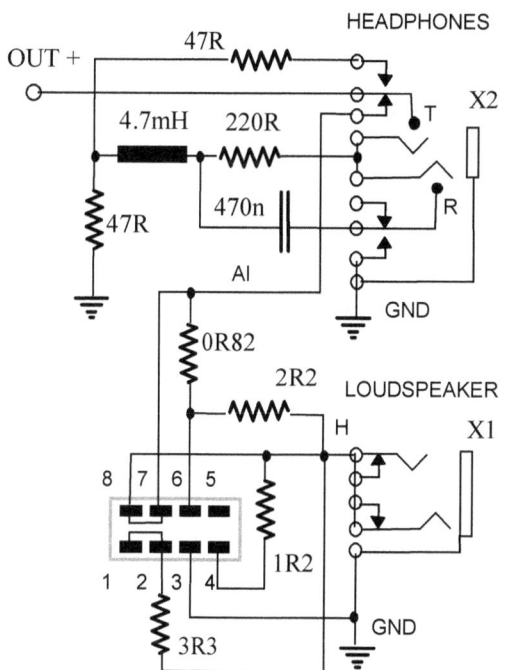

Bugera V5 output attenuator, headphones and loudspeaker connections, © The Music Group

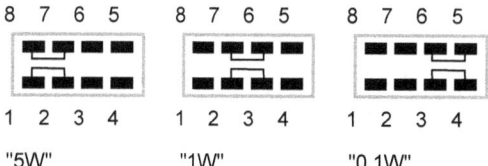

Contact connections for the three positions of the attenuator switch

The same will happen with tip "R." Its middle flap is now connected to the upper arrow, to which nothing is connected. Once the headphone jack is plugged in, the bottom V tip will hit tip "R" and push that contact down, thus connecting the middle flap to the bottom arrow, whose terminal is wired straight to GND. That will connect one end of the 470n capacitor to GND.

Now you can trace the signal from the OUT+ terminal (the HOT secondary of the output transformer from the previous page) down to terminal 7 of the sliding attenuator switch. That switch has two contacts, the upper and the lower one.

In the 5W position contacts, 7 and 8 are connected together, so the signal enters through contact 7 and exits the switch through contact 8 out to the hot tip of the X1 jack (+ of the speaker). The bottom contact in this position does nothing because it shorts switch terminals 1 and 2, and terminal 1 is not connected to anything. In that position, the three resistors (0R82, 1R2, and 3R3) are not in operation.

Before you proceed, as a practice, draw the two circuit diagrams of the output attenuator from point "AI" (Attenuator Input) down for positions "1W" and "0.1W". Remember also to draw the connected speaker, represented as a 4R resistor (4 ohms). In both cases, we are looking at the signal voltage between point "H," the speaker's + or HOT terminal, and GND or ground. The circuits are simple voltage dividers.

The output power is $P = V^2/R$, so if we calculate the voltage attenuation of the voltage divider in each of the two cases and compare it with the standard or ⅕ Watt operation, we will get the output power levels.

In attenuator switch position "1W" terminals 6 and 7 are connected, shorting out the 0R82 resistor. Point AI is connected to the left end of resistor 2R2 and the bottom slider shorts terminals 2 and 3, thus connecting the upper end of the 3R3 resistor to GND (since terminal 3 is permanently wired to GND).

So, we have a series resistance of 2.2Ω, followed by the parallel combination of the 3R3 resistor and the 4R speaker. Their parallel equivalent is $R_P = 3.3*4/(3.3+4) = 1.81Ω$!

Now, the voltage across the speaker is $V_H = V_{AI}*1.81/(1.81+2.2) = 0.4525*V_{AI}$. Since the output power is proportionate to the square of the voltage, $V_{H2} = 0.4525^2*V_{AI2} = 0.2047*V_{AI2}$, the maximum output power compared to the 5-watt case is $0.2047*5 = 1.024$ Watts, so the specified 1 Watt output is correct.

With the attenuator switch in position "0.1W", terminals 5 and 6 are connected, so now the 0R82 resistor is not shorted out anymore. The bottom slider shorts terminals 3 and 4, thus connecting the bottom end of the 1R2 resistor to GND in parallel with the speaker. The series resistance of 0.82 ohms is followed by the series resistance of 2.2Ω (a total of 3.02Ω), followed by the parallel combination of the 1R1 resistor and the 4R speaker. Their parallel equivalent is now $R_P = 1.2*4/(1.2+4) = 0.923Ω$!

The voltage across the speaker is $V_H = V_{AI}*0.923/(0.923+3.02) = 0.234*V_{AI}$ and $V_H^2 = 0.234^2*V_{AI}^2 = 0.055*V_{AI}^2$, meaning the maximum output power compared to the 5-watt case is $0.055*5 = 0.275$ Watts, so the specified 0.1 Watt maximum output is incorrect.

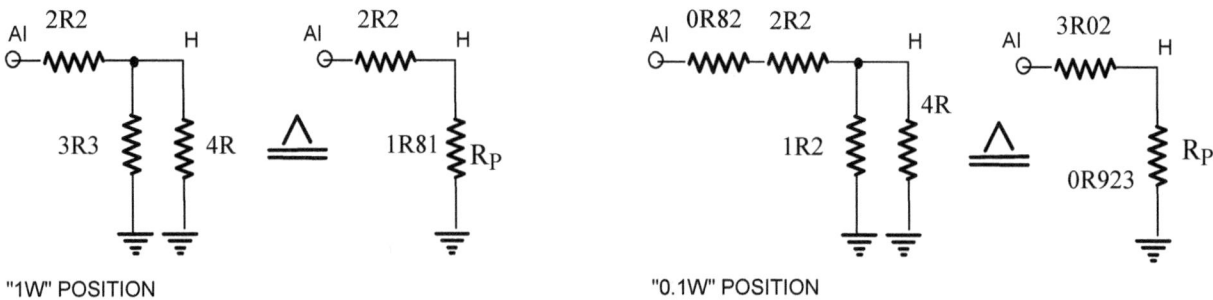

"1W" POSITION   "0.1W" POSITION

The output impedance of the attenuator and the amplifier

There are two types of output attenuators. The simple ones do not have a constant output impedance, while the more complex one keep the output impedance of the amplifier constant. If the output impedance of the amplifier isn't constant across various settings, the output attenuator switching will not only change the output power levels, but will also alter the interplay between the amplifier and the speaker, thus changing the tone and the sound of the amp-speaker combination. Does Bugera's output attenuator keep the amplifier's impedance constant? Let's find out!

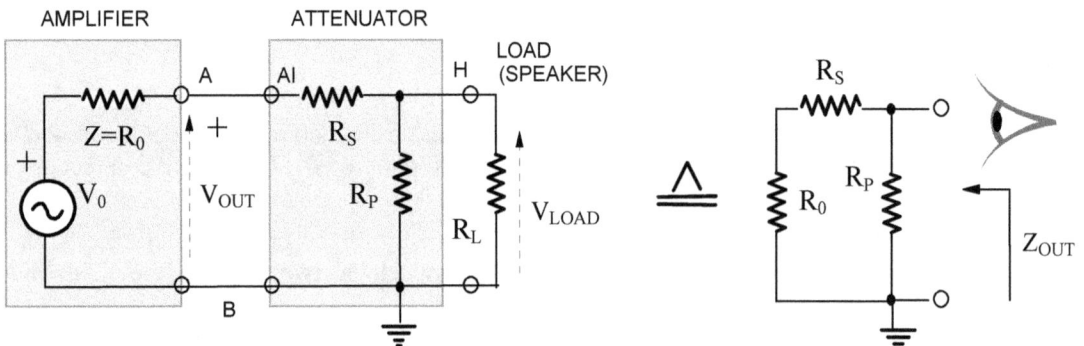

The Thevenin's equivalent circuit of an audio amplifier   The impedance that the load (loudspeaker) "sees"

A whole amplifier can be depicted as a series combination of a signal source V_0 and the output impedance Z. We will assume that such impedance is purely resistive and call it R_0 in further analysis. This simplification does not change the facts or the conclusions; it only simplifies the algebra.

If we disconnect the speaker load, so terminals H and B are open and look "inside" the amp, we will get the output impedance of the amplifier/attenuator combination, as the speaker "sees." Remember, the ideal voltage source V_0 has no internal resistance, so in calculating the output resistance, we replace it with a short circuit and get R_0 in series with RS, which is then in parallel with R_P.

When we measure the resistance across the amp's output, we get $R_0 = 0.6$ Ω. Again, let's assume that is the amp's impedance and that it is constant with frequency. That assumption is not strictly speaking correct, but it is true in the first approximation. With "1W" attenuator in the circuit, $R_S = 2.2$ Ω and $R_P = 3.3$ Ω, so $Z_{OUT} = (0.6+2.2)||3.3 = 2.8*3.3/(2.8+3.3) = 1.515$ Ω.

In "0.1W" position R_S=3.02Ω and R_P=1.2Ω and we have Z_{OUT}=(0.6+3.02)||1.2=0.9Ω. Therefore, when used as an output attenuator, as in Bugera's case, the simple voltage divider does NOT keep the amplifier's output impedance constant!

The headphone circuit

Now, let's redraw the headphone interface circuit, but this time simplified, without the 1/4" jack and with headphones plugged in. Remember, the top c/o (changeover) contact T moves up, and the bottom c/o contact R moves down.

Those of you more experienced in this kind of mental gymnastics will be able to do it in one go, but this is an educational book, so we will split it into two steps here. We'll keep the relevant terminals and retain the position of the components in the first step, and then "clean it up" by repositioning the circuit elements in step two.

It is now evident that the signal from the secondary of the output transformer goes through a resistive voltage divider first, comprised of the two 47W resistors, then through the low-pass LC filter, and finally through the series 220 resistor into the headphones. This resistor forms another voltage divider with the headphone impedance. Assume low impedance dynamic headphones of 30 ohms impedance. Since 30/(220+30) = 0.12, only 12% of the signal will reach the headphones; the rest will be wasted on the series resistor. Likewise, with, say, 600 ohms headphones, 600/(600+220) = 0.73, so in that case, almost 3/4 of the signal will reach the headphones.

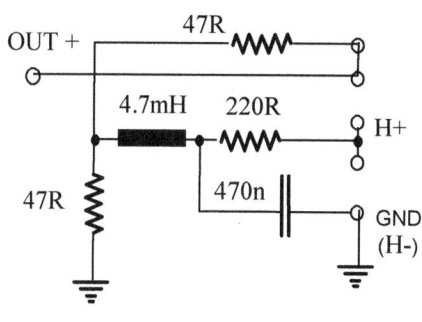

STEP 1: "Cleaned up" circuit

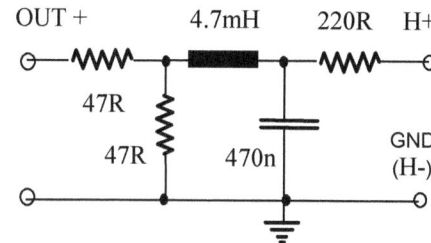

STEP 1: Redrawn circuit

So, the first issue is that the output power and loudness will significantly vary with headphones of various impedances.

Furthermore, the load on the OT secondary is now much higher than the nominal 4 ohms of a loudspeaker, so when reflected onto the primary, that will change the operation of the EL84 output stage and thus the sound!

Finally, the RLC filter formed by the 47R resistors and the LC filter is a resonant circuit with natural angular frequency $\omega_N=\sqrt{1/LC}$ or, in terms of frequency $f_N=1/2\pi\sqrt{(LC)}$, because $\omega_N=2\pi f_N$. So, with L=0.0047H and C=470*10^{-9}F, in this case we have $f_N=1/2\pi\sqrt{(LC)}$ = 3.39 kHz.

This means there will be a resonant peak around that frequency, although the resonant frequency will be slightly different from that natural frequency due to the impact of the source resistance and load resistance. In other words, headphones of different impedances will shift the resonant frequency up and down, and will also change the damping factor ξ (ksi), which can be calculated as $\xi=0.5R\sqrt{(LC)}$. Q is the quality factor of the resonant circuit, which is related to the damping factor Q=1/(2ξ), or $Q=\sqrt{(LC)}/R$.

Again, that means that upper midrange frequencies of between 3 and 3.5 kHz will be emphasized, while high frequencies will be significantly attenuated by the 2nd order filter. In essence, that is akin to boosting the midrange tone control and turning the treble tone control way down.

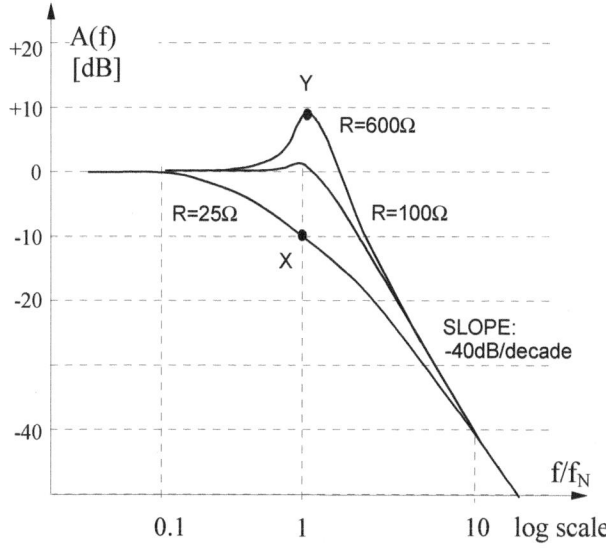

The attenuation or amplitude curves for RLC filter are shown on the left for three different values of R. In our case $\sqrt{(L/C)}$=100, so R_1=0.25*$\sqrt{(L/C)}$=25Ω, R_2= 1*$\sqrt{(L/C)}$=100Ω, and R_3=6*$\sqrt{(L/C)}$=600Ω. The higher the impedance of the headphones, the higher the peak. The lower the impedance of the headphones, the more this circuit will attenuate mid and high frequencies, and the earlier such attenuation will start!

For instance, with R=25Ω at f/f_N=1 (meaning at the frequency of 3.39kHz), the signal is attenuated 10dB (point X), which is a significant cut. Remember, the - (minus) sign means attenuation, the + sign means amplification or boost.

So, with R=600Ω at f/f_N=1, the signal is actually boosted by almost 10 decibels (point Y).

Ibanez Tubescreamer TSA5TVR-S

Let's look at the headphones and line-out circuitry of a similar small combo amp, Ibanez Tubescreamer TSA5TVR-S (analyzed in Volume 1 of this book). To trace the circuit on the power PCB, we unbolted it from the chassis, as we did with the "output" PCB. Since the power PCB is not flat on the copper side (due to the tube sockets protruding), we could not scan it in the flatbed scanner. It would be too complex and time-consuming to use it for this example, anyway, so let's trace the output PCB and reverse-engineer it into a circuit diagram.

This approach can be used to draw a circuit diagram from any PCB-built amplifier, both tube and solid-state.

When you mark the soldered components on the scanned PCB together with their values, pay special attention to the fact that the component side and the copper side (shown here) are mirror images of one another. This means it is very easy to make a mistake with pin numbers of semiconductor components such as transistors, voltage regulators, and integrated circuits.

There are two connections coming from the secondary of the output transformer, TX+ (blue wire) and TX COM (black wire). The COM terminal is duplicated and another black wire comes out of it and connects it with GND terminal on the power PCB.

To protect the output transformer, without a speaker plugged in, the shorting contact connects a 8Ω 7Watt resistor as a dummy load across the output transformer's secondary winding, between TX+ and TX COM terminals.

The two small caps, 100n and 1n5 are in parallel and permanently connected across the speaker terminals.

The headphones (HP) jack is a stereo type, and since there is only a mono signal, both of its tips are strapped together. The PCB track also shorts the two shorting terminals to the same point.

The HP out signal is obtained from TX+ terminal through two RC stages, basically two low pass filters (33Ω+1µF to GND) each with 4.82 kHz upper -3dB frequency.

The line out signal is processed in a similar fashion, through two low pass RC filters, 1k+100nF and 1k+50nF, with -3dB frequencies of 1,592 Hz and 3,183 Hz respectively.

Notice that plugging in headphones does not disconnect the internal speaker, which is a very strange choice. The whole purpose of using headphones is to practice without disturbing others. You could unplug the speaker from the back, but then the 8Ω dummy loud would automatically be switched in and since the headphones have a much higher impedance (30-600Ω) very little sound would actually reach the headphones.

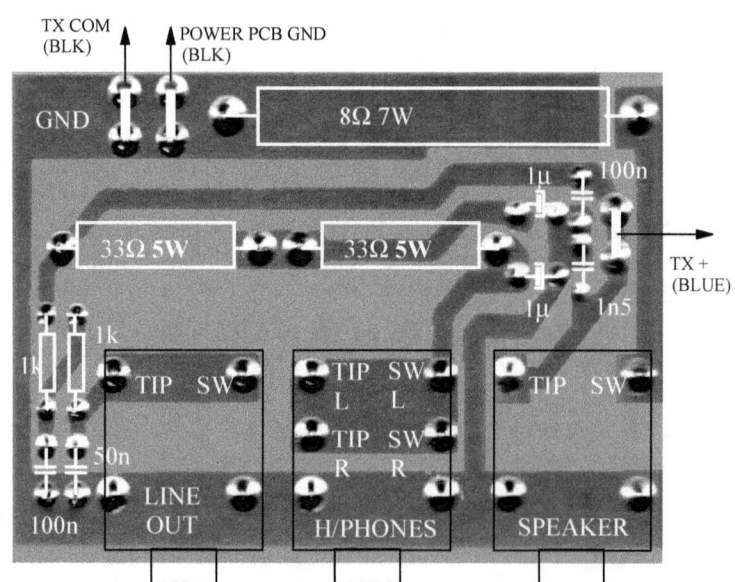

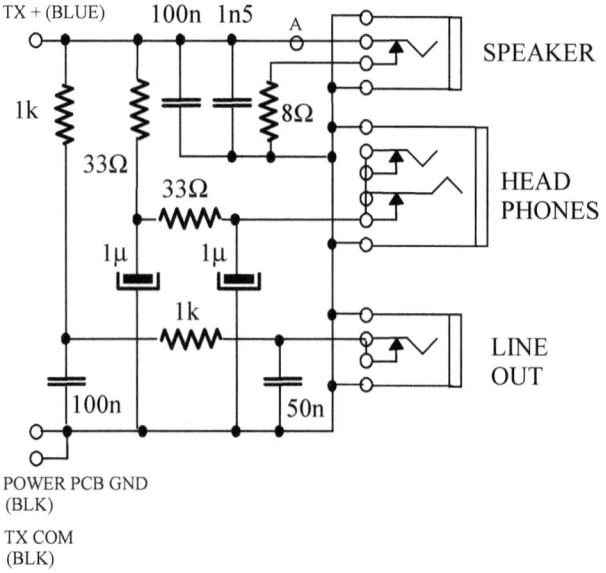

The scanned output PCB with marked components and terminals (ABOVE) and its schematics (BELOW).

TROUBLESHOOTING AND REPAIRING TUBE GUITAR AMPLIFIERS

- TYPES OF FAULTS AND COMMON CAUSES OF FAILURES
- FAULT LOCATION AND SIGNAL TRACING METHODS
- POWER SUPPLY TROUBLESHOOTING
- CAPACITOR TROUBLES
- VARIOUS COMMON SYMPTOMS AND MOST LIKELY CAUSES
- ESSENTIAL TOOLS AND TEST GEAR
- AC (SIGNAL) TESTS USING A VOLTMETER (MULTIMETER)
- TESTING & MATCHING TUBES FOR GUITAR AMPS
- CHECKING AND MATCHING TUBES WITHOUT A TUBE TESTER

If you understand the physical principles behind tubes and other electronic components and the operational principles of tube amplifiers, you possess about 80% of the knowledge required to successfully test, troubleshoot and fix tube guitar amps. The final 20% is the more specialized knowledge and know-how (the practical aspect of amplifier repair), and since those last 20% are so important, we devote a whole chapter to them!

TYPES OF FAULTS AND COMMON CAUSES OF FAILURES

The first step towards locating and fixing faults is the understanding of their causes and types. Some faults have a singular cause, others are caused by two or more contributing factors. Often one such factor would not be sufficient to cause trouble, but two factors together certainly do. In no particular order of importance or commonality, here are the major reasons amplifiers fail:

- POOR DESIGN: tubes operating beyond their maximum rated voltage, currents or power levels, capacitors' voltage ratings exceeded, leading to a breakdown in insulation, resistor power ratings exceeded, causing overheating and burning out, hum caused by poor topology or parts positioning.
- POOR WORKMANSHIP: contact problems, loose connections, cold solder joints
- COMPONENT FAILURE: aging, drifting values, burnout
- WEAR & TEAR + ROUGH HANDLING (vibration, rubbing, loose connections)
- IMPROPER PREVIOUS REPAIRS, MODIFICATIONS OR "IMPROVEMENTS": it is always faster, easier, and cheaper to fix amplifiers in their original vintage condition than those butchered by ignorant "experts"!
- DAMAGE IN TRANSPORT

The prime suspects amongst the components

As a "rule of thumb," the main suspects among the components are tubes and capacitors, especially electrolytic capacitors, which age and dry out, becoming leaky and lose capacitance.

Due to their nature (moving parts and poor contacts due to dirt, grime, and oxidization), switches are also a common cause of problems. Many current made-in-China tube sockets are of poor design and/or construction. Socket pins lose contact with tube pins, break off or even fall out (!) Corrosion and oxidization seriously affect even the best sockets of yesteryears.

While checking DC voltages or AC signals, if you get intermittent readings, consider replacing a suspect tube socket straight away; otherwise, you could be wasting lots of time trying to find the fault around the socket connections.

Resistors and transformers are less likely to be the source of problems. Old carbon composition resistors (the molded type) drift significantly in value, causing improper operation of the circuits that require precise matching of resistors (for instance, phase splitters). However, the title of the Prime Suspect goes to wires, sockets, and contacts. Let's look at them in more detail.

Contact issues and intermittent troubles

Intermittent faults are the most frustrating of all. They suddenly appear for a while, a moment, a few seconds or minutes, and then all is well again. That unpredictability makes them difficult to trace and locate.

The cause behind all intermittent faults is contact loss between two points in the circuit that should be connected or unwanted contact or short between points that should not be joined together!

Quite a few factors can cause the loss of contact. Bad or "cold" solder joints are probably the most common issue, especially in poorly constructed DIY equipment and vintage kits. Although a joint of two or more wires passes the visual inspection, it is uncertain what is happening inside it - mechanical stress or corrosion could have caused a break in a wire or a fraction in the solder joint itself.

If in doubt, re-solder all the joints in the affected area, for instance, around a tube stage that does not work correctly. This is often faster than signal troubleshooting.

In amplifiers that use printed circuit boards, especially double-sided, there could be a poor solder joint or a break in the metal rivet that joins two sides via the plated-through hole. A quick ohmmeter check between two copper layers would confirm such an issue, which is illustrated in a).

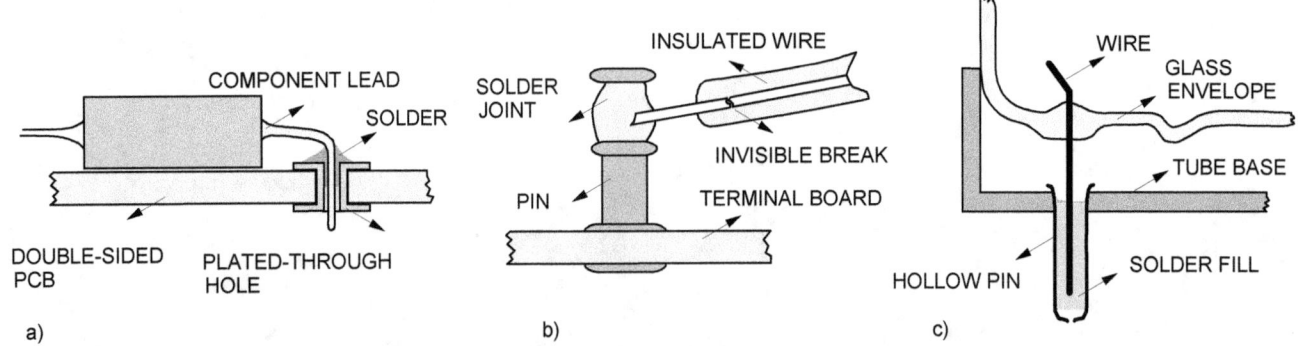

a) b) c)

Illustration b) shows a situation that also passes a purely visual inspection. The hookup wire has an invisible break under the insulation. That is the reason why visual inspection should always be performed together with the mechanical check. Use a small insulated screwdriver, tweezers, or a dental probe to jerk wires and component leads around. 3-piece dental tool kits are cheap, around $5-6 online, including a probe, tweezers, and a mirror. A mirror and a magnifying glass are very valuable to see what is happening in hard-to-reach places!

This issue usually goes back to when wire-strippers or cutters were used to strip the insulation off the wire before soldering. If too much force was used, the wire would be partially cut-through and, if subjected to vibration or mechanical stress, can break off days or even years later.

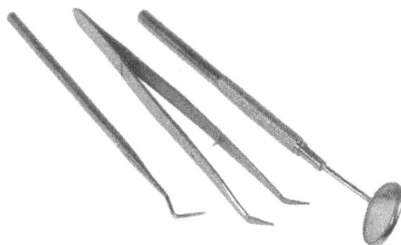

Some amplifiers use Snap-On and other types of connectors. These should always be on your list of suspects if a connection problem develops. A contact defect can also happen inside a tube. If a loss of contact occurs inside a glass envelope, nothing can be done, and such a tube should be discarded. A common problem, illustrated in c), is poor solder joints inside the tube pins, especially in the current Chinese-produced tubes.

These cheap basic dental kits are useful in identifying "cavities" in your solder joints and other contact problems.

In octal and other larger tube sockets, the wires that bring tube elements out to the base must be soldered inside the hollow pins.

Due to mechanical stresses and thermal expansion and contraction, a poorly soldered contact will soon be lost. So, if a tube tests "bad" or dead, don't throw it away in haste; re-solder all the pins first and then test it again. It could be just a poor contact, and a tube could be perfectly healthy otherwise.

Many currently manufactured tube sockets suffer from contact problems. The metal contacts are often poorly designed and make contact only in one or two points (instead of enveloping the tube pin around its whole perimeter). Some of the materials used are very brittle, so the contacts deform or even break off after a dozen or so plugging and unplugging of the tube.

> **HOW TO PROPERLY PULL TUBES OUT OF THEIR SOCKETS**
>
> When pulling tubes out of their usually very tight sockets, be extremely careful. Never jerk a tube sideways. This can bend or break delicate pins and loosen the socket contacts. Some tubes have a bottom glass pinch, which can be snapped off, and even the glass envelope can be cracked due to excessive force. Both situations would render a tube useless. Always pull the tube straight up, with minimal lateral movement!

Another kind of contact problem is where two points in a circuit should not be joined together but are short-circuited due to a mechanical, thermal, or construction fault. It is very easy to make a soldering mistake on smaller (7- and 9-pin) tube sockets or when soldering components onto a printed circuit board. Too much solder can make a short circuit between two PCB tracks or tube socket lugs.

Another cause of shorts during the construction or repair phase is loose metal objects, such as blobs of solder or wire or component leg cutoffs which got stuck inside the amplifier and were not noticed during the cleanup. When you finish the soldering stage, inevitably with an amplifier upside down, turn it the right way up and shake it gently so that the loose particles drop out. The ones that got stuck inside have to be found and removed manually. The probing tool is again invaluable here.

A short can develop during the postage or transport of vacuum tubes due to shock, excessive vibration or sagging of the grids. That is why you should always test tubes after transport before you plug them into an amplifier! Permanent or intermittent internal shorts can happen during the tube's operation due to small mechanical tolerances or overheating, which would cause thermal expansion of the tube's elements.

FAULT LOCATION AND SIGNAL TRACING METHODS

Checking order

The checking order is ALWAYS the same. Skipping one or more steps increases the chance of the fuse blowing or the amplifier going up in smoke, resulting in expensive damage to one of the transformers or power tubes!

1. VISUAL INSPECTION: Inspect the internals for any overheating, burned-out, missing, broken, loose or disconnected parts, links, and components. Identify non-original parts and previous repairs and/or modifications.

2. MECHANICAL CHECKS: Using a dental probe or a similar tool, jerk wires and component leads and check soldered connections for any loose or cold joints, broken or detached wires, and components, anything that could not be identified by a visual inspection only.

3. COLD & COMPONENT CHECKS: With an ohmmeter and LCR meter, check connections, continuity in various circuits (grid, cathode, anode), and components (inductors, capacitors, resistors, chokes, transformers) without powering the amplifier up.

4. HOT CHECKS: Check AC and DC voltages in critical points, such as the heater, cathode, screen, and anode voltages.

5. SIGNAL CHECKS: Measure hum, noise, distortion, output power. Study the waveforms on an oscilloscope.

Fault location methods

SUBSTITUTION: The most commonly used method, insert a known good tube or replace a component with a new one. Of course, this approach does not fix secondary faults since the replacement component will also fail if the cause of the fault is somewhere else.

COMPARISON: Used when one channel or an identical stage is working well to compare its DC and AC voltages with the faulty channel or stage.

SEPARATION: Break the signal chain by pulling out a tube, disconnecting loads from a transformer, or disconnecting a coupling between two stages. The aim is to isolate and identify the offending stage.

BRIDGING & BYPASSING: Bridge the suspect wire, connection, or component or bypass the whole stage and see if anything changes.

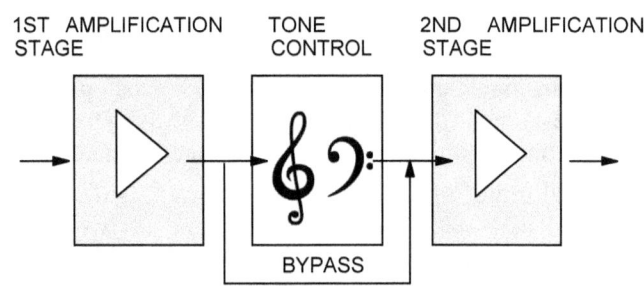

ABOVE: Bypassing a resistor that is suspected to be burned out or an elco that may have lost its capacitance

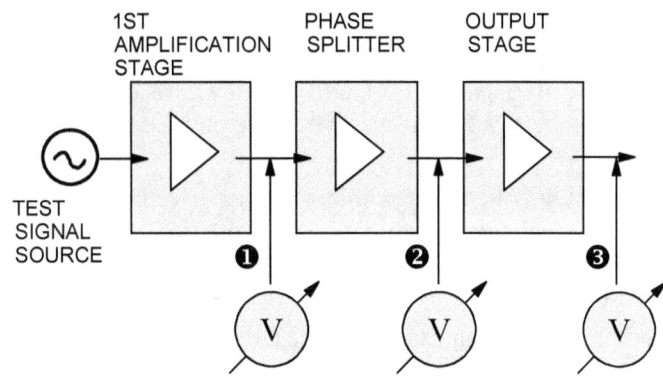

ABOVE: Bypassing a tone control stage, which often creates trouble due to a high number of capacitors and two or more potentiometers!

Signal tracing options

There are two ways to signal trace an amp or preamp. In FORWARD TRACING, the audio oscillator is set to the required input voltage and left connected to the input. The input to the test instrument (voltmeter, oscilloscope, or signal tracer) is moved from the output of the first stage down the signal path.

In BACKWARD TRACING, the test instrument is connected to the amp or preamp output, set for a required (expected) voltage, and not touched again. Instead, the function generator's output is constantly adjusted for the required voltage and moved from the input of the final stage back towards the input.

In most cases, audio oscillators cannot provide high enough voltage to drive the output stage of amplifiers, so for that reason alone, forward tracing is more practical. Backward tracing can be used with preamplifiers, though.

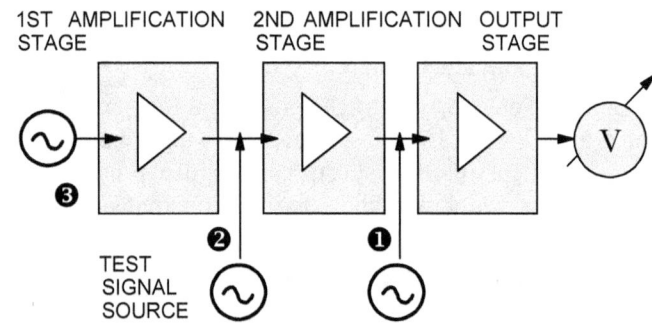

RIGHT: Forward and backward signal tracing.

POWER SUPPLY TROUBLESHOOTING

Powering up

Once you have completed the visual and mechanical inspection and performed checks of all the major components (transformers, chokes, tubes, coupling, and filtering capacitors), you are ready to power the amplifier up.

Variac power-up. Bring up the voltages gradually up with a variable autotransformer. This would give the power supply electrolytic capacitors some time to reform the electrolyte. Most autotransformers can exceed the rated mains voltage; use that test to see if an amplifier will cope with wide power fluctuations and if the power supply capacitors have enough voltage margin.

Light-bulb power-up. Wire a 40-60W incandescent light bulb in series with the amp's AC cord. Fluorescent, LED or halogen types are not suitable. Turn the amp on. If there are no shorts in the power supply, you should see the globe flash briefly to full brightness and then go down to dimly lit. If the lamp stays at full brightness, there is a short circuit within the amplifier. By wiring the lamp in series with the amp's AC supply, you limit the current that can flow through the circuit to safe levels and prevent any damage if a short is present.

Staged power-up. Remove all tubes except the tube rectifier. Disconnect high voltage after the CRC or CLC filtering. Power the amplifier up. If the fuse does not blow, measure power supply DC voltages. They should be higher than nominal due to the lack of anode loads.

Plug preamp tubes in and turn the amp back on. If the fuse does not blow, measure DC voltages on grids, plates, and cathodes of preamp tubes. If within +/- 10% of nominal, turn the amp off again and plug the power tubes in. Turn the amp back on. If the fuse does not blow and power tubes' anodes do not glow red, measure DC voltages at main points again.

ABOVE: Variacs of 500W-1kW power rating can be used to gradually increase the supply voltage to a suspect amplifier.

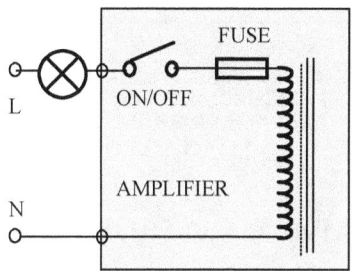

ABOVE: The globe acts as a load in series with mains transformer's primary winding. if the primary has shorted, the globe will glow brightly.

Replacing selenium rectifiers

Selenium rectifiers age, regardless of their use. So, even NOS (New Old Stock) rectifiers sitting on a shelf for forty to sixty years have aged and should not be used as replacements. Gradually, their forward voltage drop increases, and the reverse leakage current increases. Their DC rating reduces as they age; they run hotter and may even catch fire. The smoke is highly toxic.

This low current selenium stack in the amplifier (1) was used to provide the negative bias voltage for the output 7189 tubes. The bigger the size of the plates, the larger the current capacity of the selenium rectifier. The more plates in a stack, the higher the rated voltage of the selenium stack.

Each plate can take about 30V, so by counting the number of plates you can determine the voltage rating of the selenium diode!

When you replace selenium rectifiers with silicon diodes, the DC voltage will increase by 5 to 20 Volts. The forward voltage drop of silicon diodes is much lower, only around 0.6 Volts.

This increase may be useful or may present a problem, depending on the circuit and the application. In the bias circuit mentioned, the new voltage will be higher (more negative), and the anode current through the output tubes will be lower, so you may have to re-bias the output stage.

If such a voltage rise is unwanted, add a suitably sized (in terms of power rating and resistance) resistor in series with the silicon diode(s).

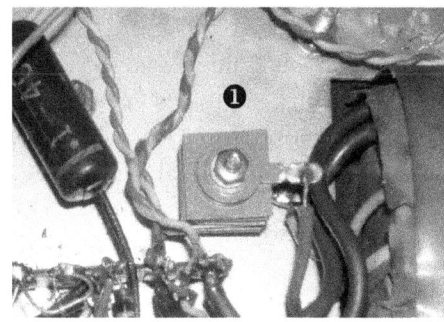

ABOVE: Selenium rectifiers used in vintage guitar amps should be replaced even if working.

Symptom: the mains fuse blows

When a fuse blows once, there is nothing to be concerned about. Fuses age and deteriorate, so a slightly higher inrush current on power-up may be enough to melt their inner wire. However, if a replacement fuse blows, something is causing the higher than normal current draw.

If output tubes are unplugged, that effectively isolates the output stage. Once you power the amp up without the power tubes, if the fuse does not blow, the fault was either in one or more power tubes (internal short-circuit) or the related components (cathode resistors, bypass capacitors, or output transformers). If the fuse still blows, the problem is most likely in the power supply.

By unplugging a tube rectifier, the filtering circuit is disconnected from the mains transformer, and if the fuse still blows, the fault is almost certainly in the power transformer, which needs to be thoroughly checked and most likely replaced. A shorted winding section (between two winding layers) and often even a few shorted turns may be enough to increase its no-load current to very high levels, enough to melt the primary fuse.

If the fuse does not blow after unplugging the rectifier tube(s), check the filtering components: elcos, chokes, and resistors. If all are OK and there is no short circuit between the positive and negative (ground) rails, check and/or replace the rectifier tube since the fault will almost certainly be due to a faulty rectifier tube.

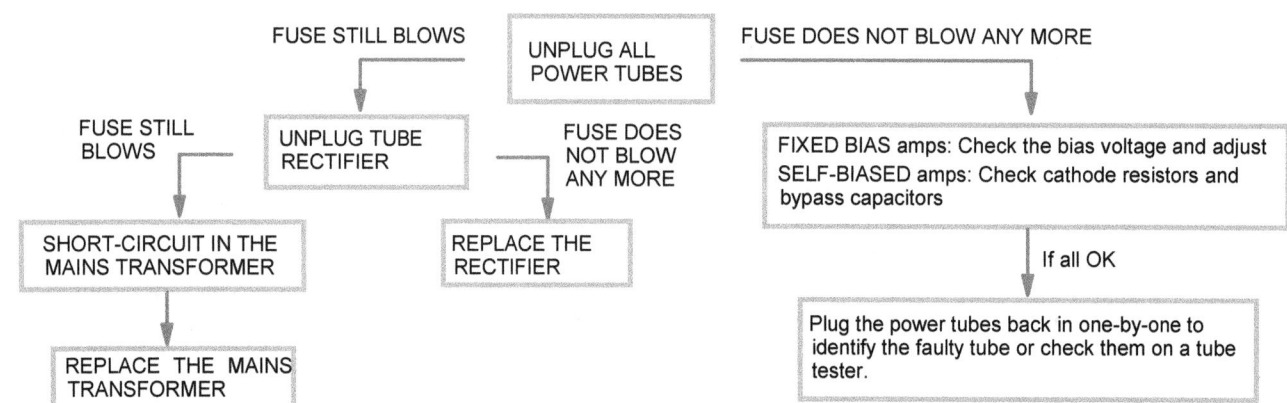

The troubleshooting chart for a very common symptom, an amplifier repeatedly blowing a fuse

CAPACITOR TROUBLES

Capacitor leakage is a common cause of all sorts of problems, from improper DC conditions (bias) to distortion and even tube destruction. When a cathode bypass capacitor becomes leaky (allows a small DC current to flow), the consequences are not serious, and the condition may not be discovered at all. Say the leakage resistance is 300kΩ. Together with the 2k2 cathode resistor (V2 in the circuit below) it will form a new parallel combination of 2k2*300k/(2k2+300k) = 2.184kΩ As a result, the cathode DC bias of V21 will not change at all.

What about the same leakage resistance of 300kΩ in parallel with the coupling capacitor CC?

The VB DC voltage of 250V will be divided according to the voltage divider formed by the anode resistor of 47kΩ, the leakage resistance of 300kΩ, and the grid resistor of 330kΩ. Even without the exact calculation, since 47k+300k is roughly equal to 330k, you can conclude that the DC voltage in point X (the grid of V2) will be slightly less than half of the 250V supply voltage, around 120V!

The nominal cathode voltage is 4.4V, so V2 will be so positively biased that a full anode current will flow through it. The anode will overheat, and the tube will be destroyed.

Even a relatively small leakage, the equivalent of 3MW leakage resistance, would result in the positive grid bias voltage of $V_G=250*330/(330+47+3,000) = 250*0.1 = 25$ V, which is still more than enough to destroy V2.

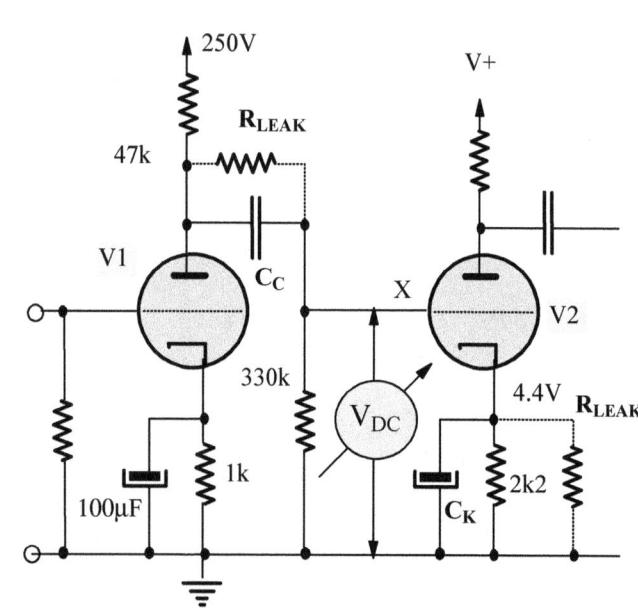

In-situ testing of coupling capacitors

Testing a D-factor on digital LCR meters is not conclusive because such meters use a very low test voltage (less than 9 Volts). The real test is under the operating voltage in an amp. One way to do that is illustrated above.

Before powering up a suspect amplifier, unsolder the coupling capacitor C_C in the point marked X (the grid of V2), then power the amp up without any signal. If you don't want to do unsoldering, the other option is to pull the V2 out; R_{G2} (330k) can stay in the circuit.

Use a good quality FET voltmeter or VTVM (vacuum-tube voltmeter) with high input impedance and measure DC voltage between point X and ground if you unplugged V2 or between the unsoldered end of C_C and ground if you unsoldered it. You should get no DC voltage at all; the higher the measured voltage, the leakier the capacitor is!

VARIOUS COMMON SYMPTOMS AND MOST LIKELY CAUSES

High frequency oscillations

In contrast with low-frequency oscillations (motorboating), HF oscillations are not always obvious. The human ear cannot detect ultrasonic instability (above 20,000 Hz), but instruments can.

Method #1: High-frequency oscillations are indicated on an oscilloscope by widening or smudging the sine wave (signal voltage) at the output of an amplifier.

Method #2: Too much negative feedback can cause instability and oscillation. Sustaining such oscillations (no matter if they are of the low- or high-frequency kind) requires energy from the amplifier's power supply. Hence, the presence of oscillations increases the total power draw from the amp's power supply.

Turn the amplifier off. Insert an mA meter (DC) in the high voltage supply CT (Center Tap). Turn the amplifier on. Without any signal at the input, measure the mA current draw. Momentarily disconnect the feedback loop. Compare the mA reading. If the mA reading goes down, the amplifier oscillated with the NFB applied.

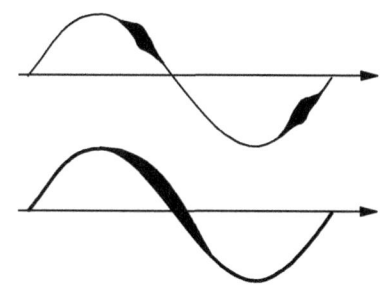

ABOVE: The "thickening", spreading or fuzziness of the oscilloscope waveform indicates HF oscillation and/or noise

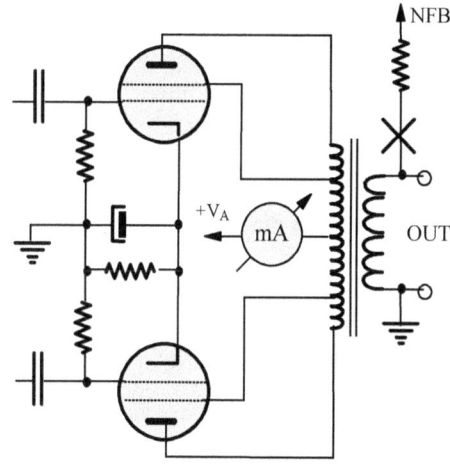

ABOVE: NFB can cause HF oscillations, which can be detected by disconnecting the NFB loop and observing any drop in anode current draw.

Installing grid-stopper resistors

While amplifying audio-band signals is desired, we don't want guitar amps to amplify radio frequency signals or oscillate at such high frequencies. The task of grid-stopper and anode-stopper resistors is to prevent such oscillations from taking place.

Their action rests on simple filtering of high frequencies and their diversion to ground. The input capacitance of a tube C_{GK}, together with the grid-stopper R_G forms a 1st order low pass filter with the -3dB frequency of $f_U = 1/(2\pi R_G C_{GK})$

Assuming 40pF grid-cathode capacitance and a 2k grid-stopper, the upper -3dB frequency would be around 2MHz.

However, increase the grid-stopper's resistance to 100kΩ, and the upper -3dB frequency of thus formed low-pass filter drops down to 40kHz. By increasing the value of the grid stopper resistor or by adding a low-value capacitor between the grid and ground you can deliberately limit the bandwidth of an amplifier should you wish to do so.

Grid stopper resistors should be placed as close as possible to their grids. Since that is not obvious from a circuit diagram, where the line between the resistor symbol and the socket pin can be interpreted as a length of wire, sometimes a dot is placed on the grid to act as a reminder to the constructor to solder it *directly to the tube socket pin*.

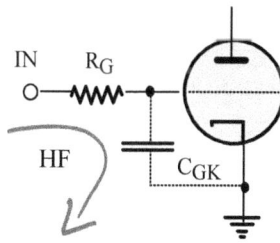

ABOVE: The low-pass filter formed by the grid stopper R_G and tube's input capacitance diverts interference and HF noise to ground.
BELOW: Solder grid stopper resistors directly to their tube socket pin!

Reducing RF interference with small ferrite rings

A very effective RF interference measure is to wrap shielded cables and ordinary hookup wires a few times around a small ferrite ring or bead. This creates a tightly-coupled in-line transformer. Adding series inductance increases impedance for high radio frequencies (common-mode signal) without attenuating the lower frequencies (differential audio signal). The impedance of most ferrites is limited to around 100W, making them most effective in the low impedance line, such as power supplies.

DC-DC converters, computer power supplies, and most other modern electronic gear have this measure implemented, so ferrite rings can be salvaged from such devices and reused.

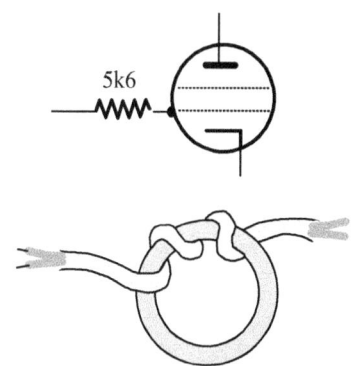

ABOVE: Wrapping the offending section of hookup wire around a small ferrite ring is often enough to prevent radio frequency interference or high-frequency oscillations in a guitar amp

Plates (anodes) of power tube(s) glowing red

Improper bias (control grid DC voltage not negative enough, say -5 V instead of -15 V) or a total loss of bias will cause power tubes to draw too much DC current. This will, in turn, cause anodes (plates) to overheat. This glow can be localized in one or two spots, or, in severe cases, the whole metal plate structure may glow bright red.

When you power up a newly bought, just constructed or repaired amplifier for the first time, even if everything seems fine (no burning smell, no crackling sounds), switch the lights off or darken the room and observe the power tubes in the dark. If there is no plate glow, go back to your voltage checks. If ordinarily black or gray anodes (plates) start glowing, turn the amp off immediately and check the biasing circuit.

Blue or purple glow inside a tube

The blue-purple glow located around the inside surface of a glass bulb is called fluorescence and is more common in power tubes. It results from electrons hitting the glass and has no adverse effect on the tube's operation. However, if the glow is of lower intensity and located inside the tube's metal structure, uniformly distributed between the cathode and the anode (plate), it indicates a gassy tube.

High energy electrons moving from cathode to anode hit gas molecules and turn them into positive ions, producing a glow. The positive ions, much heavier than electrons, are attracted by the "negative" cathode (relative to the anode at high positive potential) and hit the cathode's surface, causing "pitting" and wear out the oxide layer on the cathode's surface. Since that oxide layer supplies electrons for the electron cloud, with time, the emissive capability of the cathode deteriorates, and the gassy tube wears out prematurely.

Milky white coating inside a tube

This cramped internal layout (photo on the right) is from a faulty Fender Champ amp. The amp powered up but produced no sound at all.

The problem is immediately apparent. While the top getter of the 5Y3 rectifier is dark and shiny, even reflective (1), as it should be, the top getter of the 6V6 output tube is milky white (2), a sure sign that the tube has lost its vacuum and must be replaced.

RIGHT: Milky white coating on the tube's top or sides (depending on its getter location) means the getter reacted with outside air and turned white. The tube has lost its vacuum and is destroyed.

Low frequency oscillations (motorboating)

In the chapter on power supplies, we discussed the need for proper decoupling of amplification stages to minimize the effect of common power supply impedance, which can propagate unwanted signals between stages and provide negative feedback to some and positive feedback to others. If there is enough gain in the circuit, this positive feedback can cause high- or low-frequency oscillations.

The low-frequency instability usually happens a few times a second and sounds like an outboard boat engine, put-put-put, giving the phenomenon its popular name, "motorboating". The cause is almost always in the power supply. The original electrolytic capacitors, usually of low values (typ. 22-33mF) have dried out, and even the meager filtering they provided is now nonexistent. This is why the first symptom in these amps is usually a loud hum due to a huge AC ripple on DC lines.

Another consequence is a significant increase in the power supply's internal resistance. This elevated internal resistance is causing the motorboating, so replacing all elcos with higher values and lower ESR modern units will cure the instability.

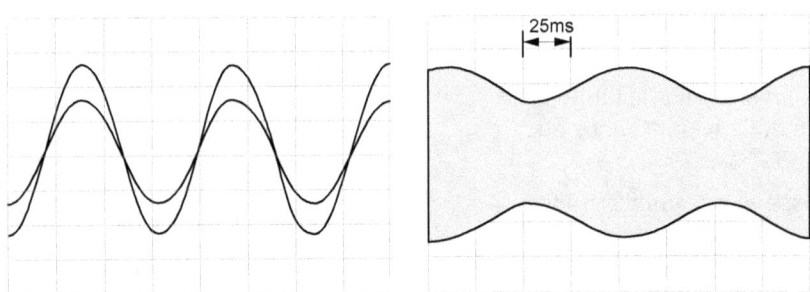

FAR LEFT: The amplitude of 1 kHz sine wave at the output of an amplifier changes a few times a second

LEFT: Reducing the time base of the oscilloscope reveals an amplitude- modulated 1-kHz signal. The LF modulating signal has a period of 5*25ms = 125ms, so its frequency is 8Hz!

Hum: Replacing a shorted choke (Fender PA100)

The filtering choke that supplies screen grids of the output tubes and preamp/driver stages in the Fender PA100 amp is a tiny affair. When we measured its resistance, we got only 7.7W, which seemed way too low (it should be 100-300 W). Likewise, an LCR meter indicated an inductance of only 0.18mH, practically zero, so both tests pointed out a partially shorted winding. Since there was no choke filtering action of any kind, that explained the presence of a slight hum on the amp's output.

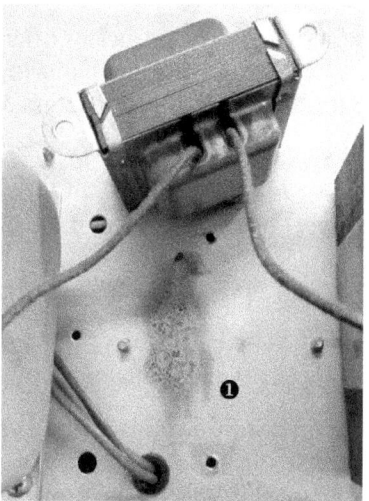

Once removed from the chassis, melted varnish traces were visible (1). There was a short circuit downstream, most likely one of the output tubes' screen grids shorted and was drawing an excessive current. The current wasn't high enough to fuse (melt) the wire the choke was wound with (thinner screen grid wire fused first), but it cooked the choke, so the varnish leaked out. Since varnish is the only insulation around the magnet wire, the now bare winding wires inside the choke fused and created a virtual short circuit.

We had a spare choke rated at 70mA, larger in both the physical size and inductance, around 7H. Although no data was specified for the original Fender choke, either on the circuit diagram or on the choke itself, our best guess is an inductance of only 1-2H. The wire was quite thick so very few turns could fit on such a small core, and since inductance varies with the square of the number of turns, it must have been very low.

The baked-in dark brown stains (1) on the chassis are a common sight in vintage amps. They were caused by boiling and leaking varnish (lacquer) around magnet wires, indicating an overheating transformer or choke.

The amp's chassis and even knobs are too hot to touch: Crate Palomino V32

Crate Palomino V32 owners complain that the amp gets very hot; even control knobs become so hot they are almost too hot to touch. This is a sure sign of either a poor electronic design or poor ventilation (mechanical design and construction). In any case, it is not a good sign.

V32's circuit diagram shows a common cathode resistor of 60Ω (10Watts) and a cathode voltage of +10.0V_{DC}. There are four EL84 pentodes in that output stage. The total cathode current is 10V/60Ω = 167 mA or 41.6mA per tube. The power dissipated into heat on that resistor is 10*0.167 = 16.7 Watts, yet it's rated at only 10 Watts! To add insult to injury, it is soldered onto the amp's circuit board and placed right next to its bypass elco. That elco will have a short and stressful life. No wonder Crate amps suffer from bad reputation, at least when reliability is concerned.

At least two 120Ω 10 Watt rated resistors must be used (in parallel, to get 60W total), and they must be bolted onto the metal chassis, so the chassis acts as a heatsink. Even 20 Watt resistors dissipating 16.7 Watts will be stressed out, so a better option would be to use three 180Ω 10 Watt resistors in parallel!

Alternatively, add a small 10V_{DC} power supply as a fixed bias supply for the output stage. All cathode resistors and their losses (heat) will be gone!

Amplifier "eats' tubes (tubes last a very short time, weeks and months instead of years)

If an amp "eats" or "chews" through tubes very quickly (meaning tubes' life is short), there are two possible causes; either the heater voltage inside the amp is way too high (or too low), or tubes are working above their maximum voltage or power levels. The second case is more common with output or power tubes.

The first thing to check is heater voltage. Most commonly, it will be 6.3V, usually AC, although some amps use DC heating. It should not be higher than 6.5V or lower than 6.0V. Some designers underheat preamp tubes (5.9 -6.1V instead of 6.3V) to reduce hum and noise. That is OK. If the voltage is too high, lower it using the method we've described in this book's Joyo Sweet Baby case study.

2. Check the approximate anode power dissipation. Don't worry about screen current. To get the current in mA, measure the DC voltage on the cathode and divide it with the cathode resistor's resistance. Then measure the DC voltage between the anode and cathode and multiply it with the cathode current. Compare the result with the maximum allowed anode dissipation for your output tubes (from data sheets). If it's higher, as it is in too many amplifiers, reduce it by using one of the methods described.

However, changing the voltage and current through the output stage will change the amp's voicing and tone! If you love its tone and don't want to change it, be prepared to spend a small fortune on replacement tubes (and possibly on rebiasing the amp) so often.

Distortion

Paradoxically, when an amplifier is dead, it usually isn't difficult to find the culprit. On the other hand, distortion is one of the more difficult faults to troubleshoot, mainly because it is the symptom of a problem, not the primary fault or problem itself. Many factors can cause it:

- Leaky coupling capacitors
- Gassy tubes (grid current flowing)
- Over- or underbiasing of the output stage
- Low anode voltages
- DC or AC imbalance of push-pull driver and output stages
- Unmatched phase inverter or output tubes
- One faulty phase inverter or output tube, so a push-pull amp is working in a single-ended mode

The main cause of distortion is improper bias of one or more stages. This can happen due to one or more of the problems listed above. A leaky coupling capacitor will allow DC current to flow through the grid resistor to the ground. This voltage drop on RG will bias the following stage's grid positively, and that stage will distort. If a tube is pulling too much anode current, the voltage drop on the anode resistor will increase, and the anode voltage will drop, again causing the stage to distort.

A cathode resistor in a self-biased stage, especially in output stages, has high currents flowing through it. It gets hot, and due to those thermal stresses, its resistance may drift and change significantly over time. In turn, that would upset the bias of the stage and increase distortion.

DC imbalance as one possible cause of distortion

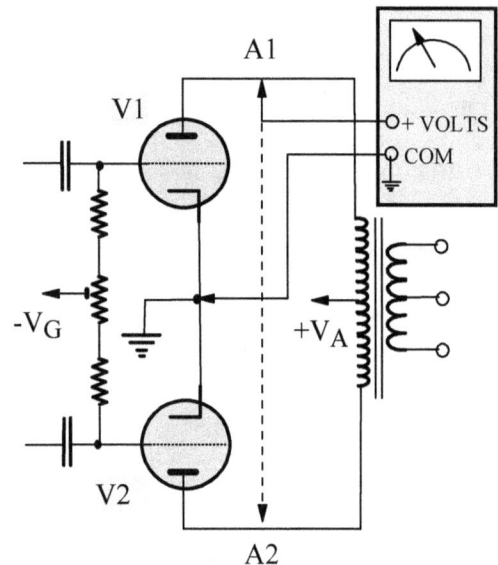

Depending on what type of voltmeter you have, you'll have to perform either one or two measurements to test a push-pull stage for DC balance.

Power the amp up with a speaker or dummy load connected, but bring no signal in. If neither of the terminals of your voltmeter is grounded (both are fully "floating"), as is the case with battery-powered digital multimeters, measure the voltage difference between the two anodes. Ideally, the voltage will be zero or very low. The higher the voltage measured, the greater the DC imbalance.

If the negative terminal of your voltmeter is earthed or grounded, measure the voltage between each anode and ground and subtract the lower voltage from the higher. Again, ideally, both voltages should be the same, and their difference should be zero.

WARNING: Make sure you observe the correct polarity! Connect the meter's + (RED lead) to the anode, and the - (BLACK lead), which is grounded, to the ground. Reversing their polarity will ground the anodes and create a dead short circuit across the amplifier's high voltage supply!

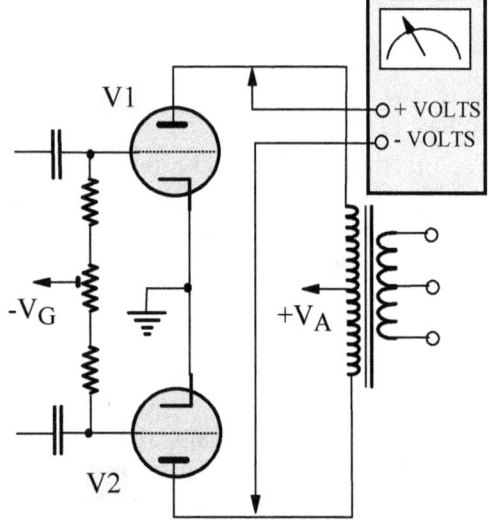

In most cases, you will detect some DC imbalance, even if you use matched tubes! The reason is that most output transformers use a faster and cheaper winding method for the primary (as mentioned earlier), where sections are wound on top of each other.

ABOVE RIGHT: If one terminal of a test instruments such as a voltmeter or oscilloscope is earthed (grounded), then two measurements must be performed, between A1 and GND and between A2 and GND.

RIGHT: One measurement is all it takes to detect DC imbalance between A1 and A2 if a floating voltmeter is used, such as a battery-powered digital multimeter.

ESSENTIAL TOOLS AND TEST GEAR

Soldering irons

Many DIY constructors ask me if they should buy an expensive Weller soldering station or a cheaper alternative? We have never had a Weller, so we cannot comment on their quality or longevity. This WER 927D soldering station (pictured) and similar budget stations served us well. You even get a set of interchangeable tips of various shapes and sizes, a convenient bonus (1).

The sponge (for cleaning the tip) is about 1mm thick, an utter joke (2), but everything else seems fine. The only irritating aspect is the fact that the plastic holder (3) is too light, so whenever you put the iron back into its "holster" (4), the whole thing falls off. A heavy metal holder would be much better. Alternatively, bolt it down to the top of the main unit (5).

For large or rough desoldering jobs for which 40-60 Watt regulated soldering stations would be too good or not powerful enough, you will need a high power soldering iron, 100-120 Watts.

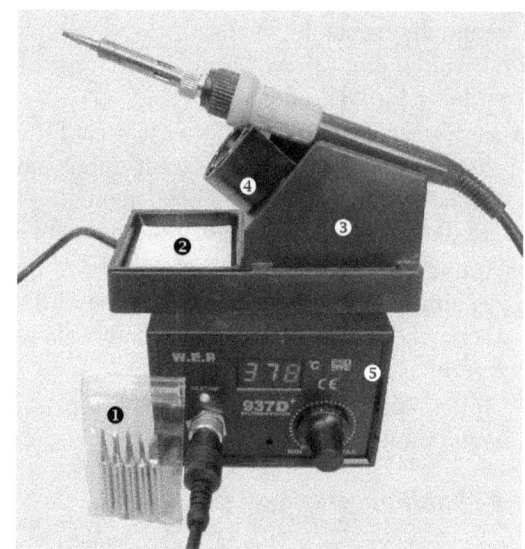

ABOVE: Temperature-regulated soldering iron
BELOW: High power soldering iron

One such job is unsoldering components from a metal chassis, which is such a large heatsink that even 80 Watts of heat would not be enough, and it would take a few minutes to get the solder hot enough to remove such components. Automotive parts retailers and craft shops also sell these high-power soldering irons since they are also used for leadlight soldering.

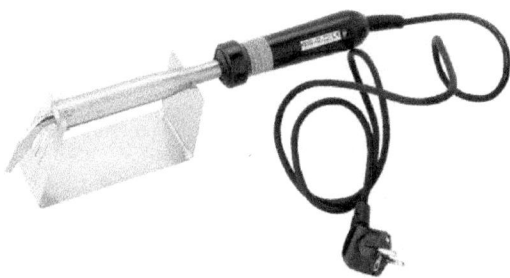

Test adapters

Opening an amplifier up and removing its chassis from the cabinet is often a tedious and risky job. Some combo amps have a timber "lip" or shelf on which the chassis sits, so even when you remove the bolts that hold it in place, the chassis cannot drop to the bottom of the cabinet and damage the tubes and the speaker in the process.

However, with most amps, once you remove the last bolt, you need to hold the chassis with the other hand to prevent such damage.

BELOW: These DIY test sockets are the most useful tool of all! Build one for all commonly used socket types (7-pin mini, Noval, Octal, Magnoval, RimLock, etc.) Pin 1 is marked with a plastic sticker.

These test sockets or similar adapters make it quick & easy to measure DC voltages and AC signals on tube pins without opening the amp up and exposing yourself to dangerously high voltages and hot tubes. Each pin on the tube socket is brought out to its own test point, so measuring the voltage between pins or each pin and GND is a breeze.

Plug the male adapter into the tube socket and then plug the tube into the jig; power the amp up, and you are ready to measure!

Multimeters

A multimeter is the most-used piece of test gear, so buy the best one you can afford. Most modern ones are digital and auto-ranging, and thus very easy and quick to use. However, many amplifier builders prefer analog meters, not for their romantic and old-fashioned "feel," but for the smooth behavior of the analog indicator.

For quick checks, it isn't even necessary to read the exact figures on the scale; one glance at the needle is enough to ascertain that the voltage is "in the ballpark," usually 1-5V for cathodes and 100-200V for anodes. That cannot be done with a digital readout.

Also, when testing potentiometers for tracking and smooth resistance change (to detect any sudden jumps or breaks), analog ohmmeters are the only way to do it! Plus, many digital meters take a very long time to settle; the digits annoyingly bounce around for a few seconds. So, test-drive a multimeter before you buy it.it.

Types of multimeters

- Passive analog non-RMS meters: very basic and cheap, manually-selectable ranges, low input resistance (affects the measured circuit), cannot measure frequencies above 1-3 kHz, not suitable for DIY audio.
- Active analog non-RMS meters, VTVM (Vacuum Tube Volt Meters) or FET meters: high input impedance, can measure frequencies way above 20 kHz, manually-selectable ranges, a good but old-fashioned choice, for die-hard analog fanatics.
- Active digital non-RMS meters: a wide range of quality and capability levels, from cheap and cheerful to so-so. No point buying them when quality True-RMS meters are more affordable than ever before.
- Active digital True-RMS meters: the best choice for the digital generation.

FET multimeters

FET multimeters replaced their VTVM predecessors in the early 1970s. The black case of this Micronta model makes it look dated, and it is dated; it hails from the eighties. Nevertheless, it is a good quality test instrument with a very large and precise mirrored scale. It can be used for audio work since its upper -3dB frequency limit is way above the 20kHz minimum required.

The JFET differential amplifier is at the heart of Micronta Dual FET analog multimeter

True-RMS digital multimeters

Escort's model EDM-89S is a typical example of the mid-late 1990s handheld digital multimeters. We have been using it for more than 20 years now, without any trouble at all. Let's look at its specifications:

- 5,000 count digital display + 53 segment analog bargraph
- 0.1% basic accuracy, True RMS, Auto/Manual ranging
- Sampling rate: Digital 3.3 times per second, Analog 20 times per second
- Selector: DC V, DC mV, AC V, µA, mA, A, Ω, diode check, audible continuity, frequency, capacitance
- V_{DC} high range up to 1.0kV, ACV 50Hz-20kHz bandwidth
- dBm measurement with 20 selectable impedances
- Frequency measurement 1Hz-10 MHz, with resolution of up to 10mHz
- Relative mode, Tolerance mode, ZOOM mode
- Dynamic recording with time stamp
- "Auto Power Off", "Sleep Mode", mA & A ranges overload fuse protection
- Capacitance measurement from 5nF to 50µF, with up to 1pF resolution

Digital LCR meters

A digital LCR (inductance-capacitance-resistance) meter is a "must-have" instrument for transformer and amplifier builders. The Taiwan-made Escort LCR meter and the identical Tenma model 72-960 (next page) are the mid-1990s technology.

An auto-ranging instrument, it measures parameters at two frequencies, 120Hz, and 1kHz. The 120Hz frequency was chosen since it's double the 60Hz mains frequency in the USA, and that is the frequency of the fully rectified AC ripple in a power supply. This LCR meter displays two values simultaneously, capacitance C and D-factor for capacitors and inductance L and Q-factors for coils and transformer windings. It also measures "resistance" on its R-range, but that is somewhat misleading. The "resistance" is measured with an AC signal, so it's the impedance modulus at the test frequency.

The B&K Precision LCR meter can also measure parameters at 10kHz, a handy feature when measuring transformer leakage inductance, which is of interest only at high frequencies.

Both have slots into which component leads can be inserted. Alternatively, you can use test leads terminated with crocodile clips, included with both instruments.

The Escort unit chews through 9V batteries very quickly. Luckily, there is a provision for an external DC supply connection to save you a small fortune on replacement batteries.

Function generators

A good multimeter is sufficient for measuring DC and AC voltages at important points within an amplifier. However, something has to provide a test signal of various waveforms, amplitudes, and frequencies, something a guitar cannot do. Such instruments are called function generators or oscillators.

General-purpose function generators aren't of a low distortion kind and, as such, aren't suitable for precise distortion measurements but are fine for daily testing and repair work.

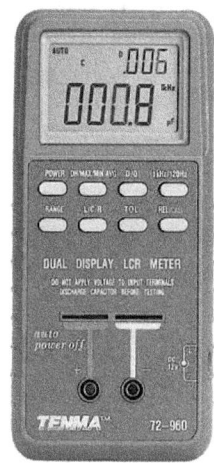

Other instruments: oscilloscopes, THD meters and IM (Intermodulation) distortion meters

An oscilloscope is one test instrument any serious constructor or fixer of tube amps should have and know how to use. However, an oscilloscope is such a complex instrument that it would take us far too much space to cover them here. That was done in Volume 2 of my book "Audiophile Vacuum Tube Amplifiers" and many specialized books dealing exclusively with scopes or CROs as they are also called (from "Cathode Ray Oscilloscopes").

Other instruments such as THD meters and IM (Intermodulation) distortion meters are also covered in the same book, but in guitar amp work, we don't usually measure distortion; the more, the merrier!

R-, C- and RC-substitution boxes

You can fix and build amps without substitution boxes; the same job can be done by swapping fixed resistors and capacitors and testing or listening to the amp you are repairing or building, but these small jigs make it all so much faster and easier.

Except for the two coupling capacitors at the output of phase inverters in push-pull amps, where you'd need two of these boxes, one box is enough to do voicing tests on guitar preamps. Change the coupling capacitor's value from, say, 10nF to 22n to 47n and listen to the changes in the amp's tone.

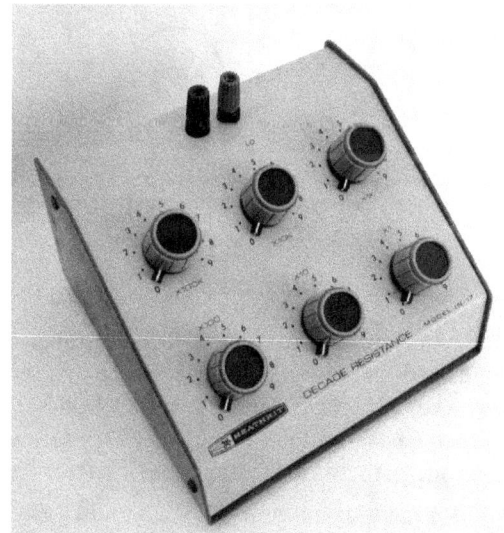

Laboratory decade resistors and capacitors can be very expensive, but there are still vintage units by Heathkit, Eico, and a few other budget test equipment makers of the 1950s and 60s.

Two 18-position switches to select one of the common resistor values, and a changeover switch to select one of the two main switched banks of resistors, and you have Eico's RTMA resistance box, model 1100. It is much cheaper to buy a used one than to make it yourself.

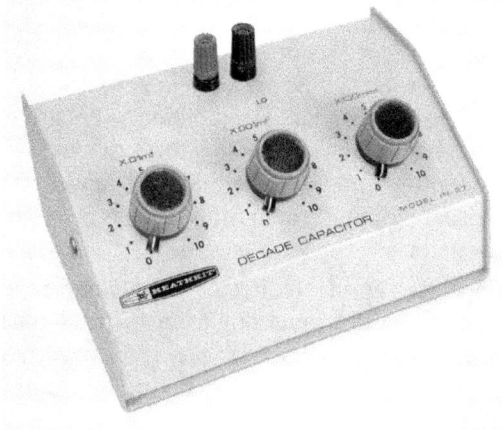

Heathkit IN-17 is a 6-decade resistance box with six 10-position rotary switches, marked X1, X10, X100, X1k, X10k and X100k. The markings denote the value multiplier, so when X1 switch is in position 7 that is 7Ω, when X100 switch is in position 4 that is 400Ω, and so on. The minimum value is 1Ω, and the maximum is 999,999Ω (for all intents and purposes 1MΩ). Ten 1% resistors are used per decade.

Heathkit IN-27 is a 3-decade capacitance box with three rotary switches, marked x.01mF, x.001mF, and x100mmF. Heathkit and many other American manufacturers used "m" improperly, it does not indicate "milli" or 1/1,000 part of a Farad, but "µ" ("micro") or 1/1,000,000 part! So, the first decade is in steps of 10nF ("nano"), the second in steps of 1nF and the smallest values are in steps of 100pF ("pico") for which the Americans use the "mm" prefix.

COMPONENT TESTING

Now that we understand the operational principles behind the most common test instruments, let's see how we can use them in audio work. As always, we start with "cold" checks using an ohmmeter. Usually, this is not a separate instrument but a multimeter set on one of the "ohms" ranges.

Testing solid state diodes

Since analog ohmmeters use their internal DC battery to push current through an unknown resistor, a diode should conduct in one direction only when the + of the battery is connected to its anode. The measured resistance should be very low, 20-50 Ω. When the + of the battery is connected to the cathode, the diode should not conduct at all, and the ohmmeter should indicate a very high, almost infinite resistance.

The cathode is the negative electrode, marked with a band or a line on one side of the diode's body, corresponding to the line on the symbol.

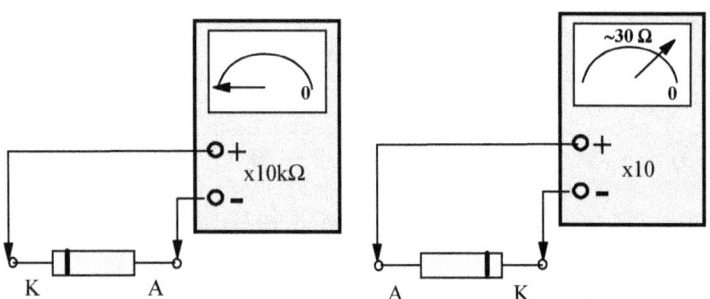

ABOVE: Ohmmeter diode check must show very low resistance one way and a very high resistance the other way.

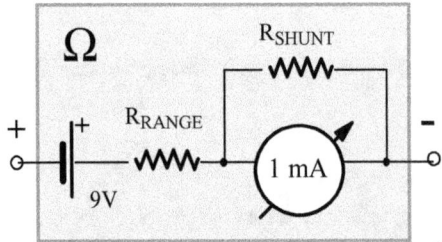

ABOVE: The internal circuit of some analog ohmmeters. Notice the - end of the battery marked as + end on the outside of the meter!

In some multimeters, the + lead is connected to the - (negative) pole of the battery, so you need to mentally reverse the connections when testing with such ohmmeters. Your - lead will have positive DC voltage on it, and its + lead will be the negative side!

In this case, it seems that everything is the opposite. That diode conducts when it's reversely polarized and doesn't conduct when it's directly polarized, which would not make sense unless you understand the internal battery connection!

Testing bipolar transistors

In the first approximation, a bipolar transistor is comprised of two PN junctions (P stands for "Positive," N for "Negative") or "diodes." The first is the base-collector junction (test #1), and the other one is the base-emitter junction (test #2).

Using an analog ohmmeter, we should get a low resistance when the base-collector junction is positively or forward polarized, so the diode is conducting and its internal resistance is negligible. When we swap the two test leads, we should get a high resistance. The PN junction is negatively or reversely polarized, so the diode is not conducting, and its internal resistance is high. These readings will be the opposite for a PNP transistor.

Since the emitter-collector path comprises two "opposing" PN junctions (one forward- and the other reverse-polarized), the resistance reading must always be high for both NPN and PNP transistors.

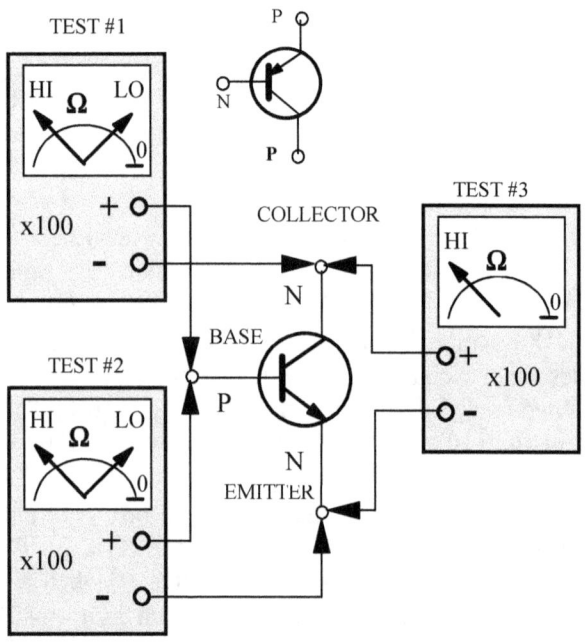

ABOVE: Since there are three electrodes (terminals) in a transistor, three tests need to be performed. The simple test will show reversed readings for NPN and PNP transistors.

What "low" and "high" mean will depend on the transistor; power transistor resistances will generally be lower than those of low power transistors. However, keep in mind that these ohmmeter tests are not conclusive. Even if a transistor passes these rough tests, it may still be faulty (a false negative). Transistors that don't pass this quick check are definitely faulty, so there are no false positives. It is not possible for transistors that show "bad" on this test to be good and work in a circuit.

Testing Zener diodes

Zener diodes work in the reverse quadrant, not in the forward-biasing arrangement as ordinary diodes. This means that the cathode of the Zener diode needs to be more positive than its + electrode (anode)!

To determine the Zener voltage V_Z of a diode, you'll need an adjustable DC power supply and a series resistor R_S. If your DC power supply (or a battery) is not adjustable (most are), connect a 1 kΩ potentiometer across it so you can adjust the test voltage. Increase the voltage until there is a sudden current jump, meaning you have reached the V_Z point (breakdown when the diode suddenly starts conducting current in reverse), and read that voltage V_Z on the voltmeter.

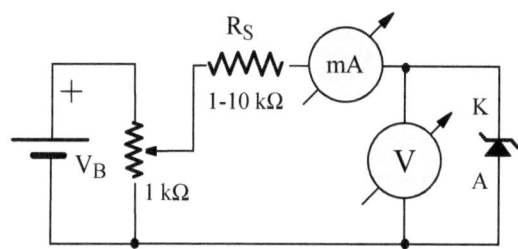

ABOVE: How to find out the Zener voltage of a Zener diode and check its operation at the same time

Testing electrolytic capacitors

To test electrolytic capacitors, use an analog ohmmeter on x1kΩ (x 1,000) range. If the capacitor is open circuit, there will be no meter indicator movement. For a good capacitor, the meter needle should move quickly towards the low resistance side of the scale as the battery inside the ohmmeter charges the capacitor. Then it should slowly drop back towards infinity. The final resting point of the needle will indicate a very high resistance, the insulation resistance of the capacitor.

If the capacitor is short-circuited, the needle will move towards zero Ω and stay there.

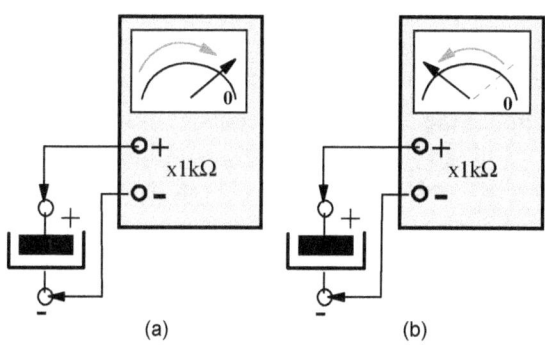

ABOVE: If ohmmeter meter needle moves quickly towards the low resistance side of the scale (a) and then drops back towards the very high resistance (b), the elco is most likely good.

Measuring the input resistance of an amplifier

LCR meters are great for measuring input and output impedances of various devices such as filters, amplifiers, and attenuators. Since these are usually of capacitive-resistive nature, you could first measure the input capacitance (LCR meter on "C" range) and then change the LCR meter's range to R or "resistance" range and measure the device's input impedance.

Remember, despite calling this mode "resistance," digital LCR meters measure AC impedance! You can do the test while the device-under-test is powered up since LCR meters inject an audio-frequency signal into the input (usually of 1kHz frequency) and measure the circuit's response, so you cannot damage anything.

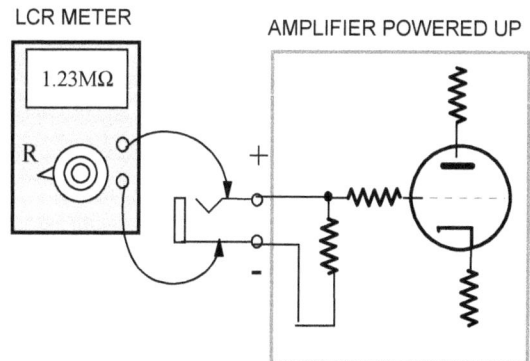

ABOVE: Digital LCR meters make checking input impedance of an amplifier or a single internal stage fast and simple.

AC (SIGNAL) TESTS USING A VOLTMETER (MULTIMETER)

A voltmeter, multimeter, FET of vacuum tube voltmeter can test most parameters of interest to amplifier builders, such as the maximum output power and the frequency range (upper and lower -3dB frequencies) of an amplifier. While faster than oscilloscope checks, these tests are "blind", meaning they don't reveal the waveform of the voltage signals, only their effective or RMS values (assuming you are using a true RMS multimeter).

Checking the maximum output power with an AC voltmeter

Assuming an 8Ω amplifier's output, terminate the amp with an 8Ω dummy load (R_L). With 1 kHz frequency and 1 V_{RMS} at the amp's input, measure the maximum output voltage at the speaker terminals of the amplifier. The maximum power is then $P = V^2/R$, where R=8Ω. Say you measure 12.3 V_{RMS}. The power is then $P = 12.3^2/8 = 18.9$ Watts.

Obviously, the dummy load must be a resistor of a suitable power rating. Instead of messing around with various resistors, buy a 100 Watt 8Ω resistor which should be fine for all but the most powerful (100W+) amps.

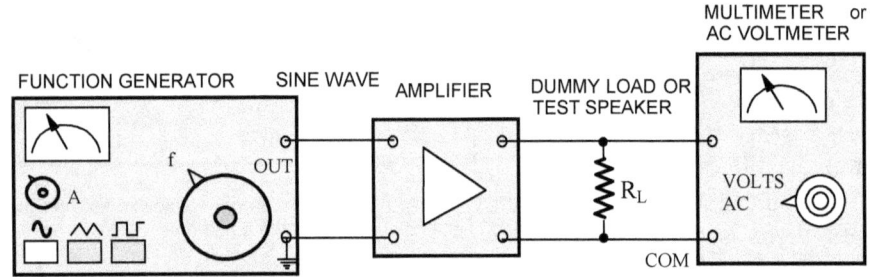

The basic test setup for voltmeter measurements

Determining the frequency range of an amplifier using an AC voltmeter

Using the same test circuit, start with the frequency of 1 kHz on the function generator (sine wave) and adjust the reading on the voltmeter or scope to be 2.83 V_{AC} on the voltmeter or 4 V_{PP} on the oscilloscope.

Finding the lower -3dB frequency f_L: Reduce the frequency until the amp's output voltage drops to 2.00V_{AC}, or until the peak-to-peak voltage on the CRO drops to 2.8V. This is the lower -3 dB frequency, or f_L. The voltage dropped to approximately 71% of its value at 1 kHz.

Finding the lower -3dB frequency f_U: Increase frequency back to 1 kHz and recheck the indication on the voltmeter or the scope, it should return to the original 2.83 V_{AC} on the voltmeter or 4.0 V_{PP} on the CRO. Now increase the frequency until the voltmeter's reading again drops to 2.0 V_{AC} (or 2.8 V_{PP} on the scope). That is the f_U frequency.

Why 2.83V or 4.00V? These voltages were chosen simply because of the easy calculation. 71% of 2.83 is 2.0 and 71% of 4.0 is 2.8! You can choose any voltages you like; just go down and up until the output voltage drops to 71% (-3 dB) of its original value A_0 at 1 kHz!

Two necessary preconditions must be met for this test to be accurate and meaningful. First, make sure the output amplitude of your function generator or oscillator stays constant. If it varies with frequency, you must adjust it to the same level as at 1 kHz during this test. Second, the AC voltmeter should have its own upper-frequency limit f_U higher than the f_U of the tested amplifier. This eliminates cheap analog and even some digital multimeters, which only go up to a few kHz!

Cheap multimeters have f_U of around 2-3 kHz and are thus inadequate for these tests. You'll need an true-RMS multimeter whose f_U is up around the 200kHz mark or higher.

Most modern oscilloscopes go up to at least 20 MHz, and cheap vintage ones (with tubes!) such as those made by Eico and Heathkit go up to 500kHz or 1Mhz, so even they are fine for audio measurements.

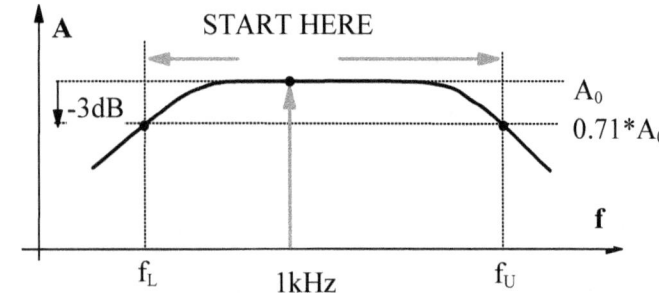

Dummy loads for amplifier testing

High power resistors used instead of loudspeakers for amp testing convert audio power into wasted heat, which is probably why they are called "dummy" loads. One such made-in-China wirewound resistor (100 Watts 8 Ω) is pictured (next page).

Alternatively, resistors of lower power rating (7 Watt and 10 Watt resistors are common) can be connected in any combination (series, parallel or series + parallel), to get the resistance and power rating you need. Three of many options are shown in DIY PROJECT frame. With some values you will not end up with exactly 8Ω, but that is fine as long as you know the resistance of your dummy load and can alter your calculations accordingly.

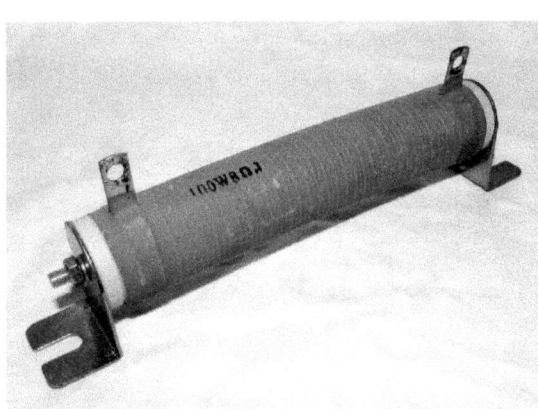

8 Ω wirewound resistor rated at 100 Watts.

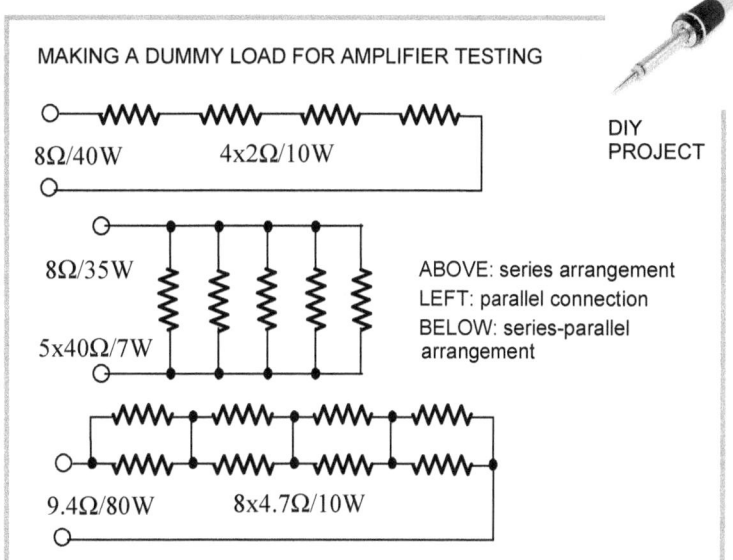

Other tools needed

Surprisingly few other tools are needed for amp repair and construction work. A set of small-to-medium screwdrivers (both flat head and Philips head range) is mandatory, as are two hand tools, wire cutters, and long-nose or "needle nose" pliers. Don't buy these in electronic shops; such universal tools are usually much cheaper in general hardware stores, automotive parts retailers, and even in fishing sections of supermarkets!

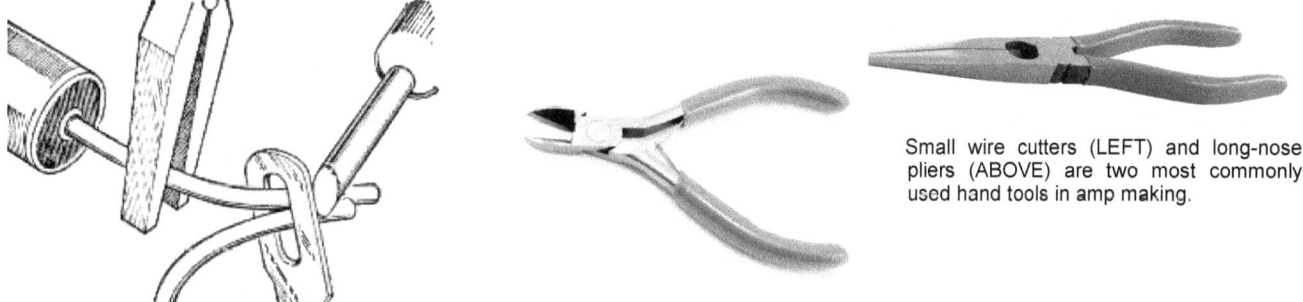

Small wire cutters (LEFT) and long-nose pliers (ABOVE) are two most commonly used hand tools in amp making.

ABOVE: Long-nose pliers are usually used together with a soldering iron. Apart from physically holding a resistor or capacitor's lead in place, they also serve as a heat sink during the soldering of sensitive components such as silicon diodes and transistors.

Hole punching

Amplifier builders find metalwork the most tedious and most challenging part of making any amp or preamp. Drilling or cutting large holes is a tedious and time-consuming job. Plus, tube sockets require neat holes, and the best (the only?) way to achieve that is to use hole punches.

Manual or screw-type punches are OK for softer and thinner materials. You drill a pilot hole, usually 10mm in diameter, and tighten the punch with an Allen key until it breaks through the chassis. However, hydraulic hole punches are a much better option. Even if you make only one or two amps, the investment of a hundred or so dollars will quickly pay off.

TESTING & MATCHING TUBES FOR GUITAR AMPS

Tube testers seem easy to operate but are highly complex instruments (apart from the most basic tube checkers). As with all other test gear, you need to understand their operating principles, capabilities, and limitations; otherwise, you will not benefit much from using them. However, tube testers are a specialist topic, and to cover them here would take quite a few pages so that we will focus only on the most critical issues.

Emission "checkers" connect all tubes (triodes, pentodes, etc.) as diodes and impress some low AC voltage across them (20-50V). The tube rectifies the AC voltage into pulsating DC peaks. The analog moving coil meter displays some figure on its scale, supposedly calibrated in % of the nominal emission or cathode current. These quick checkers can only identify some serious defects inside tubes and their results are questionable, and as such should be of no interest to a serious amplifier builder.

In *dynamic conductance* testers, AC voltages of adjustable amplitude are brought to the grid (lower voltage) and anode/screen grid (higher voltage). The instrument indicates a composite of cathode emission and transconductance (since the amplifying action of the grid has an impact on the reading). Examples include Eico 666 and 667, Sylvania 219, 220 and 620, Jackson 648, and Simpson 1000.

The test circuit inside *proportional* mutual conductance testers is similar to dynamic plate conductance testers, the differences being the addition of the AC grid signal source and the AC meter in the anode circuit. Note that the biasing AC voltage is out of phase with the rest of the secondary voltages. These are called "proportional" Gm testers because proportional AC voltages are used (proportionally higher, screen voltage is lower than the anode voltage, for instance).

The tube-under-test acts as its own rectifier; the control grid rectifies its grid bias voltage, the screen grid rectifies its voltage, etc.

When a small AC signal is applied to the grid, an AC component of anode current is measured. The meter is calibrated to read mA/V or micromhos directly.

Mutual conductance testers

TThe testing speed and low production costs were the main design aims when this family of testers was conceived. Most of these very common testers used the Hickok method (Hickok, B&K, Precise, Mercury, Stark, military testers by Hickok).

Many (such as later B&K models 700, 707, and 747) don't display true gm reading; the scale is "relative micromhos" marked 0-120 or 0-130, presumably in percentage points of the "nominal" value.

Testing is "blind" - no bias voltage, grid signal, or anode current is displayed, and the bias voltage cannot be adjusted! Gm is tested in only one arbitrary chosen point unknown to the operator (unless voltages on tube pins are measured).

The circuitry is simple: unregulated voltages and the mains frequency used for the grid signal (no separate grid signal oscillator). All feature "sensitivity," "meter," "shunt," or "calibration," or similar controls which directly affect the meter reading. Examples (other than Hickok and Stark branded testers): Mercury 1000, 1200, 2000, Sencore MU140, and MU150.

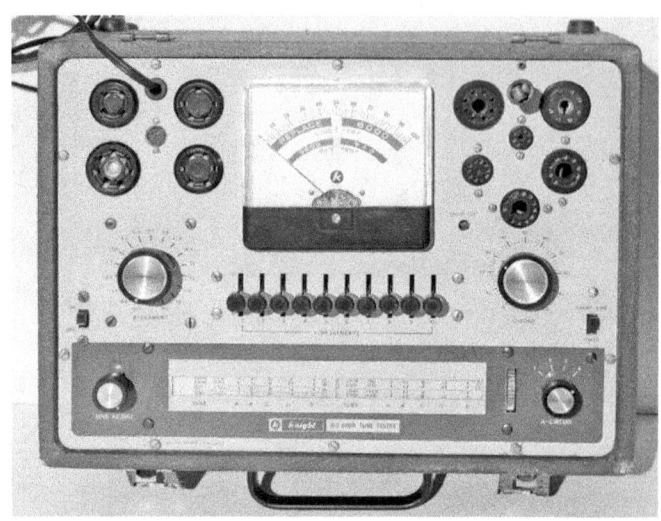

ABOVE: Knight KG-600B is a typical vintage emission tester, ten levers (one for each tube pin) in up or down position (all electrodes connected either to cathode or anode) and a choice of four very low test voltages.

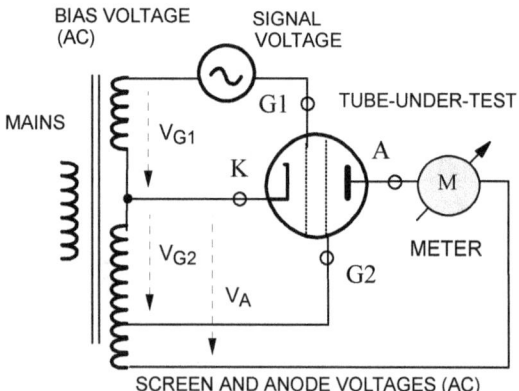

ABOVE: Proportional" mutual conductance testers also use AC voltages brought to their control grid G1, screen grid G2 (if used) and anode (A).

BELOW: Matrix LX310E is a very capable albeit rare French tube tester. "Polarisation" means bias. Notice the selector switch; the X1 range means that bias is adjustable between 0 and -10 Volts. When the switch is in the "X5" position, the bias voltage is adjustable from zero to -50V.

Better mutual conductance testers approach the laboratory standard. The bias, screen, and anode voltages are DC and are adjustable if not continuously (that would be ideal), then at least in steps. A separate AC oscillator supplies grid signal voltage, which can be adjusted in steps; the smaller signal voltages are used for preamp tubes and the larger ones for power tubes.

For instance, Triplett 3444 tube tester uses four grid signal levels, 33mV, 100mV, 0.333V, and 1.0 V. The test frequency is 4 kHz. There is no range selector switch. The two bias ranges (0 to -5V and 0 to -50V) are automatically selected depending on the position of the "Plate voltage" switch or switch "C". Since the bias DC voltage is continuously adjustable (0 to -10V) or 0 to -50V) you can test a tube in any point along their transfer curve, not just in one, as preselected by the tester's designer, as with basic gm testers.

Experiment: How to plot a tube's transfer characteristics on your tube tester

The worst way to "match" tubes is based on anode current in one steady point only, yet this is what tube sellers and even manufacturers do. The next best option is to match them for both anode current and mutual conductance, albeit in only one operating point (one anode voltage, one bias voltage). Short of plotting the whole family of anode characteristics (which requires a range of anode voltages), the quickest way to match tubes is to plot their transfer curves (by varying the bias control on your tester) and superimpose them on top of each other.

This assumes that the bias voltage on your tube tester is variable. If it isn't, your tester tests tubes in only one point, predetermined by the tester's designer. Even if they belong to the family of mutual conductance testers (much better than the rather primitive emissions testers), these are not serious testers, only "quick tube checkers." Mercury, Sencore, B&K, and Seco transconductance testers belong to this category.

A transfer curve shows the relationship between DC plate current IA as a dependent variable (Y-axis) and negative grid bias voltage as an independent variable on X-axis at a fixed plate voltage. As we select 3-5 typical plate voltages, we get a family of transfer curves manufacturers specify in their

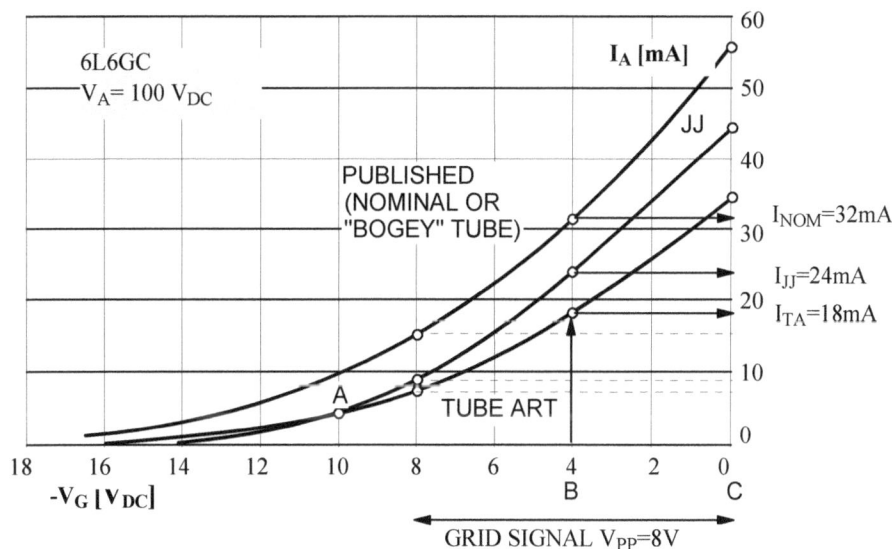

In this experiment we measured transfer curves on Triplett 3444 tube analyzer of two brands of 6L6 tubes, the current production JJ and NOS Chinese production from the 1960s, labeled "Tube Art".

A $100V_{DC}$ plate/screen voltage was used in order to stay below the 50 mA current limitation of the tester. With $250V_{DC}$ anode/screen voltages the anode currents would very quickly shoot toward 100mA or even higher.

The results are superimposed on the graph from GE catalog for the 6L6GC tube to see how much variation there is between the two brands and the vintage datasheet specification. Notice the difference between the published curve by GE and the two tubes tested. Would these tubes perform differently in an amplifier? Absolutely!

Both the Tube Art (TA) and JJ tubes pull significantly lower currents than the nominal, average, or "bogey" tube from the GE datasheet. Also, below point A (where two transfer curves cross), the TA tube pulls more current than the JJ tube. JJ's TX curve is steeper in that region than TA's, meaning JJ tube has a higher mutual conductance.

To the right of point A, JJ's TX curve is still steeper (the slope of the tangent at any point) than that of the TT tube, and the anode currents increase faster, so JJ's transconductance is higher than TA's across the whole range of operation.

What would happen if we plugged each of these tubes into an amplifier? For instance, let's assume that our amp is biased in point B ($V_G = -4\ V_{DC}$). While GE datasheet specifies that their average tube should pass around 32mA of anode current in that point ($V_G = -4\ V_{DC}$ and $V_{AK} = +100\ V_{DC}$), the JJ tube only draws 24 mA and TA pulls even less, 18mA!

And finally, let's look at the maximum anode currents reached in point C, the zero bias point ($V_G = 0\ V_{DC}$). We assume a symmetrical grid signal around point B or -4V bias, a sine wave with its peaks at $V_G = -4\ V_{DC}$ and $V_G = 0\ V_{DC}$ or 8V peak-to-peak. The TA tube's anode current will vary from a low of 8 mA to a peak of 34 mA, a swing of 34-8 = 26 mA. JJ's tube will swing between 9 mA and 44 mA, a total of 35mA. Finally, the bogey GE tube should swing between a low of 16 mA at -8V bias to a high of 57 mA at zero bias, a total of 41 mA.

You probably realize now where this discussion is heading. The TA tube will produce the lowest audio power of the three tubes. Assuming identical anode voltage swings, due to its larger current swing JJ's tube will generate a 35/26=1.35 or 35 % higher power output. However, it will still fall significantly short of the vintage GE tube that would produce 41/35 = 1.17 or 17% higher output still!

However, you may protest, we cannot make meaningful conclusions based on such a tiny sample, a single tube from each maker, and you'd be right. We cannot proclaim that all Tube Art tubes are "weaker" than all JJ tubes or that all JJ tubes are "weaker" than the vintage GE-made tubes. This was to illustrate that variations between different brands and even between tubes made in the same factory can be significant, and swapping such tubes around in your amp will change operating conditions, output power, and the tonal voicing of your amp.

Investigation: The Groove Tubes performance rating system

Groove Tubes is a tube reseller/rebrander based in San Fernando, California. They source tubes from manufacturers in Russia, Slovak Republic (JJ), and China, but in the past also rebranded some Yugoslav tubes (Ei - Elektronska Industrija). Since GT literature contains a notice that Groove Tubes is a registered trademark of Fender Musical Instruments Corporation, it seems that it is no longer an independent business but is now owned by FMIC.

According to their website and info gained by us from various photos, after the 2-hour burn-in and stabilizing period, preamp tubes are tested for hum and microphonics in guitar amps. This is followed by the gain test in their customized testbeds, where the signal frequency is swept over the audio range, and the gain-vs-frequency curve is plotted using PC-based "Audio Precision" hardware and software.

Tube parameters are measured using the Amplitrex AT-1000 tester. Based on the results, the best (most balanced) tubes are selected for the "Special Applications Group," further tested on Hagerman Vacu-Trace curve tracer. The tubes with dynamically matched halves are named "Matched Phase Inverter" (MPI) tubes and sold at a premium.

The photo on the GT website shows only one curve used for matching, not a whole family of curves. It is not clear if that is one of the anode curves or a transfer curve at a specific anode voltage since transfer curves also have the same shape (for triodes only).

Apart from the similar basic tests, power tubes are "dynamically energized" and matched in terms of distortion levels or "gain-to-distortion ratio" and graded on a 1-10 scale. The Crate Palomino V8 amplifier tag hints at the differences between the three categories of tubes but does not explain what that means.

The "early distorters" have the least amount of headroom, meaning they start clipping the peak of the signal early, and, as we know, a clipped signal is severely distorted. Their "gain-to-distortion ratio" is the lowest of the three groups. These are marked as grades 1-3. "Normal" or average tubes are graded as levels 4-7, and late distorters (our names) have the widest headroom and thus distort at much higher signal levels.

Groove Tubes amplifier tag
©Groove Tubes

Experiment: testing a pair of Grade 4 and a pair of Grade 7 GT-6L6C tubes

In our stash of various used tubes, we found two pairs of Groove Tubes GT-6L6C, one pair marked as "Test 7", the other as "Test 4", so we plugged them into Triplett 3444 analyzer. The results for 2V jumps in bias voltage are shown in the table (next page). The test was done with 250V on the anode and 100V on the screen. Notice that all test points do not perfectly fit the smooth curves, but again that is due to the manual and error-prone (subjective reading on an imprecise scale) nature of the test.

The two Grade 4 tubes were well matched, with almost identical test results, certainly within the margin of error (manual setting of the bias pot). The two used Grade 7 tubes were not identical, perhaps they were matched when new, but now they have aged differently.

BIAS [V_{DC}]	0	-2	-4	-6	-8	-10	-12	-14	-16	-18	-20	-22
Grade 4 tube #1	41	31	25	20	16	11.5	8	4.6	2.3	1	-	-
Grade 4 tube #2	41	31	25	19.5	15.5	11.2	8	4.6	2.3	0.8	-	-
Grade 7 tube #1	49	39	33	28	22	16	12	7	3.7	1.7	0.6	-
Grade 7 tube #2	46	36	29	24	20	15	10.5	6	3.2	1.35	0.3	-

Groove Tubes claim that their tubes will age identically, but that is questionable. Tube aging is a complex process and depends not just on tubes but also on the amplifier they are used in. Even if two tubes were perfectly matched initially, they are unlikely to stay matched after a few years of operation!

Of course, the test results shown here and our conclusions are still valid. Assuming a fixed bias of -6V, in idle state (operating point Q4), G4 tubes will pull 18mA, and G7 tubes will be biased much hotter, pulling 28mA.

Assuming a maximum input grid signal of 12VPP (peak-to-peak), centered around -6V bias, the anode current of the G4 tube will swing between the minimum of 8mA to a maximum of 41mA, a total peak-to-peak value of 33mA. The anode current of the G7 tube will vary from 12mA to 49mA, a total of 37mA. So, a G7 tube will amplify the signal more and distort at a higher power level (later) than the G4 tube.

Notice that the positive halves for both tubes are identical, around 21mA (identical positive headroom), but the negative peak of the G4 tube is smaller, 12mA compared to 16mA for the G7 tube (smaller negative headroom for the G4 tube). Since there is a bigger difference between its positive and negative halves, the G4 tube will distort more. Remember, if + and - halves were identical, the harmonic distortion would be zero. So, although GT groups these two pairs into the same 4-7 category, there are noticeable differences in their characteristics and behavior.

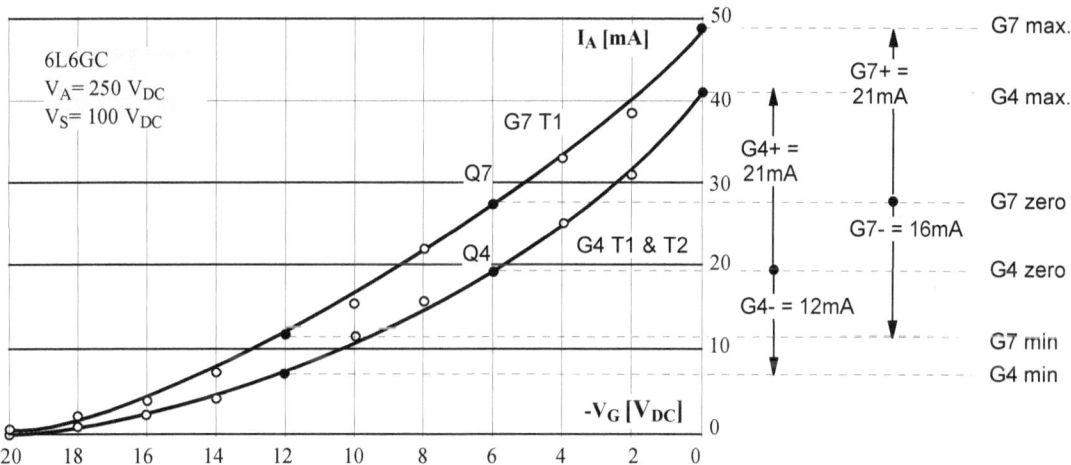

HOW TO CHECK AND EVEN MATCH TUBES WITHOUT A TUBE TESTER

The ultimate test of any tube is inside its amplifier. Tube testers, even the more elaborate ones, suffer from both false positives (a tube tests "bad," but it works in the equipment) and false negatives when a tester passes a tube as good, but the tube then fails to work properly in an amp.

So, unless you buy & sell tubes commercially, you don't really need a tube tester for amplifier repair and construction work.

These three tests range from the simplest, using ohmmeter only, to the more complex tests with the tube in a hot state (heaters energized).

Ohmmeter checks in the cold state

These cold checks are necessary but not sufficient to proclaim a tube to be OK. An analog ohmmeter should show a very low reading (a few ohms) when measuring the resistance of the tube's heater.

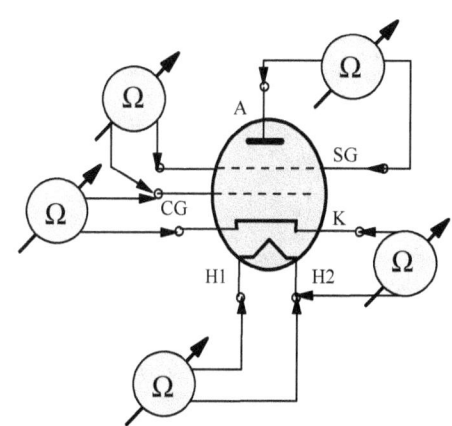

ABOVE: Apart from the very low ohm reading across the heater pins, all other ohmmeter checks should show infinite resistance.

If the ohmmeter shows infinite resistance, the heater has burned out, and you can throw that tube away. Between all other electrodes, the ohmmeter should show infinite resistance (no shorts).

Pairs of electrodes physically close to each other are more likely to show a short circuit, for instance, heather to the cathode, control grid to the cathode, and screen grid to the anode. However, even anode to cathode test can show low resistance due to foreign material stuck in the mechanical structure of the tube.

Hot ohmmeter checks

A similar "hot" test is used by most cheap tube emission checkers. Connect a small mains transformer to supply the rated heater voltage. If the heater needs 6.3 V @ 3 amps, a 20 or 30 VA transformer will be sufficient. You can use a universal adjustable DC power supply instead.

Connect an analog ohmmeter between the cathode and the control grid. You don't even have to connect the anode to the grid, as shown. Connect the + lead to the anode and/or control grid and the negative lead to the cathode. The ohmmeter should show some deflection; how much will depend on the type of tube tested, its internal resistance, and other parameters. If you get no indication, reverse the ohmmeter's leads.

Use the highest range on your meter, for instance, 10k or 100k. Most multimeters use an internal 9V battery for one or two of the highest resistance ranges, but only a 1.5 V battery for lower ranges. That may not be enough to check a tube. Digital ohmmeters and multimeters, which work on a different measurement principle, are not suitable for this test.

Hot checks with a function generator and a multimeter

The low voltage power supply provides a DC heating voltage (adjustable and readable on the power supply voltmeter). A function generator provides a sine or square voltage that is connected between the cathode and the grid/anode. A transformer of suitable voltage and current rating can be used for AC heating instead of the DC power supply.

A DC milliammeter or a multimeter set on DC current range displays grid/anode current. At such low test voltages (most function generators provide 7-10 Volts RMS signal) the tube's grid should be safe. Some higher-spec models may even go up to 30V. If you are still worried about damaging the tube-under-test due to too much current flowing through the grid, strap the grid to the cathode instead.

FURTHER READING: "How to Use, Calibrate, Repair and Upgrade Vacuum Tube Testers"

For a more in-depth look at tube testers, tube testing and matching, please consult my book "How to Use, Calibrate, Repair and Upgrade Vacuum Tube Testers", available on Amazon and all good online bookstores.

Design, functionality, calibration and modifications of vintage testers by B&K, Hickok, Triplett, Mercury, Sencore, Weston, Simpson, AVO, Taylor, RCA, Precise, Precision, Eico, Jackson, Sylvania, Knight, Heathkit, Seco, Sico, Conar, Metrix and other brands are discussed.

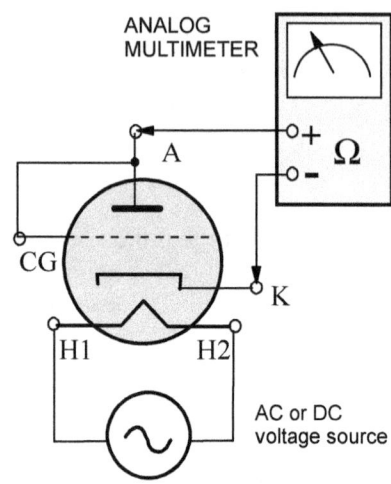

ABOVE: 9V from the ohmmeter's internal battery is enough to cause a current to flow through a hot tube, which indicates emission levels from its cathode.

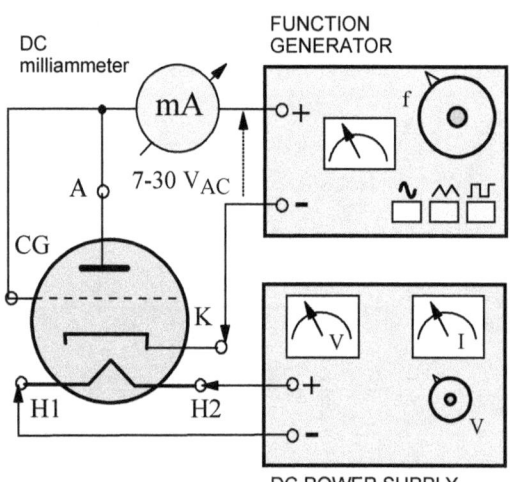

ABOVE: A function generator as a signal source plus an AC or DC heater supply, and you have a simple yet effective emission tester!

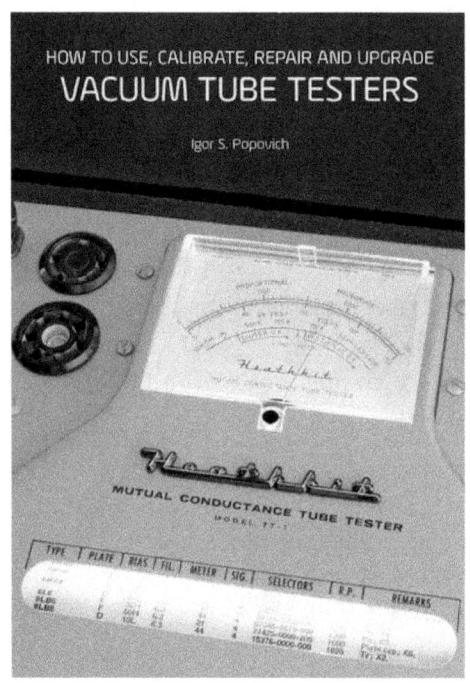

WIRING, SOLDERING & MODIFICATION PRACTICES

- GENERAL GUIDELINES FOR MODIFICATION SUCCESS
- SAFETY RULES AND PRECAUTIONS
- WIRING OF THE AC CIRCUITS
- SOLDER & SOLDERING
- REDUCING HUM: ISSUES AND SOLUTIONS
- HOW TO POWER AN AMPLIFIER UP FOR THE FIRST TIME

Designing, building, modifying, and repairing tube guitar amps involves both right- and left-brain activities, and very few people are equally good at both or all of those tasks. Some excel at designing and conceptualizing; others have superior building and wiring skills but could not design an amp if their life depended on it.

The good news is that practice and experience lead to improved performance, so even if your wiring and soldering skills aren't that good, you will improve over time. In this section, we overview some important principles amp builders and fixers need to understand and follow at all times, including the most important of all, safety rules!

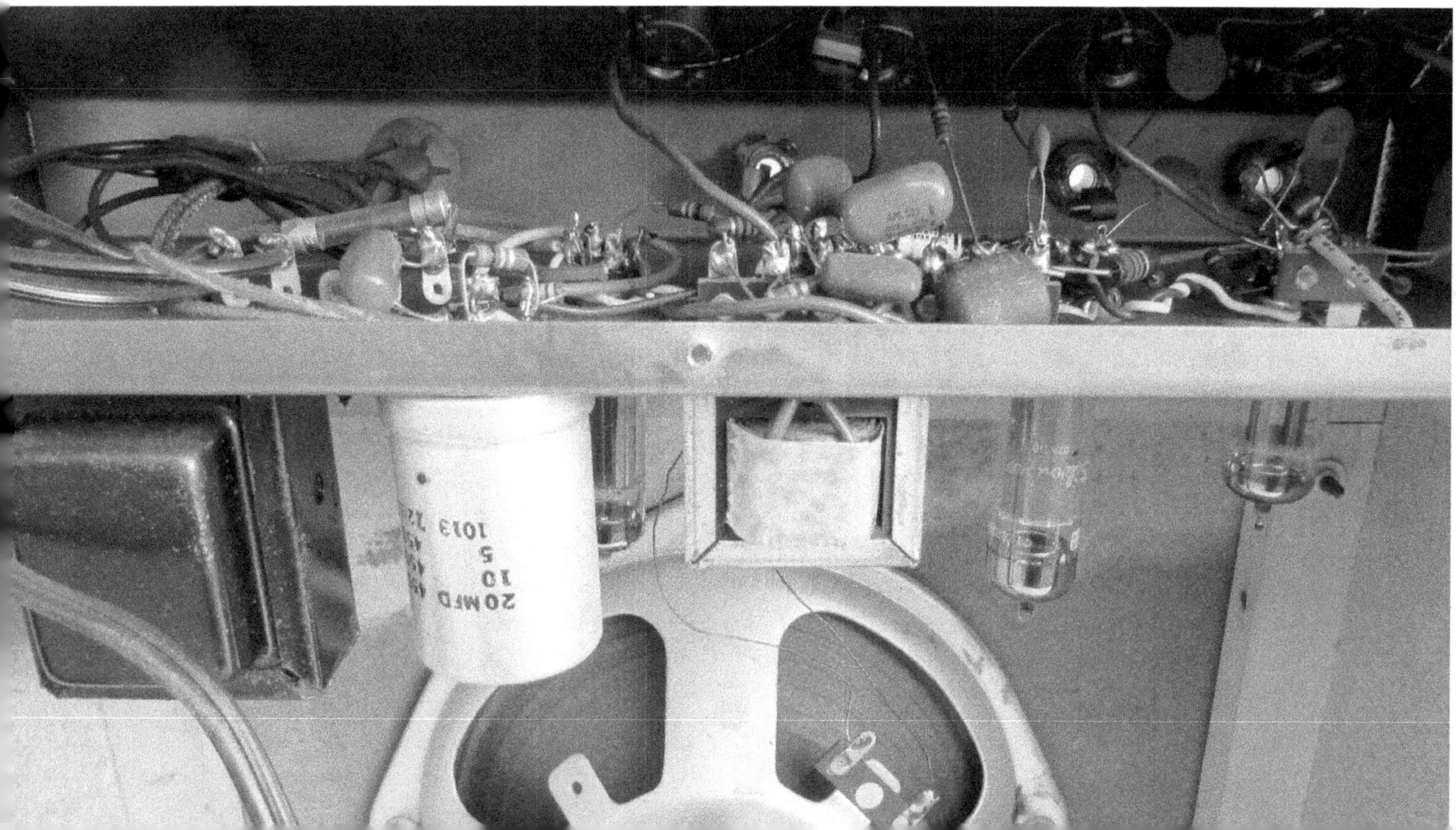

GENERAL GUIDELINES FOR MODIFICATION SUCCESS

The first question you should answer is, "Why are you modifying a particular amplifier or preamplifier?" If you are a collector or want to resell desirable amps for profit, you should not change anything apart from components needed to repair the amp. Even then, to preserve the value of the collectible original, the replacement components (capacitors, resistors, etc.) should be the same type and, if possible, the same brand as the original ones. Some, usually more expensive amps, will fetch much higher prices in their original condition.

Keep in mind that many types of repairs can also be considered improvements. As soon as you change the brand, the type, or the rating of a component, that is a modification. However, there is no point replacing an old and leaky Sprague "Black Beauty" or Marbelite "Good-All" vax capacitor with an identical NOS part, which is also likely to be leaky. So, prudent judgment is needed.

To restore and upgrade an old amp or to build a new one?

Another option is to buy a cheap new amplifier (usually made in China, Vietnam, and other Asian countries) and do it up. Ideally, look for amps with large chassis, built point-to-point or on a terminal board, and with large power and output transformers. You are effectively buying the cabinet, the chassis with the control panel, and the magnetic components (transformers and chokes). The resistors and capacitors are often cheap and nasty and are to be upgraded anyway.

RESTORE & IMPROVE	DESIGN & BUILD
✓ Compromised interiors (rusted chassis, poor contacts on tube sockets, components drifted too far from the original values, leaky capacitors (-)	✓ Complete freedom of choice (design, topology, the looks, components) (+)
✓ Limited availability of quality vintage guitar amplifiers and high price of "in demand" models (-)	✓ Lots of manual labor (metal work and soldering) required (-)
✓ Battered exterior, torn Tolex, dings, scratches (-)	✓ Low resale value, unless you are a known, reputable designer & builder (-)
✓ A higher resale value, especially for "in demand" models (+)	✓ New is new and old will never be new again (+)

Common mistakes inexperienced tweakers make

1. Using a replacement component of a lower rating. If the original designer specified a 600 VDC coupling capacitor, you should not install one rated at 400VDC in its place.
2. Using a replacement component of a different type. Replacing a burned-out non-inductive resistor with an ordinary one may fix one problem but cause two new ones.
3. Implementing changes that are unsuitable for a particular design. Generally, increasing the value of coupling capacitors will improve the bass response of an amp and improve the sound. Still, it should not be done on some designs or done indiscriminately without considering other design issues.
4. Not considering the impacts of a particular modification. For instance, replacing tube rectifiers with solid-state diodes will raise plate voltages by 15-40 Volts in all stages. That will shift the operating points, change the output power and distortion of the amp, and may exceed the rating of filtering capacitors and cause them to explode. It will definitely change the sound of the amp.

Choosing an amp to buy & upgrade

The choice of the amp depends on your ultimate purpose. If you want to keep it for your own use, the criteria will be different from the situation where you want to profit by restoring and/or improving the amp for a resale. Some criteria apply to both cases:

- The larger the power transformer, the better. If you need to add a tube or two, a larger trafo will have enough spare capacity to accommodate the additional load on the heater winding and the increased plate current draw.
- Look for amps with lots of space around components, which makes them easier to work on.
- The condition of the exterior is a significant factor. You can fix the insides of an amp, but if the metalwork, the cabinet or the speaker is damaged, corroded or compromised in any way, stay away from that amp.
- Smaller amps are cheaper to post or send by couriers, especially internationally. If restoring for profit, just as with real estate, you don't make a profit when you sell it; most profit comes from low acquisition costs, buying it cheaply, and saving on postage.

The Ten Commandments for improving vintage amps and preamps

1. SAFETY FIRST. Bin the old and illegal 2-core USA mains cables and two-prong reversible plugs. Install modern 3-pin plugs and approved 3-core mains cables. Earth (ground) the metal chassis. Remove all capacitors on the AC side of the mains transformer. Add a fuse-holder if the stingy manufacturer did not install one.
2. POWER SUPPLY. The heart of an amplifier. The first and most important improvements start right there.
3. ENERGY. The more stored energy in the power supply, the deeper the bass and the better the amp's dynamics.
4. ELCOS. Electrolytic capacitors are bad for the sound. Replace them with film types wherever possible.
5. PENNY-PINCHING ATTITUDE. Most vintage tube amps were made with cost savings in mind. If you are restoring or improving them, all cheap & nasty components should be replaced with more reliable and better-sounding modern ones. Buy the best quality components you can afford.
6. HUM. The enemy of focus, clarity, and detail. Reduce hum, and your amp's tone will improve too.
7. NEGATIVE FEEDBACK. Reducing negative feedback is a matter of changing only one resistor. This simple & easy change can result in significant audible improvements.
8. SIMPLICITY. A more complex circuit does not necessarily result in more superior sonics. To save time & money and to preserve the "character" of the original, changes should be kept as simple and as minimal as possible.
9. GOING OVERBOARD. Completely redesigning and rebuilding a vintage amp makes no sense. You'd be better off designing and constructing a new one from scratch, on larger and better-looking chassis, with better topology and layout, using better quality switches, tube sockets, potentiometers, and transformers.
10. SYNERGY. For best results, all the changes (add-ons, improvements) should work together. It's the combination of changes that improves or ruins an amp.

The art of salvage - sourcing quality parts

If you plan to keep fixing, modifying, or building tube amps, the best long-term investment is to get as many vintage TVs, tape recorders, oscilloscopes, and any other piece of electronic gear you can get. A large stash of used parts will save you not just many hours you would otherwise spend online or going through electronic shops looking for a specific component but also thousands of dollars.

Compared to their trade prices (what manufacturers pay), modern components sold in electronic shops to DIY enthusiasts are grossly overpriced. Many are of poor quality, bought cheaply, and sold at huge markups (hundreds and even thousands of percent)!

When you look inside a vintage Tektronix tube oscilloscope or any other quality test instrument, you can see that it was designed and build properly. In comparison, vintage tube guitar amps look cheaply made (and they were).

The components used in test instruments were first class: tubes, trimmer capacitors, trimmer resistors, carbon composition resistors, precision resistors, tube sockets, potentiometers, extension shafts, porcelain soldering strips and quality hookup wire can all be reused!

Vintage gear will be dusty & dirty, so clean it first with compressed air. Small air compressors (for pumping tires) are available in auto-parts shops.

The internals of a vintage Tektronix tube oscilloscope

Make sure you wear a good quality dust mask; the dust may not be just full of bacteria and viruses but also may contain toxic inorganic particles. Alternatively, use a water hose. Don't soak the transformers; a gentle removal of the surface dust is all that is needed. Then, dry them in the sunshine or a very slow oven, 50-70 degC.

Don't just snap the component leads off - try to unsolder them first. That way, the reused components will have slightly longer leads which may prove critical in your project. There is nothing worse than when the leads of a particular resistor or capacitor are a few millimeters short, so you have to spend ten or more minutes searching for another one in your stash. Unfortunately, in some vintage gear, the component leads were first wrapped around a terminal or tube socket lug and then soldered, so removing them is a tedious and frustrating task.

Not all vintage parts are worth salvaging. None of the electrolytic capacitors will be suitable for reuse, and many of the film capacitors will be leaky, but most other components will be fine.

SAFETY RULES and PRECAUTIONS

Electric shock - how it happens and how to avoid it

Just as people can drown in 10cm of water and in 10m deep water, they can get a lethal electric shock from 100V, which can be as deadly as 1,000V! The severity of an electric shock depends on many factors, for instance, on the age, gender, and physical condition of the victim. The main factor, however, is the level of current flowing through the body.

The threshold of perception is around 1mA. Currents up to 5mA produce tingling but not severe pain. Muscular contractions start at around 10mA, while 100mA of current would start the fibrillation of the heart muscle, preventing it from pumping blood and causing death unless the fibrillation is stopped. Above 300mA, the heart's contractions are so severe that fibrillation is prevented, so if the electric shock is halted quickly, normal heart rhythm will probably resume. Thus, 100-300mA is the most dangerous range.

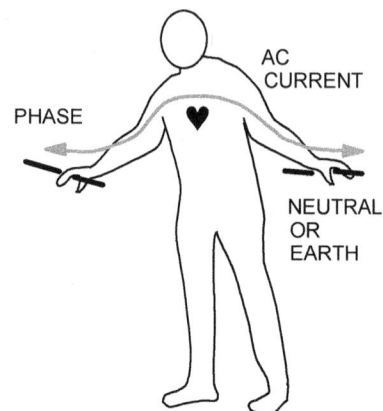

The worst situation: current flowing through the upper body (torso) and the heart

The lethal voltage level depends on the resistance of the current-conducting path through the body, which in turn depends on skin resistance and how the contact is made. Skin resistance, in turn, depends on its moisture level. Dry skin can have as much as 500kW of resistance, while the wet skin's resistance can be as low as 1kW! As electric current flows, it punctures and breaks down the outer layer of the skin, and skin resistance falls rapidly. This is why it is of paramount importance to break the contact with the live conductor as quickly as possible.

The most dangerous situation involves a voltage between two hands or arms, as illustrated when the current flows through the upper torso where the heart is located. That is why you should only use one hand when measuring voltages. The other hand should not be touching anything - keep it in your pocket!

Checklist: Electrical hazards and required safety measures

Tube amplifiers involve high voltages that may be lethal, so numerous precautions must be taken to mitigate the risk of an electric shock when powering up, testing, and working on "live" circuits.

- Install a device called ELCB (Earth Leakage Circuit Breaker), ground fault interrupter, or RCD (Residual Current Device) in your house's switchboard or workshop. These are now mandatory by law in Australia, but electrical laws in many other countries are much laxer. Use an isolation transformer. That will minimize the probability of an electric shock. These protective devices may save your or your children's lives (if they poke a knife into a toaster, for instance, or into a tube amplifier!)
- Unplug your amp from the wall outlet when not in use and while working on it. Once unplugged, discharge the power supply capacitors before doing any work.
- Don't just use a straight wire for discharging; the spark and the high discharge current may damage the capacitor. Make a discharge lead with insulated crocodile clips at both ends and a $1k\Omega$ 2W resistor in series to limit the discharge current.
- Use only one hand when working on a "live" amp. Don't touch the chassis or any other part of the amplifier with the other hand. If practical, wear thin cotton gloves.
- Wear shoes with rubber soles. Never stand barefoot on a bare concrete floor while working with electrical appliances or amplifiers. Use a rubber mat. Timber flooring and carpet (wool or polypropylene) are also good insulators. Use tools with insulated grips, and don't touch their exposed metal parts.

Checklist: Other hazards and precautions needed to mitigate them

- Your workspace should be clean, tidy, spacious, well organized, well lit and well ventilated.
- Keep children and pets away from the amplifier and your working area.
- Tubes get hot very quickly. Never touch or handle hot tubes with bare hands. Use a thermally insulating glove.
- Tubes and soldering irons are fire hazards. Remove all flammable material from their immediate surroundings and always have a fire extinguisher ready. It must be of the type approved for electrical fires.
- Never wear dangling jewelry, headphones or anything else that may get caught inside the equipment you are working on and create either a mechanical or electrical hazard (short circuit).
- Metalwork is another serious risk. When cutting, drilling or grinding metal chassis or other parts, always wear eye protection and gloves. Secure the pieces you are working on (clamp them down). Have a First Aid Kit handy.

WIRING, SOLDERING & MODIFICATION PRACTICES

- Amplifiers are heavy; many weigh 30 to 40 kg! Exercise caution when lifting them to the workbench and whenever you move them around. Bend your legs (your knees), not your back!
- Building, repairing, or modifying tube equipment is a demanding activity; it requires concentration and full awareness of all your senses. In case something is wrong, you need to be able to hear when a capacitor starts frying, transformer insulation crackling or varnish smelling, or to smell a tiny whiff of smoke when a resistor begins burning. One mistake may cost you an expensive transformer, a pair of tubes, or even your life!
- Never work on amplifiers when you are tired, sick, upset, hot, sweaty when you have food, sex, or anything else on your mind. Don't do anything else while working on tube equipment: don't eat, drink, have your headphones on, or talk on a mobile phone.

The principle behind isolation transformers' operation

Isolation transformers galvanically isolate the primary and secondary windings. The secondary of an isolation transformer is not grounded, so one short circuit to ground (either live or neutral) does not change its operation, but the safety isolation aspect would be lost after such a fault.

Touching the secondary L or N terminal would not result in electric shock since these are floating (not "referenced" to the ground), so the path through the human body cannot be closed, and the circulating current cannot flow.

As indicated by the arrows, one fault in the secondary (load) side, such as a short circuit to earth (or the metal case, which is earthed), is tolerated. It will not endanger the user since it will simply "ground" (or "earth") either a phase (L) or neutral (N) on the secondary side. However, the isolation properties of the transformer would be lost since that would "reference" the secondary winding to the ground.

To have two faults simultaneously is highly unlikely, and even if it does happen, it will short-circuit the transformer's secondary and the secondary fuse (2) would blow and/or the primary circuit breaker (1) would trip.

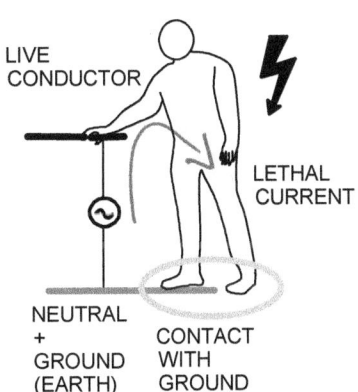

Ordinary mains installation: neutral and earth bonded together, once the body closes the circuit, lethal currents flow

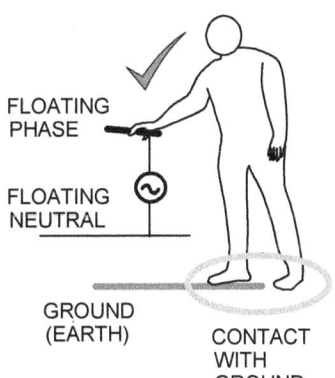

Floating system when isolation safety transformers are used. Neither neutral nor phase are referenced to ground!

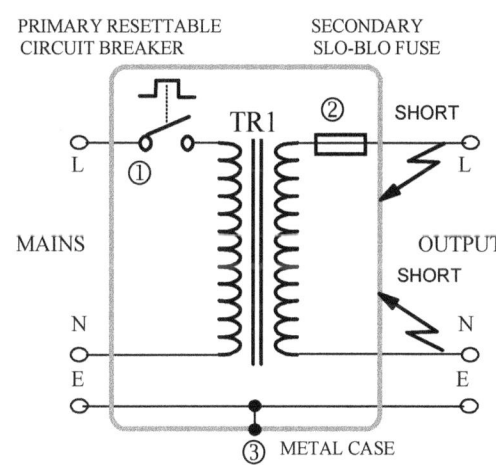

Isolation transformer's features.

Since the mains voltage (or any other AC or DC voltage in an amplifier) is or should be referenced to ground (the neutral connected to the earth connector at the house's switchboard), if a person is in contact with the ground (earth), touching a "live" conductor or "phase" closes the circuit, and circulating currents will flow through the body.

However, touching a live conductor is harmless if the voltage is floating, i.e., not referenced to ground (just like birds sitting on HV overhead power lines). The circuit cannot be closed, so the lethal currents cannot flow through the body. Remember, it's not the voltage but the currents that kill!

WIRING OF THE AC CIRCUITS

The most critical safety upgrade for all vintage and some currently made amplifiers

Most amplifiers, tube testers, capacitor checkers, signal generators, and other audio or test equipment built before the 1970s in the USA used a two-pin mains plug. There was no earth pin, so these units were not earthed (or grounded) at all. Furthermore, these 2-prong plugs are reversible, so instead of the neutral being connected to the metal chassis, you can end up with the live 115 VAC on it. In many cases, the fuse and the power switch are in different branches of the mains circuit, as in the Gibson Skylark GA-5T amplifier (next page).

In many cases, there were film capacitors connected between the phase or neutral and the chassis. Once these caps become leaky, they pass the mains voltage onto the metal chassis and have been suitably named "widowmakers" due to their lethal nature! As such, even if not leaky, they must be removed. These days only X2-rated and approved capacitors can be used in such critical positions. Short circuits cannot develop in those capacitors. Your first job, before you start repairing or upgrading anything, is to perform this safety conversion.

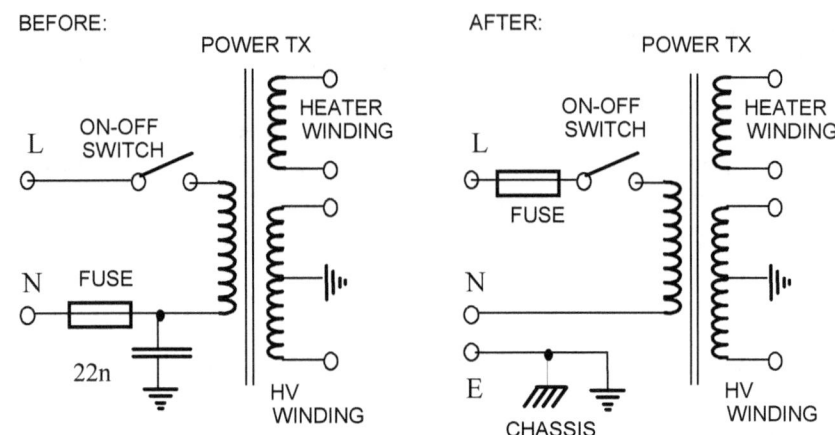

ABOVE: The mains circuit of Gibson Skylark GA-5T amplifier in its original state (LEFT) and after safety modifications (RIGHT).

Wiring the mains side of the power transformer

As for replacements, start with the power cable, to cover yourself safety-wise and legally. Use only approved power cables, mains-rated on-off switches, and approved fuse holders. Never make your own power cables. These wiring practices apply to both restorations and to building new amps from scratch.

The outer insulation should protrude pass the strain relief gland 2-3 cm (one inch) to relieve the pressure and strain caused by cable movements. Never pull the mains cable or pick an amp up by its mains cable!

There should be enough slack in the "looped" wiring; the wires should not be straight and tight, which means they were cut too short. A few modern amps had such wiring mistakes.

Notice how the loop for the earth conductor is much longer than for the live and neutral wires; if the mains cable is pulled or becomes detached and touches the metal chassis, the earth conductor will stay connected and protect you from electric shock.

The USA system uses black for the "live" or "phase" conductor's insulation and white for the neutral wire. In Australia and most other countries, the live conductor is brown, and the neutral is blue.

The neutral conductor should go directly to one end of the power transformer's primary winding, while the "live" conductor should pass through the fuse holder and continue onto the on-off switch.

All connections should be insulated by plastic shrouds or heat-shrink, so even if you or someone else accidentally touch them while working on the amp, you will not get an electric shock.

The fuse and the on-off switch must both be in the same leg, the live (L) phase. Of course, you can always use a double-pole mains switch that will have contacts in both Live and Neutral legs, but the fuse must be on the L side!

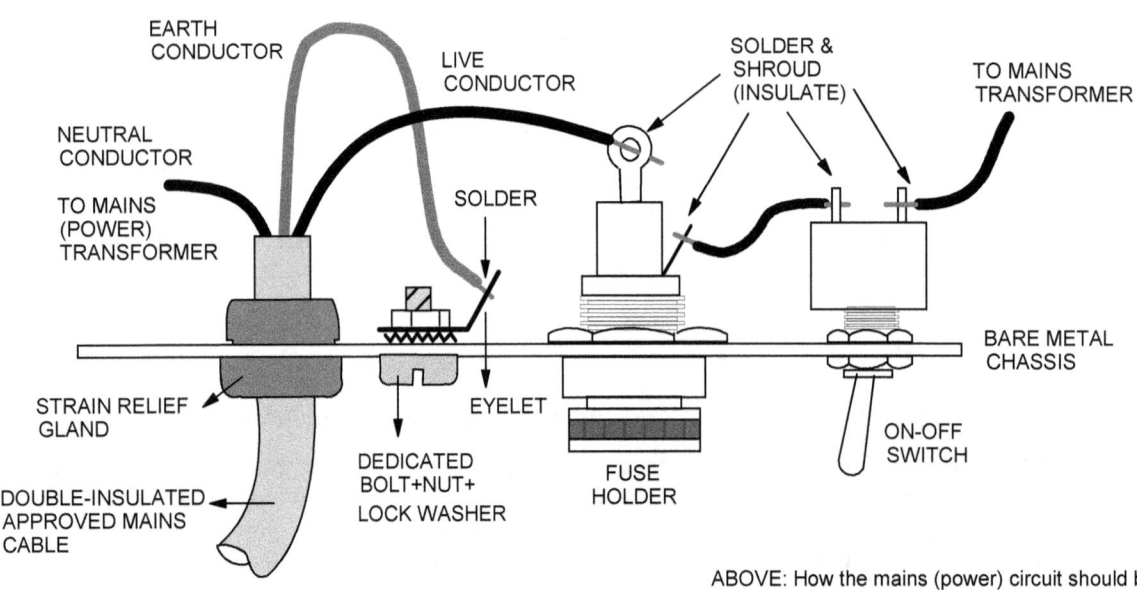

ABOVE: How the mains (power) circuit should be wired up

WIRING, SOLDERING & MODIFICATION PRACTICES

SAFETY WARNING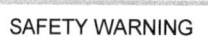

✓ In Australia and many other countries, fuse holders whose tops can be unscrewed by hand and thus live internal contacts accessed are illegal. They were ruled unsafe since even a child could open them up. Only fuse holders that require a crosshead (Philips) screwdriver for top-cap removal are approved and legal.

Wiring IEC combo inlet sockets

Vintage amps invariably had a fixed mains cable entering the chassis through a compression gland, permanently wired to a separate fuse holder and the on-off switch, as shown on the previous page. These days, most modern amps use an IEC power inlet socket, which is often combined with an integral fuse holder (2-in-1 socket). Some even have an integral on-off switch, which usually has an internal neon light - we could call such a combination a 4-in-1 power socket. This saves amp makers construction time and thus money.

Each of these types can be of a snap-in design, with plastic "lips" that snap in place once the inlet is inserted into a rectangular opening in the chassis, so no screws, bolts, or rivets are used. Of course, the chassis hole must be marginally larger than the plastic assembly; otherwise, the whole inlet would not be held in place and could even fall out. A few currently-produced amps suffered from this problem, so the Chinese factory tried to glue them in. Due to high mechanical push and pull forces when the power cable is inserted and removed, such a primitive solution does not work - sooner or later, the inlet will get loose and pop out.

The other type has two triangular lips, each with a mounting hole for bolting it to the chassis. Here your rectangular cutout does not have to be so precise.

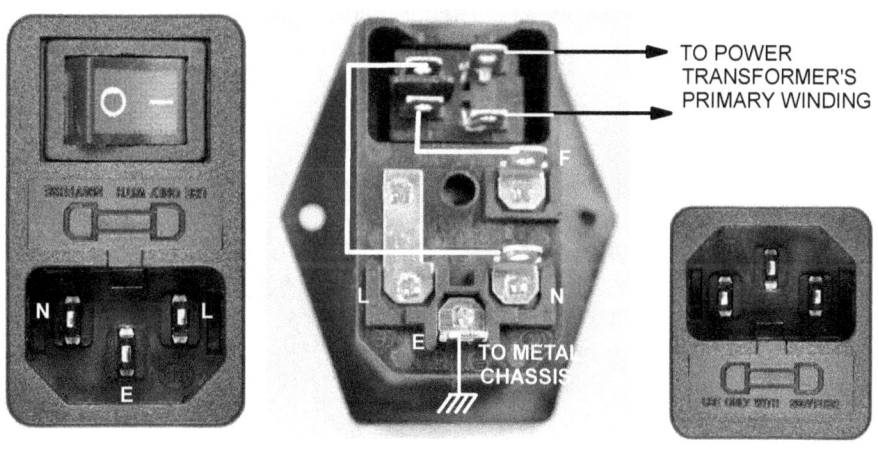

FAR LEFT: The 4-in-1 IEC inlet (snap-in type), with an integral fuse, on-off switch and pilot neon light (inside the illuminated switch)
MIDDLE: The rear view of a similar screw-in type inlet, its terminals and the way to wire them up
LEFT: The 2-in-1 IEC inlet with integral fuse, snap-in type

The on-off switch inside these combo sockets is a separate switch and can be pushed out and used by itself. You will notice that two of its four lugs have very thin wires soldered onto them. These go to the neon indicator inside the switch, meaning these lugs must be on the load side. Usually, the wires leading to the power transformer primary winding are connected there, as illustrated.

The other two lugs need to be connected to the IEC inlet, one directly to the neutral (N) lug, the other to the load side of the fuse or the "F" lug. You should shroud all these mains terminals with a heat-shrink to prevent accidental contact with them while testing a live amp (in operation).

IEC plug & socket

While older amplifiers used permanently wired power cables, modern designs mostly use detachable cables. At one end is the plug of the country the amps are sold in, and at the amplifier end is the IEC plug since the amp's chassis has an IEC socket.

Usually, a "socket" means a female receptacle, but in this case, the "male" plug actually has "female" contacts, which are recessed so they cannot be touched for the obvious reason - they carry lethal voltages if the other end is plugged into a power outlet. Likewise, the "female" IEC socket on the amp's chassis has "male" pins.

AU plug & socket

Australia, New Zealand, Papua New Guinea, China, and Argentina use a 10A plug with two flat 1.6 mm thick blades, set at 30° angle to the vertical. These are live and neutral pins and are insulated with plastic covering (white in the photo but usually black). Older versions did not have such insulation and have been illegal since 2004.

The earth pin also measures 6.3 by 1.6 mm but is longer (20 mm). It makes contact first while plugging it in and breaks the connection last when unplugging an appliance, a safety feature.

As with most other major world plugs, there is also an ungrounded version without the earth (ground) pin, but that type can only be used on double-insulated devices. If unterminated with the IEC plug, as in the photo, most cables come with the earth lug already attached to the earth wire.

EU plug & socket

Most European countries use the so-called "Schuko plug", the name coming from the German "Schutzkontakt" or "protection contact." There is no earth pin. In the vertical plane, the earth clips of different shapes and sizes are positioned on both sides of the plug, while the two uninsulated round pins in the horizontal plane are L (live) and N (neutral). The pins aren't insulated since the EU socket is recessed, a feature that prevents finger contact.

UK plug & socket

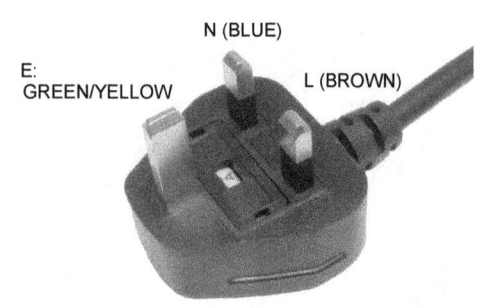

Used mainly in the United Kingdom, Ireland, Cyprus, Malta, Malaysia, Singapore, Hong Kong, and some Middle Eastern countries, this plug has rectangular pins and an integral fuse. As with the AU plug, the L & N pins are insulated, and the uninsulated earth pin is longer. The cable color-coding is identical to AU and EU cables - the "Live" core is brown, the "Neutral" is blue, and the "Earth" is the green/yellow wire.

US plug & socket

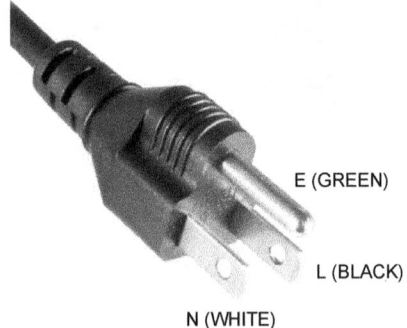

The USA or, more accurately, "North American" power cable (let's not forget Canada and Mexico) is usually rated at 15A and 125V.

As with the Aussie and British plugs, the ground (earth) pin is longer than the line and neutral flat blades, so the device is grounded before the power is connected. There are many other types, variations, and versions of power plugs used in the USA for specific applications such as high power loads, industrial applications, etc., but these are of no interest to us here.

Modifying 220V equipment to run on 240V mains voltage

Most modern tube guitar amps were designed for 115V or 230V mains operation. Some have dual primaries and can operate on either voltage. However, mains transformers in some amps have 220V primary windings. On 230V, that is not a big deal, but when used in 240Vmains countries (or on our 248-252V actual mains voltage in Australia!), all voltages in such amps will be way too high. The same problem happens with 115V amps used in some parts of the USA, where the actual voltage is closer to 125V.

WIRING, SOLDERING & MODIFICATION PRACTICES

When an amplifier burns out a set of tubes a month or two, most guitar players assume that is due to poor quality tubes and that it is normal. It isn't. It is usually caused by the heater and other voltages within the amp being too high. Heater voltages would be, say, $7.2V_{AC}$ instead of $6.3V_{AC}$. The anode voltages would be $460V_{DC}$ instead of $410V_{DC}$, meaning all filter capacitors rated at $450V_{DC}$ would be stressed out and could explode.

Screen voltages for output pentodes would be too high, so both anodes and screens would draw high currents, and tubes' life would suffer.

Another very common cause is poor design, where a manufacturer would push power tubes over their maximum allowable power dissipation limits. A higher mains voltage would exacerbate that problem (make it worse) and increase power dissipation/decrease tube life even further!

Replacing the power transformers in such amps is certainly an option, but a quality replacement power transformer would set you back US$100-200 or even more for a high-power amp. Luckily, there are two cheaper and easier alternatives.

Option 1: Step-down autotransformer

An autotransformer does not have two separate (isolated) windings. A single winding has a tap to which the load is connected (if the transformer is a step-down type). A step-up effect can also be achieved if the connections of the mains input and the load are reversed.

N_0 = total no. of turns (P_1-P_2), N_L = load turns (S_1-S_2)

The beauty of auto-transformers is that their power rating can be only a fraction of the power supplied to the load. The input and the output (load) currents flow in opposite directions through part of the winding, effectively canceling some of the magnetic flux and dissipated power.

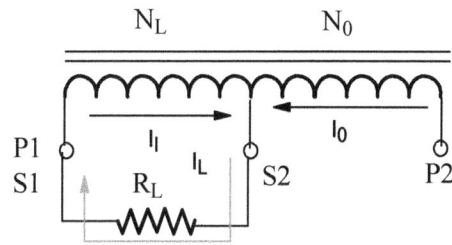

A step-down transformer. P1 & P2 and higher voltage (primary) connections, S1 &S2 are secondary taps.

POWER RATING OF A STEP-DOWN AUTOTRANSFORMER	QUICK CALCULATION

The power rating of the autotransformer is $P = P_L(1-V_L/V_M)$. The closer the step-down or load voltage V_L is to the input or mains voltage V_M, the lower the needed power rating of an autotransformer! Say your mains voltage is 240V, and you need a 240 to 220V autotransformer to power a tube amp designed for 220V, with 200 VA consumption. The power of the core needed is only P=200*(1-220/240) = 200*(1-0.917) = 17 V!

If you need a step-down isolation transformer to power 120V mains amplifiers (from the USA, for instance) on 240V mains, a transformer with separate windings would need to be rated at least at the load power (say 300VA). With a 240/120V autotransformer, its power rating would be 50% of the load rating; in this example, 50% of 300VA or 150 VA. An autotransformer is a smaller and thus cheaper solution than an isolation transformer!

Option 2: Drop-down mains transformer trick

Get a standard low voltage mains transformer of a suitable VA (apparent power) rating (minimum 30VA), and connect it as per the diagram on the right. You are dropping part of your mains voltage across the transformer's secondary winding and taking the reduced AC voltage off its primary winding.

In this case, the secondary voltages are symmetrical (15-0-15V), but they don't have to be; we only need one secondary voltage, say 0-24, 0-28, or 0-30V. Instead of the nominal 240VAC, our workshop's mains voltage is 248V, so we got 218V out, perfect!

The same trick can be applied to amplifiers designed for 100VAC mains (Japan) in 115VAC countries such as the USA.

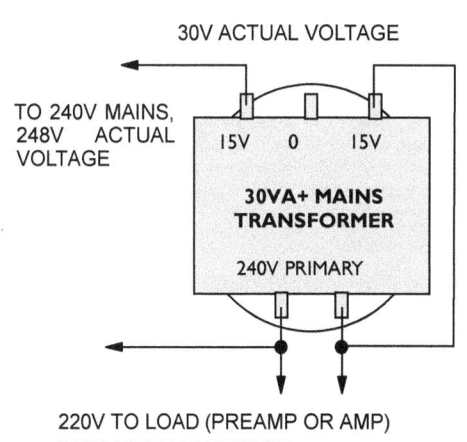

How to use a small low voltage transformer to run 220V amplifiers on 240V mains

SOLDER & SOLDERING

Types of solder

The job of the solder is to bind components, terminal strips, printed circuit boards, tube socket lugs, and wires together. As the name suggests, Flux-core solders have a core of flux, whose job is to "reduce" metal oxides on the surface of components to be soldered and improve the mechanical contact between the solder and the metal surface. To "reduce" means to "return oxidized metals to their metallic state." It also helps solder flow (spread out). The alloys most commonly used in electronics are 60/40 tin/lead (Sn/Pb) and 63/37 Sn/Pb solder.

Health hazards

In 2006 the European Union Waste Electrical and Electronic Equipment Directive (WEEE) and Restriction of Hazardous Substances Directive (RoHS) became law in the EU, effectively making lead illegal. Lead-free solders, such as Sn-Ag-Cu (Tin-Silver-Copper) and An-Ag-Cu Antimony-Silver-Copper) have since gained popularity even outside the EU. As of this writing, there are no federal laws in the USA prohibiting lead in solder.

Lead is very toxic, affecting the brain and the nerve system, and the human body cannot eliminate it. Luckily, it does not get absorbed through the skin, so touching lead solder is not a hazard, and there is no lead in the solder fumes. The most significant risk is the oral transmission of particles from your hands through the mouth, so washing your hands after touching anything electronic (new components, solder, old amplifiers, etc.) is mandatory! Just because there is no lead in solder fumes does not mean they are harmless. Antimony, bismuth, germanium, indium, and other elements are present in solder and its fumes. The primary source of fumes is the flux and not the solder. The only sources that claim that solder fumes may lead to lead poisoning are companies that sell fume extraction equipment.

Proper soldering

A brand new soldering tip must be tinned first, i.e., covered with a thin coat of solder. During use, the tip of the soldering iron will get dirty, covered in gunk. Gently clean it by rubbing it against a wetted sponge. Press the tip onto the wires/contacts to be soldered together for a few seconds. Then touch the hot joint with the solder. The solder should melt upon contact. If it doesn't, the connection hasn't been heated enough.

You should never place solder onto the tip of the soldering iron, only on the joint to be soldered! The more component leads-wires and the larger they are, the longer it will take for the joint to heat up enough to take on the solder and melt it. Knowing how long to heat the joint comes with practice.

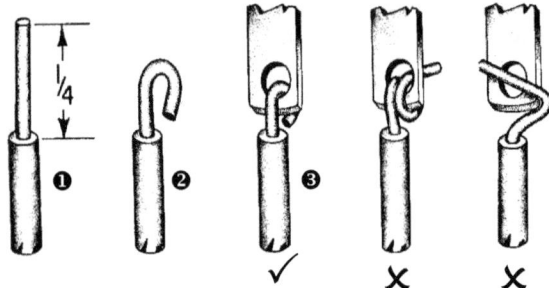

ABOVE: The three steps involved in preparing a wire for soldering onto a terminal or tube socket lug. The copper lead should only be looped once through the hole, so future unsoldering is fast and easy. The two rightmost examples are wrong, one being a double loop, the other an improper one.

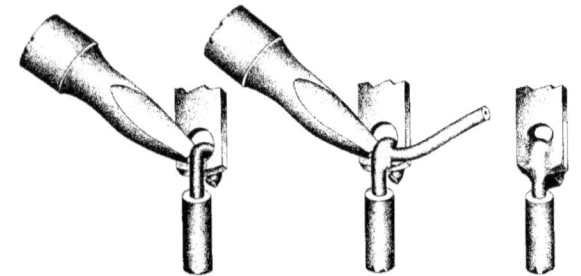

ABOVE: Once the copper lead is looped through the hole, the tip of the soldering iron should be touching both the lead and the lug, thus heating both together. The solder is then brought against the lead and the lug and melts to cover both without even touching the soldering iron's tip.

The solder should flow smoothly and freely and should cover the whole joint relatively quickly. It should not 'blob" and fall off. If that happens, the surfaces are not clean enough; they are dirty, greasy, or oxidized and must be cleaned before attempting to solder the joint again. The tip should be touching the component lead or wire and the terminal lug or pin simultaneously.

Large joints with many wires/leads may need to be heated from various sides, not just at one point. Don't just heat one lead entering the joint; try to touch two or three together. Otherwise, one will be hot, and others will be cold and not accept the solder, so "cold joints" may develop. Again, your skills will improve with practice.

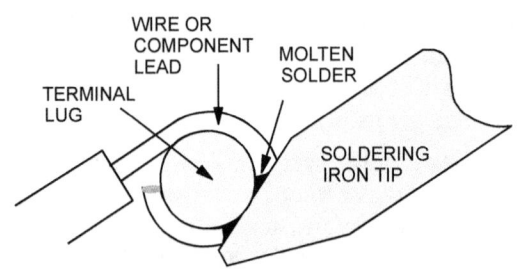

ABOVE: The tip should be touching both the lead and the lug, thus heating both together.

WIRING PRACTICES AND PRECAUTIONS

Color-coding for chassis wiring

There was a wiring code in everyday use in vintage equipment, but not all manufacturers adhered to it, and kits were often wired willy-nilly. You may follow this code if you wish or settle on your own if building a number of amplifiers.

Red and violet denote positive and negative DC supply lines, orange is the screen grid wiring, yellow is the cathode, green is the control grid, and blue are the anode connections. The AC lines are colored gray; white is used for AVC (automatic volume control) circuits and feedback paths. A red wire with blue stripes is used for regulated positive power supply lines and black for ground connections.

Of course, more complex circuits deviate somewhat. For instance, if there is a CT heater transformer secondary, the two heater wires will both be brown in color while the grounded center tap will be black. Or, in cathode followers, if there is no anode resistor, you can use a red $+V_{BB}$ wire straight onto the anode lug on the tube socket or a blue wire direct to a $+V_{BB}$ terminal.

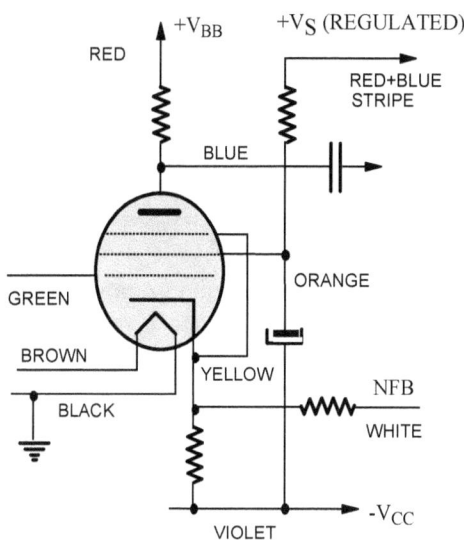

ABOVE: Not many manufacturers stick to this vintage wiring standard these days, but here it is should you decide to follow it! A pentode stage is shown but the code obviously applies to triodes as well.

The wiring order

Heater wiring should always be done first and placed against or closest to the chassis. Once the mains and heater wiring is done, you can plug the tubes in, power the amp up and check the wiring and heater voltage under load before the rest of the building/wiring is done.

Connections between adjacent terminals should be wired next. If one tag cannot accommodate all the components that must be wired to that point, two or three adjacent lugs should be strapped together. Connections to distant terminals should then go on top of the adjacent terminal links.

Resistors and capacitors that may need to be changed during the fine-tuning of the finished amplifier should be wired at the very top, so there is easy access to them. Plus, the distance D between capacitors and the metal chassis should be as large as possible to minimize the parasitic capacitances.

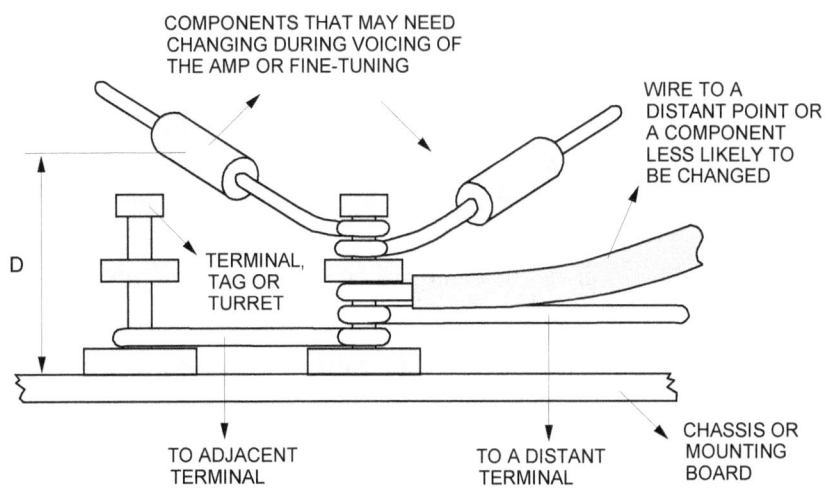

ABOVE: The stacking or soldering order for point-to-point wired terminals or lugs

All AC wiring runs must be twisted together or twisted and shielded (even better).

After twisting, these mains and heater wires should be routed as close to the chassis corners as possible and secured to the chassis by cable ties or other means, so there is absolutely no movement. No other cable must be run parallel to them or, heaven forbid, tied to them in a bundle!

The earth wire from the mains cable must be firmly connected to a dedicated bolt secured to the metal chassis body. The electrostatic shield of the mains transformer and the main star point are also connected to that point.

Shielded cables should be used for longer runs of signal-carrying wires, and their shields must be grounded at one end only. The illustration on the left shows how to terminate a shielded cable.

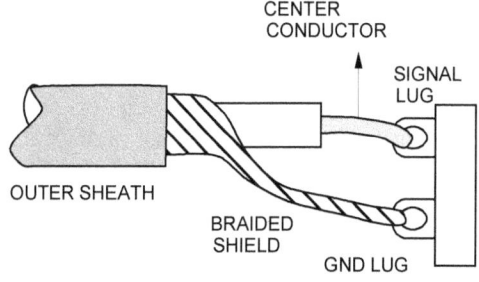

ABOVE: How to terminate and ground a shielded cable

REDUCING HUM: ISSUES AND SOLUTIONS

Getting the topology right

A quiet and reliable amplifier starts with optimal topology. The rules are simple. Keep the low signal level (preamp) stages, the input jack, and the preamp tube as far from the power supply and the speaker (in combo amps) as possible.

The on-off switch and the guitar input jack should be on opposite sides of the chassis or the control panel. The speaker output jack can be closer to the mains (power) inlet since its signal voltage levels are higher (1-20V), but still do not put them too close, right next to each other.

The rectifier tube is not critical and can be placed very close to the power transformer and the filtering choke (if any).

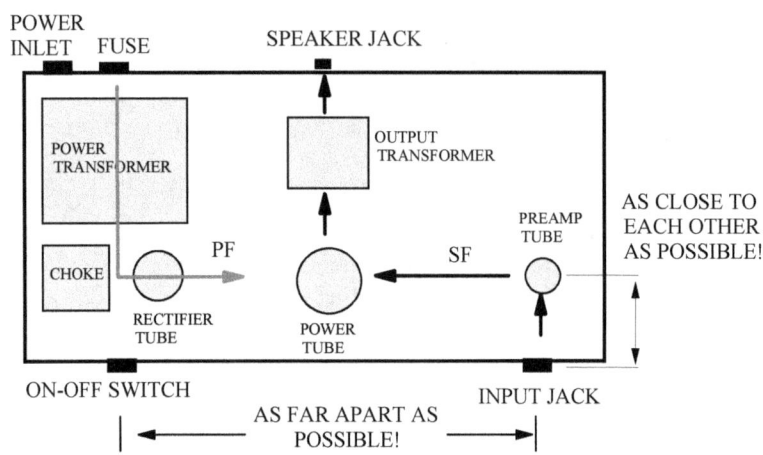

ABOVE: Optimal topology and signal flows in a guitar amplifier

Likewise, the output tube(s) can be relatively close to the power supply, but in combo amps, keep them as far from the speaker's magnet as possible.

The typical topology illustrated above shows signal flow (SF) and power flow (PF), which should progress in opposite directions. The power supply (transformer, choke, rectifier, and RC filter components) should be on the power tube's side, never on the preamp's side of the chassis.

According to the MSLP (Minimal Signal Path Length) rule, component leads should be as short as possible, and signal-carrying wiring length should be minimized. Component leads and wires act as antennas, receiving unwanted audio and radio frequency signals (interference) and transmitting them into their surroundings. The lower the signal involved, the more important that rule becomes.

Input stages are more susceptible to hum and interference than the output stage. Plus, once the unwanted hum or interference enters the audio circuit at the input, these unwanted signals are amplified together with the guitar signal. The two are impossible to separate or filter out later on! So, place the input jack as close as possible to the volume control or gain potentiometer and the input tubes.

To minimize the possibility of hiss, hum and buzz:

1. If the metal chassis used is not monolithic (in one piece, welded or machined out of a single piece of aluminium), make sure it has galvanic continuity between all its panels and parts.
2. Shield all magnetic components (chokes, power-, input-, interstage- and output transformers). Use shielded cables in grid circuits, between input jack and preamp tubes, and for negative feedback runs.
3. Shield preamp tubes (it may require changing tube sockets over to shielded types on existing amplifiers).
4. Solder grid-stopper resistors directly onto the pins (lugs) on the socket itself.
5. Route AC (mains and heater) wiring as far as possible from the preamp stage. Shield it if necessary.
6. If possible, change 6.3 V_{AC} heating to 12 V_{AC} or even 24 V_{AC}. The strength of the radiated magnetic field depends on the current, and the heater current at 24V is four times lower than at 6V!

Heater wiring

Currents flowing through conductors produce magnetic fields. Since heater wiring carries the highest currents in an amplifier, it is essential to take precautions to minimize the possibility of electromagnetic coupling with the rest of the audio circuitry. Here are the rules that should never be broken.

If you are designing an amplifier from scratch and have a choice of power transformers and tubes, use as high heating voltage as possible. That will reduce the heater currents and their magnetic fields. So, with dual heater tubes such as 12AX7 and 12AT7, 12.6V is better than 6.3V!

Keep heater wiring close to the chassis and separate from other wires, especially signal-carrying and mains (AC power) wires. Never bundle heater wires with any other wire to make the wiring look "neater"!

Twist heater wires together along the whole length. Avoid loops around tube sockets.

WIRING, SOLDERING & MODIFICATION PRACTICES

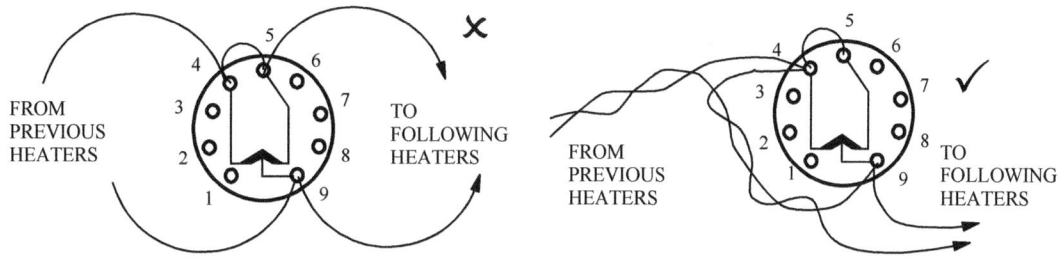

ABOVE: Never loop a heater wire around a tube socket!

There are three main types of AC heating circuits, illustrated below. If the secondary heater winding has a CT (Centre Tap), it must be grounded. However, a less known trick is connecting it to a point with a low DC voltage instead, for instance, to the power tube's cathode (s). Such elevated heater voltage helps minimize hum and increase the S/N ratio, with improvements of up to 10 dB!

Biasing to a higher voltage point, determined by the resistor ratio and B+ voltage, requires a voltage divider R_1-R_2 but allows higher voltages than those available on the power tubes' cathodes (typically 10-30 Volts).

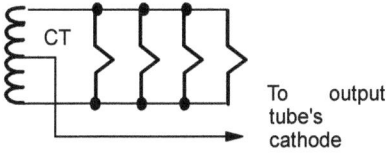

Adding a DC bias to the heaters by connecting the CT to the lower voltage on the cathode of the output tube (ABOVE) or to the high anode voltage power supply through a voltage divider (RIGHT)

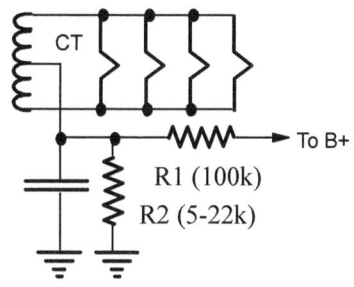

R1 (100k)
R2 (5-22k)

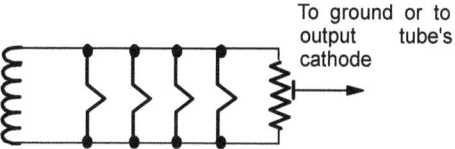

The humbucking trimmer potentiometer acts as a voltage divider to find the point of minimum hum.

The most common hum minimizing solution in vintage amps was to connect a humbucking potentiometer across the heaters. The potentiometer acts as a voltage divider to find the point of minimum hum. It can be done by measuring the AC hum voltage at the speaker terminals or by ear. It is very rarely used in modern amps.

If the heater winding has no CT, one of its ends must be grounded, or a strong hum will develop. Some designers create an artificial CT by connecting identical resistors between both heater's winding ends and ground.

Proper grounding practices

Rule #1 of proper wiring is *not* to use the chassis as the ground bus. Although some manufacturers do it, it is not recommended unless the chassis is plated in a thick layer or high purity copper, and most aren't.

Option 1: A single star point. The first option is to connect every grounded point by its separate wire to one star-point (so-called star-grounding method). That point is usually the grounded lug of the first filter capacitor in the power supply. However, in a large and intricate amplifier, connecting every grounded component to the same star point is not practical.

Option 2: Multiple star points. Each stage or section will be connected to its own star point, and then all those local star-points are connected to the main star-point (illustrated above right)

Option 3: A ground bus. A thick ground bus or "rail" is run around the amp, and all grounded component ends are soldered to it. Again, the bus is to connected (grounded) to the chassis at one point only (illustrated on the right).

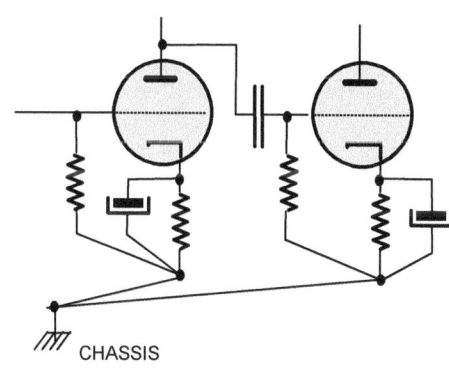

ABOVE: Each stage has its own star ground point, and then all are again connected in a global star arrangement.

BELOW: A ground bus makes component placement easier and neater looking than the star grounding method.

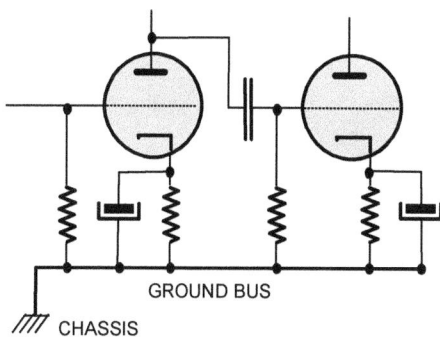

The audio common - chassis - mains ground connection

Three main grounding (earthing) methods are illustrated below. There is no connection to the chassis or mains earth in the floating audio common or audio "ground" arrangement. The most common arrangement is b), where the audio ground and chassis/mains ground are permanently connected. The "ground lift switch is optional; it is used to break the ground loops if a hum develops.

Option c) illustrates the RC connection between the audio ground and chassis/mains ground. Notice that the chassis is always grounded or earthed by a fixed and secure connection to the mains earth terminal for safety reasons!

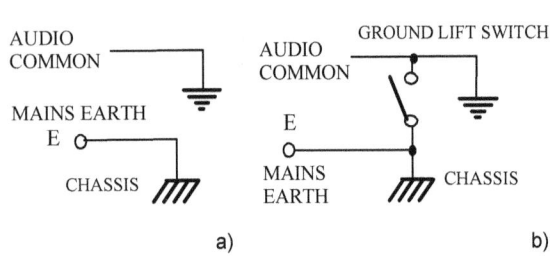
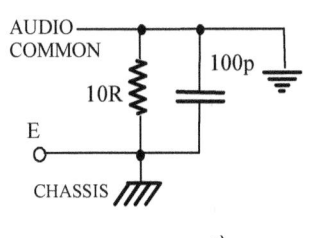

LEFT: Three main grounding - earthing methods:

a) floating audio common or audio "ground" - no connection to the chassis or mains earth

b) audio ground and chassis/mains ground connected together, with an optional "ground lift" switch

c) RC link

SAFETY WARNING

To eliminate hum, many "repairmen" and guitarists (even some ignorant manufacturers!) install "ground lift" switches in guitar amplifiers in such a way that both the audio common and the chassis are disconnected from the mains earth. Not just illegal but potentially deadly.

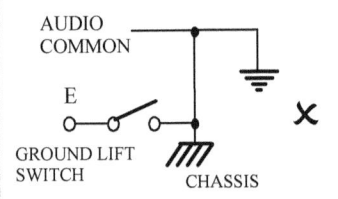

WRONG: MAINS EARTH (GROUND) DISCONNECTED FROM THE CHASSIS

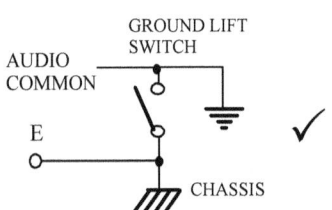

MAINS EARTH STAYS CONNECTED TO THE CHASSIS, ONLY AUDIO COMMON LIFTED

Any short circuit or malfunction can make the chassis "live" and not even RCD circuit breakers or ground-fault interrupters will operate if the earth connection to the equipment is lost or disconnected! A few guitar players were electrocuted on stage for this very reason.

Power transformer mounting

In this vintage amplifier, the whole power supply section is on a separate "floating" steel base, on rubber mounts, so the vibrations from the mains transformer are not transmitted onto the rest of the chassis (1). Ventilation holes were also punched out around the rectifier tube (2), always a good idea. Notice chassis damage, most likely from a corrosive electrolyte leaking out of the overheating filtering capacitor (3)!

The power transformer and filtering choke should be mounted vertically above the chassis. The chassis will then act as an electromagnetic screen between them and the sensitive audio circuitry underneath.

The worst way to mount transformers is to have them sit horizontally, recessed into the chassis through a rectangular hole. The shielding integrity of the chassis is seriously compromised in such a case, exposing the audio circuitry to the EMI (Electromagnetic Interference) from the recessed transformers.

LEFT: The whole power supply suspended on a floating sub-chassis

RIGHT: The worst method of mounting transformers, recessed into a chassis (side view)

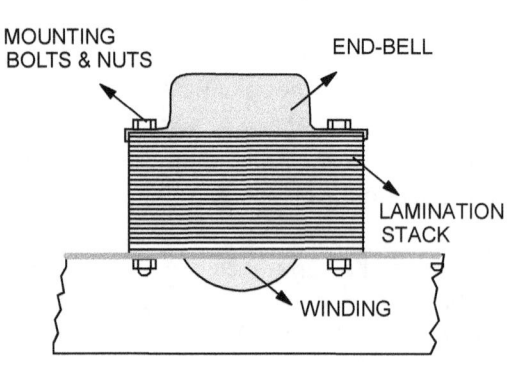

WIRING, SOLDERING & MODIFICATION PRACTICES

Transformer positioning

When open construction transformers are used (not fully enclosed in metal boxes), to minimize the unwanted magnetic coupling, position the power and output transformers at 90° angle and separate (distance) them as much as possible. Even if full metal enclosures are used, it is still a good practice to keep transformers inside them at 90° with respect to one another.

A power supply filtering choke (if used) can be close to the power transformer, but the same rules for power transformers apply to chokes, i.e., they should also be kept as far away as possible from preamp tubes and output transformers!

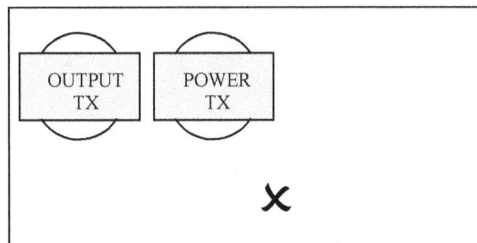
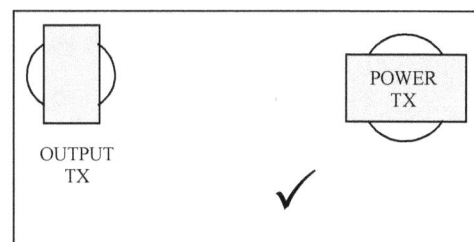

HOW TO POWER AN AMPLIFIER UP FOR THE FIRST TIME

Once you finish the wiring of the amplifier, or once you have finished replacing all the required components on a vintage amp you are upgrading or repairing, the best action is to have a break and sleep on it. Tidy up the bench and come back to the testing stage the next day. Amplifier building requires high levels of concentration, and after four or more hours, you will feel tired, not so much physically but you will feel "drained" psychologically.

Even if your wiring hasn't been completed, how can you tell if it's time for a break? Simply, once you notice that you are slowing down or starting to make mistakes. Of course, in many cases, we aren't even aware that we are making mistakes, and that is precisely the reason behind the suggestion of leaving the checking and testing for another day. It would be best if you were fresh and rested for this crucial stage.

Step 1: Cold checks

Cold checks are checks without any power applied to the amplifier. The first step is a visual and "dental" check. Just as your dentist uses a mirror and a sharp tool to poke around your fillings during your dental checkup, you should similarly check every soldering joint.

I wish I had a dollar for every cold soldering joint I found, and for every time I found one component lead not soldered properly with three or more of them going through the same lug. That is one of the drawbacks of point-to-point wiring. For instance, with three resistors and two capacitors connected to the same terminal lug, you may think all five have been firmly soldered together, when in fact only three are OK, while the other two leads are just sitting there without a proper contact. So, take small pliers and pull firmly on each component lead to check its mechanical integrity and the strength of its soldering joints.

Once you have completed the visual and mechanical joint checks, take an ohmmeter and check for continuity of all circuits. I usually start with the mains circuit. Make sure you insert a fuse into the fuse holder. With the on-off switch in the "off" position, you should get infinite resistance between live and neutral and live and ground (chassis). Turn the power switch to the "on" position, and you should get a few ohms (less than 10) between the live and neutral pins on the power cord. That is the DC resistance of the primary winding of the mains transformer.

Next, check the heater wiring for continuity and make sure it is wired to the correct pins on tube sockets. Check the ground and all components and points connected to the ground, including the shields of all shielded cables. The easiest way to do that is to use a crocodile clip on one of the ohmmeter leads (say the black one), clip it onto the star ground point, and use the red probe for testing.

Then clip the negative to +VA point (high voltage for anodes) and check continuity to all the points that should be connected to it, such as one side of anode resistors or the output transformer's primary winding.

Step 2: DC voltages check with tubes plugged in but with no HV on anodes

If all seems OK during cold checks, plug the tubes in (if you haven't done so during the cold checks), disconnect the main HV supply wire (+V_A output of the power supply) and power the amplifier up. First, measure heater voltages on the rectifier tube socket (if used) and all other tube sockets.

If heater voltages are within +/-5% (for instance, from 6V to 6.6V if 6.3V heating is used), measure the DC voltage at the output of the power supply filtering section. Since there is no load (you've disconnected its output in the previous step), it should be 5-15% higher, say 440V instead of 400V. This elevated voltage will also be a good test for the first power supply capacitor (the following ones are still de-energized). If it's marginal (old, dried-out, leaky, or in any way compromised), it may blow up, so stand aside and don't put your face too close to the amplifier. Of course, you've tested all the components before installing them in the amplifier, haven't you?

However, a multimeter or LCR meter test is one thing, and a full 400+ Volts across a capacitor is something quite different!

Measure DC bias voltage (if fixed biasing is used) for output tubes. If there is an adjustment (trimmer potentiometer), it should work and change the voltage up and down. Adjust it for maximum negative voltage on the grid pins of output tubes.

Step 3: AC & DC voltages checks

Now turn the amp off, reconnect the main HV supply wire (+VA output of the power supply) to the rest of the circuit, and power the amp back up. Again, stand away from the chassis. If any electrolytic capacitors were wired the wrong way, they would explode at this point, and the smoke and gunk from their insides will spew all over the chassis. You don't want that toxic stuff in your face or eyes. If nothing smokes or explodes, that means that there are most likely no catastrophic wiring errors or shorts.

After a minute or so, after the tubes have warmed up, measure all DC voltages again. The cathodes of preamp tubes should be at around 1-4 V_{DC} unless DC coupling is used. In that case, a voltage of around +100V_{DC} will be present on one or more cathodes, and there should be a voltage drop across their anode resistors.

For instance, there should be 250 V_{DC} between one end of the 100k anode resistor and ground and 150 V_{DC} between the other end (anode pin) and ground, or 100 V_{DC} across it (meaning the current is 1mA).

The easiest and safest way is to use a crocodile clip, fix the multimeter's black lead to the ground point (common), and use only the right hand for measurements between various points and ground! Please don't touch anything with the left hand; it should only be used in emergencies, for instance, to switch the power off quickly should anything go wrong.

For amps using a fixed bias, slowly decrease the negative bias voltage by adjusting trimmer potentiometers until the output tubes reach their nominal anode current. Bias cannot be adjusted in some fixed bias amps (the bias is well and truly fixed!), so there will be nothing to do but measure the bias and confirm that it is within the prescribed range of values.

For amps with self- or cathode bias, measure the DC voltages between the cathodes of the power tubes and ground, and calculate the quiescent currents to make sure all are within safe limits.

POWER SUPPLY MODIFICATIONS AND IMPROVEMENTS

- CLEANING UP THE POWER SUPPLY
- SIMPLE POWER SUPPLY MODS & IMPROVEMENTS
- MAKING TRANSFORMERLESS AMPS (WIDOW-MAKERS) SAFE & LEGAL
- HOW TO FIX DC VOLTAGES: JOYO "SWEET BABY"
- IMPROVING POWER SUPPLY FILTERING: PANAMA CONQUEROR
- THE OUTPUT TUBE OPERATED WAY ABOVE ITS MAXIMUM DISSIPATION LEVEL: EPIPHONE ELECTAR TUBE 10
- VARIABLE POWER CONTROL ADDED: BLACKHEART LITTLE GIANT

While contemporary amp designers and builders focus on audio circuits of guitar amps, many tend to overlook their power supplies. A power supply is a heart of an amplifier, and just as with humans, a weak heart affects the rest of the body and does not lead to a powerful performance.

Inadequate filtering (high ripple) of high voltage DC supply is a common issue in budget amps which, to reduce costs, do not use filtering chokes. This could result in audible hum, which can also be caused by poorly designed (no electrostatic shield) and/or positioned power transformers.

The heater and anode supply voltages in many amps are way too high, so tubes operate above their maximum allowable voltage and power dissipation limits. This manifests itself through overheating, poor amp reliability (various component failures), and very short tube life, requiring frequent tube replacement.

Luckily, most of these design and construction bloopers can be quickly and easily fixed. In this section, we show you how.

CLEANING UP THE POWER SUPPLY

Overvoltage protection using varistors

There are two basic types of passive protective circuits. Voltage clamps, typically using Zener diodes or varistors, are overvoltage protectors; they only reduce the level of a voltage spike without changing the rate of change or oscillation of the waveform.

Snubbers perform both functions; they reduce the magnitude of the spike but also change its spectral content and reduce the radiated RFI (Radio Frequency Interference). Although varistors are classified as semiconductors, they have no PN-junctions. Varistors are made from silicon carbide mixed with a ceramic binder. The material is then pressed or extruded into the desired shape (disc, rod, washer) and sintered, resulting in hard ceramic-like material.

Zinc oxide varistors have superior properties, a sharper I-A characteristic. Polycrystalline zinc oxide is mixed with molten bismuth oxide and sintered. The bismuth oxide forms a rigid coating around the zinc oxide grains, so the varistor acts as an open circuit at low applied voltages. When the voltage across it exceeds the value of its "clamping voltage" V_C, the varistor suddenly changes its properties from a very high to very low resistance and conducts.

This makes it the most straightforward, cheapest, and most effective overvoltage protective device and a transient suppressor. Due to its symmetrical I-V characteristic, it works on both DC and AC circuits, in parallel with the load. The voltage ratings range from 12 to 1,000V, 275V-rated varistors are most often used on 220-240V mains circuits. Of course, the voltage spike or transient energy has to be dissipated somewhere, so varistors are also rated in terms of their energy-absorption capabilities, up to 160 J (Joules). The peak current-handling ratings range from 10 to 2,000A!

Snubbers

Snubbers are circuits used to reduce voltage spikes caused by the effects of circuit inductance when a mechanical or semiconductor switches open. The sudden interruption of current flow through an inductive element causes a sharp induced voltage across the current switching device. Such voltage transients are not only a significant source of electromagnetic interference (EMI) but can damage or destroy the sensitive semiconductor switches (transistors) if their voltage ratings are exceeded during the pulses.

Protecting the power switch and eliminating RF interference

When an amplifier is switched off, the mains transformer's primary current and its magnetic field suddenly collapse. Due to self-inductance, the energy stored in the magnetic field tries to keep the current constant by preventing it from collapsing. The induced voltage is the first derivative of current, $V=-LdI/dt$, and can exceed 1kV.

This high transient voltage causes a spark between the contacts of the on-off switch, which will get damaged in this oft-repeated process and eventually burn out.

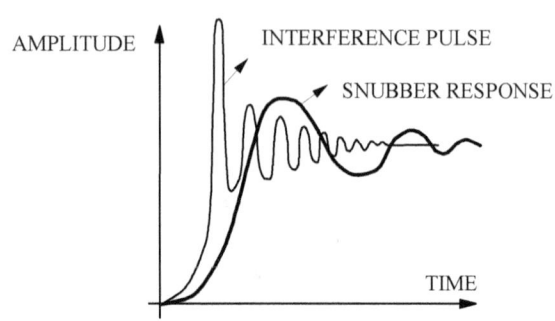

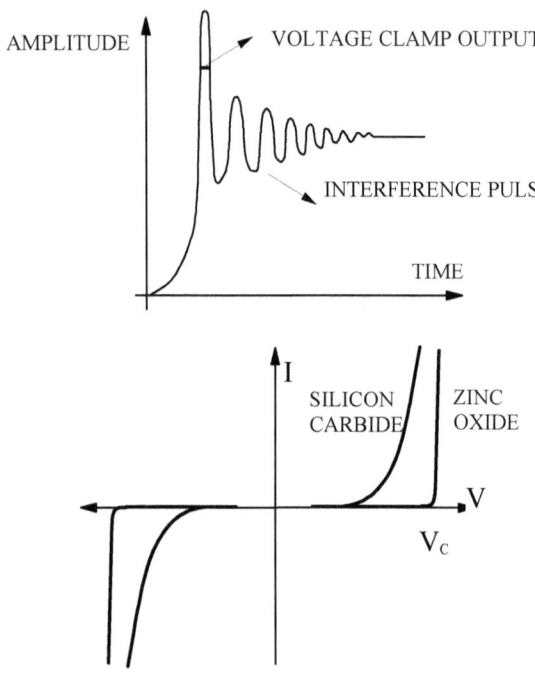

Snubber's oscillatory response to an interference pulse (ABOVE) compared to the output from a voltage clamp (BELOW)

ABOVE: The current-voltage characteristics of silicon carbide and zinc oxide varistors

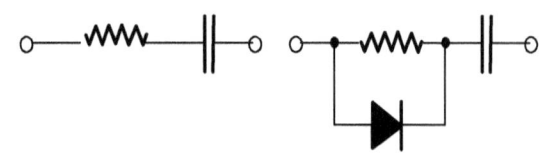

ABOVE: The RC snubber and RCD snubber

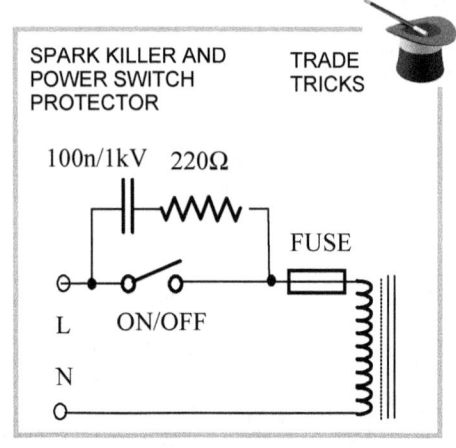

Furthermore, the spark created during this turnoff period is of oscillatory nature, a dampened sine wave, producing a rich cocktail of high-frequency harmonics known as radio interference.

The spark killer is a series RC network across the mains switch contacts, also known as a "snubber." Most manufacturers don't include snubbers in their amps so they can save 50 cents on these two components, but as an amateur, you know better. Snubbers resemble film capacitor, with both components insude a plastic case. The printing on the case will specify both values, for instance, "100n + 330R".

Rectifier diodes are often bypassed by low-value ceramic capacitors, which suppress HF and VHF radiation caused by diodes switching on and off at the mains frequency rate.

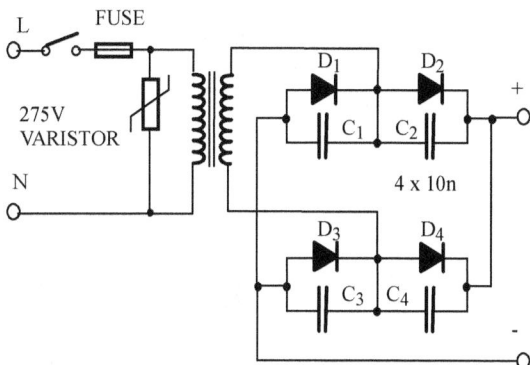

ABOVE: A typical linear power supply with varistor protection and diodes bypassed by RFI reduction capacitors

SIMPLE POWER SUPPLY MODS & IMPROVEMENTS

Using NTC thermistors for high voltage "slow start"

When a tube amp is switched on, there is a current spike through the heaters of all tubes, which are cold (tube filaments have a PTC (positive thermal coefficient) characteristic, meaning their resistance goes up as they heat up. The heating voltage also drops down due to poor regulation of power transformers (unless stabilized DC voltage is used for heating). These current surges stress out the filaments and shorten the life of the tubes.

NTC (negative thermal coefficient) thermistors ("thermal resistors") have a high resistance in a cold state and a very low resistance in a hot state, a property that makes them ideal as limiters for inrush currents.

An NTC thermistor connected in series after the rectifier will limit the no-load voltage rise and the inrush current spike, and in turn reduce the stress on all components. The "cold" resistance value of a thermistor is typically referenced at 25°C (abbreviated as R25). For most applications, the R25 values are between 100Ω and 100 kΩ. Typically, thermistors for this application will have a cold resistance of 3,000Ω and a "hot" resistance of 30Ω.

Two spots where thermistors could be used. In 1), connected between the CT (Center Tap) and ground, the thermistor provides a "soft start" feature, delaying the HT voltage from a directly heated rectifier tube, giving indirectly heated preamp and power tubes enough time to warm up before the full plate voltages are applied.

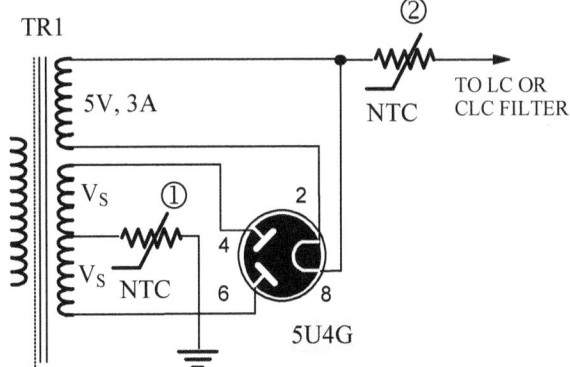

In 2) the thermistor limits the switch-on current surge. This arrangement can be used with any tube or solid state rectifier, but is not fail-proof. Let's say you are carrying out listening evaluations of various preamp or output tubes. When you turn the amp on at the beginning of the session, the thermistor is cold and limits the rush-in currents.

When you turn an amplifier off to change the tubes, the thermistor will be in its warm state and its resistance will be low. If you turn the amplifier on again after a minute or two, the thermistor will still be warm and will not limit the inrush current!

Motor start and audiophile metallized polypropylene high voltage capacitors

One of the cheapest and easiest upgrades you can make to any vintage or new amp is to replace electrolytic power supply filtering capacitors with film caps. If you like brand names, the 47mF "Solen Fast Cap" is of metalized polypropylene construction, rated at 630VDC, and sells for US$22.95 each. Since it has exposed axial leads, it cannot be mounted above the chassis, but there are polypropylene film caps that can -motor start (MS) capacitors. Some have plastic cases; others are in metal housings. The same metal clamps for securing electrolytic capacitors can be used for these bolted to the chassis from underneath.

Most of the currently produced units come from China, and higher capacitance units (30-47mF) generally cost under US$10. However, beware, not all motor start capacitors are film type; many are bipolar electrolytic caps, and these aren't sonically any better than unipolar ones.

Since the sellers do not emphasize this fact, two methods can be used to identify them. If the item description does not mention either "polypropylene" or "film," it is most likely a bipolar elco. Secondly, they pack higher capacitances into the same size cases, as in the photo below, typically three times higher, 150mF versus 30mF for polypropylene film.

Motor start caps are always rated in AC volts, but you can easily convert that to DC volts; simply multiply the AC rating by 2.82 (or approximately three times, which is easier to remember)! So, a 450 V_{AC} rated MS cap is good for up to 1,270 V_{DC}!

Four typical made-in-China motor start capacitors. L-R: polypropylene film with snap-on terminals (450 V_{AC}), polypropylene film with integral leads, bipolar electrolytic with snap-on terminals, and bipolar electrolytic with screw-type terminals.

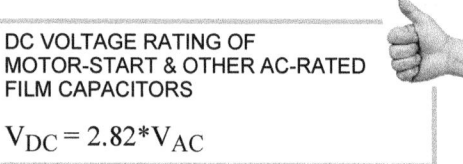

DC VOLTAGE RATING OF MOTOR-START & OTHER AC-RATED FILM CAPACITORS

$$V_{DC} = 2.82 * V_{AC}$$

Reducing the high voltage with a Zener diode

Since the voltage across a Zener diode does not vary with an increased power draw of the amplifier, inserting it between the CT and GND reduces the high voltage supply by a fixed amount (nominal Zener voltage) and thus reduces the maximum output power. The illustration shows silicon diodes as rectifiers, but the principle works with tube rectifiers as well.

The current flows from GND to CT, so the Zener diode's cathode (K) must be grounded. Normal polarity Zener diodes in DO-5 type metal cases have their anode connected to the case.

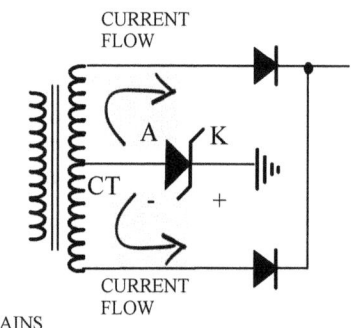

The easiest way to mount the diode would be to bolt it onto the metal chassis, acting as a heatsink. However, the amp's chassis is grounded, meaning the anode would be grounded instead of the cathode, so the Zener diode would not work as a Zener diode at all but would work as a standard diode, with forward bias. So, a reverse polarity Zener diode is needed, in which, you've guessed it, the case is the cathode.

Zener diodes, like all components, have maximum power ratings. We multiply the rated Zener voltage, say 50V, by the power draw of the high voltage circuit. For a small single-ended amp that may be 50mA in idle and 100mA at maximum power. So, the power dissipated on the Zener diode would be P=I*V= 0.1*50=5 Watts. Of course, we have to allow some thermal margin, so a 10 Watt diode would be the absolute minimum, a 20 Watter would be better.

For a large push-pull amp, the power draw could be 100mA in idle and 250mA at full power, so assuming a 50V drop required P= 12.5 watts, so a 50 Watt Zener would be perfect.

NTE5257AK (15V), NTE5268AK (30V), and NTE5275AK (50V) are just some examples of 50 Watt reverse Zener diodes made by NTE Electronics. With this manufacturer reverse polarity Zeners have a suffix AK at the end of their part numbers, while "A" denotes normal polarity. Other manufacturers use different naming conventions. For example, 1N3331RB is a reverse 50W 50V Zener diode, just like NTE5275AK. "B" is the standard suffix, and "RB" indicates reverse polarity.

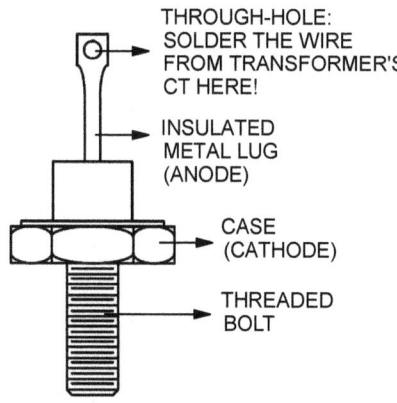

Remember that you may get a higher or lower voltage on the Zener diode since tolerance is usually +/-5%. For a 50V Zener, the tolerance is +/-2.5V, resulting in the Zener voltage range of 47.5V to 52.5V.

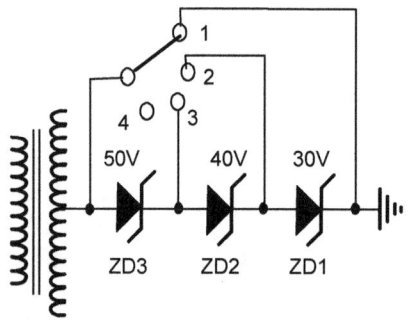

ABOVE: The 4-step power reduction circuit using three Zener diodes and one 4-position switch.

POWER SUPPLY MODIFICATIONS AND IMPROVEMENTS

You can string as many Zeners in series as you need, and they don't have to be of the same voltage rating. It's possible to get multiple power reduction steps. All Zener diodes are bypassed in position 1 of the power reduction switch, so this is the full power position. In position 2, ZD1 is active, and ZD2 and ZD3 are bypassed. In position three, only ZD3 is bypassed, ZD1 and ZD2 are in series, so the voltage is reduced by 30+40=70V. None of the diodes are shorted out in position four, so the HV is reduced by 120V!

Converting a cathode-biased output stage into a fixed-biased one

Adding a resistor between the HV center tap of the mains transformer and ground creates a voltage drop on that resistor due to the amplifier's current draw. That voltage can be used to bias the output tubes. In this case, the total DC current draw of the amplifier is 7.5V/47W = 160mA.

As the output power of the amplifier increases, the current through the resistor and the negative voltage in CT will increase too.

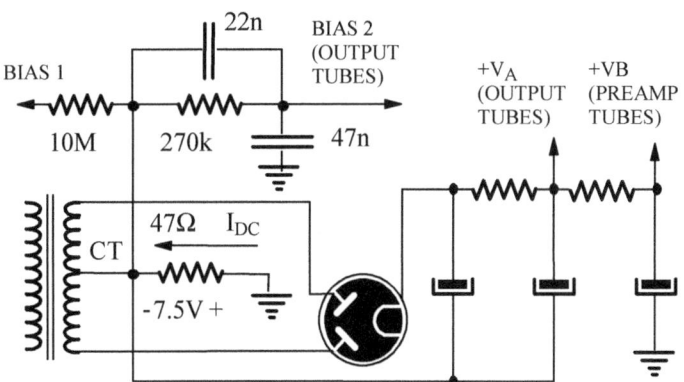

The output stage bias will increase, reducing output tubes' anode currents and the output power. In class A circuits, where the power draw of the audio circuit is constant, this circuit works well. You can use it to reduce the anode voltage a bit in amplifiers that operate their output tubes above their maximum dissipation.

In class AB or B amplifiers, where the current draw varies significantly, the circuit will also work as a dynamic limiter and reduce the amp's maximum output power.

MAKING TRANSFORMERLESS AMPS (WIDOW-MAKERS) SAFE & LEGAL

Brief history

The name Kay Musical Instrument Company appeared in 1931 when Henry Kay Kuhrmeyer bought Kay Kraft, which originates from C.G. Stromberg company that manufactured Mayflower guitars and banjos. Kay started selling their wares through the famous Sears, Roebuck and Co. in 1940, with Kay tube amps sold for a decade, from 1952 through to 1962, when a lineup of solid-state amps was introduced.

Sydney Katz bought the company in 1955 and sold it ten years later to Seeburg, famous for its jukeboxes, who offloaded it only two years later to Valco, a guitar and amp manufacturer, which survived for only a few years afterward. By the mid 60's the guitar and guitar amp market was overcrowded, and only the largest and most reputable businesses survived the tough market competition.

Topology and the choice of tubes

This is the simplest possible tube amp, a single pentode input/preamp/driver stage with very high gain (690k anode resistor), and a single-ended pentode power stage. The 50L6's power tube heater draws 150mA only, as do heaters of the two other tubes, the 12AU6 pentode, and 35Z5 diode. Since the heaters of all three tubes are in series, they must be drawing the same current. Indeed, that is why the designer chose them together all those years ago.

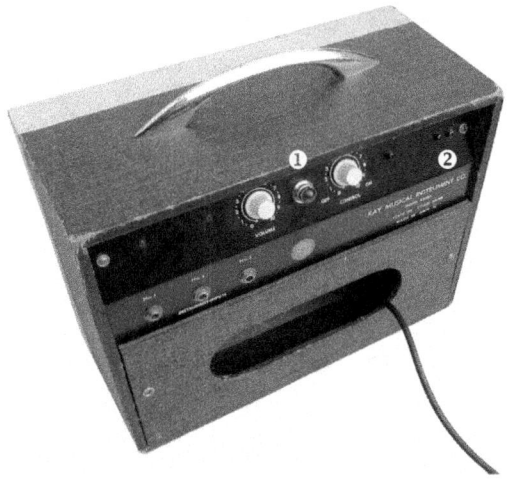

RIGHT: The front and rear look of Kay 5303 after restoration. The blue pilot light was added between the two knobs, and a fuse holder is right under the top corner, close to the added isolation transformer.

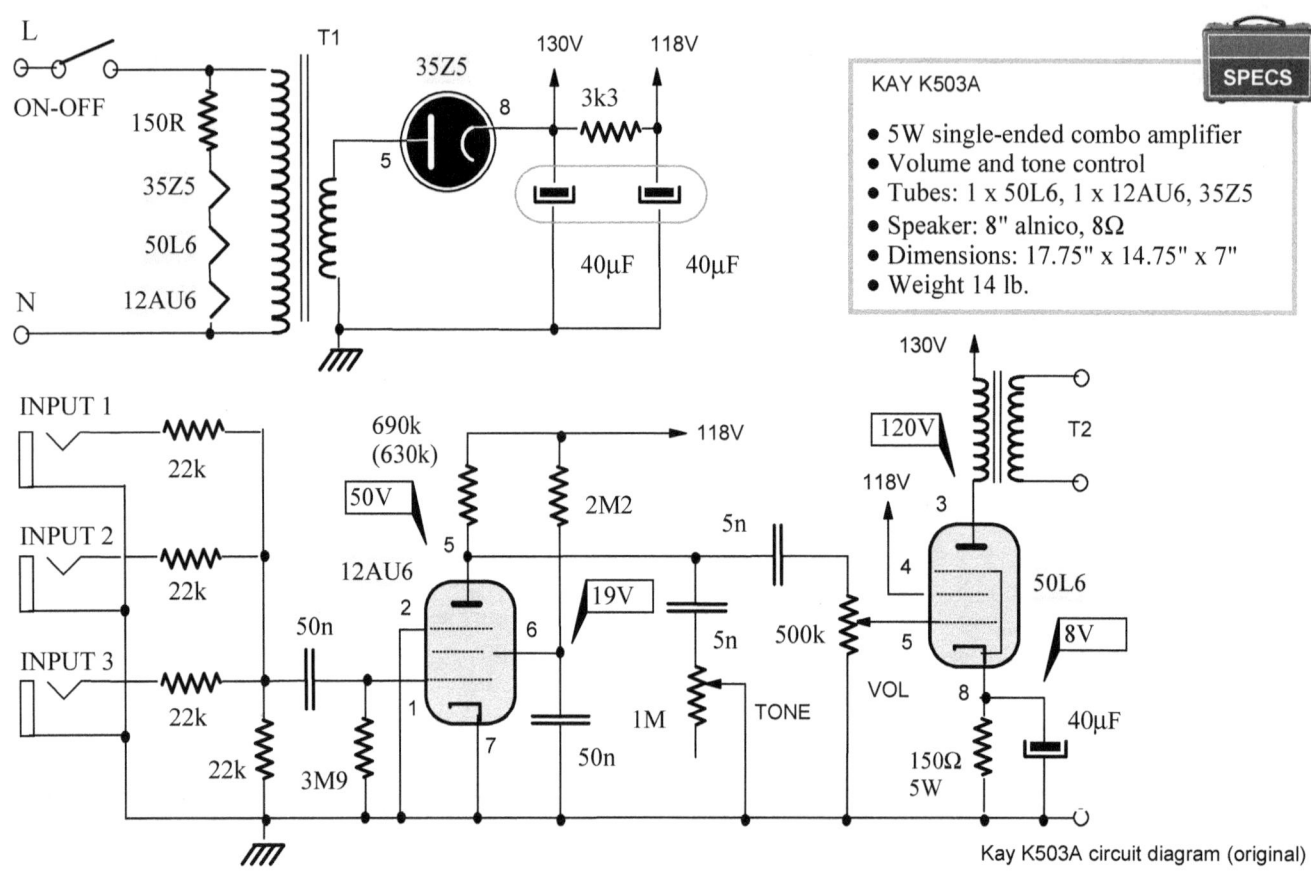

Kay K503A circuit diagram (original)

Adding the three voltage drops together, 50V+35V+12V=97V, means the designer had a few volts to "kill" on the 150Ω series resistor. Since 0.15A x 150 = 22.5 Volts, 97+22.5 = 119.5 V_{AC} mains voltage.

The heater string is connected to the primary side of the power transformer, directly across the mains voltage. This dangerous practice was allowed in the 1950s and 60s in the USA but is now illegal, so an isolation transformer must be installed. Alternatively, a completely new transformer with two 110V secondaries should be substituted.

Half wave rectification is used, so the ripple on pin 8 of the rectifier tube (fed directly to the output stage - V1 going to the output transformer) is significant. A simple 3k3 and 40μF RC filter is used for the screen grid and preamp stage anode voltage (V2).

All three inputs are identical, 22k series and 22k resistor to ground, meaning the input voltage is halved (a simple voltage divider). This lends itself to changing the values of these resistors and thus making some inputs "hotter" and some "cooler."

The amp was powered up, and DC voltages in the five critical points were measured as per the values indicated on the previous page. With 8V on its 150Ω cathode resistor, 50L6 was pulling a cathode current of I_K=8/150=53mA.

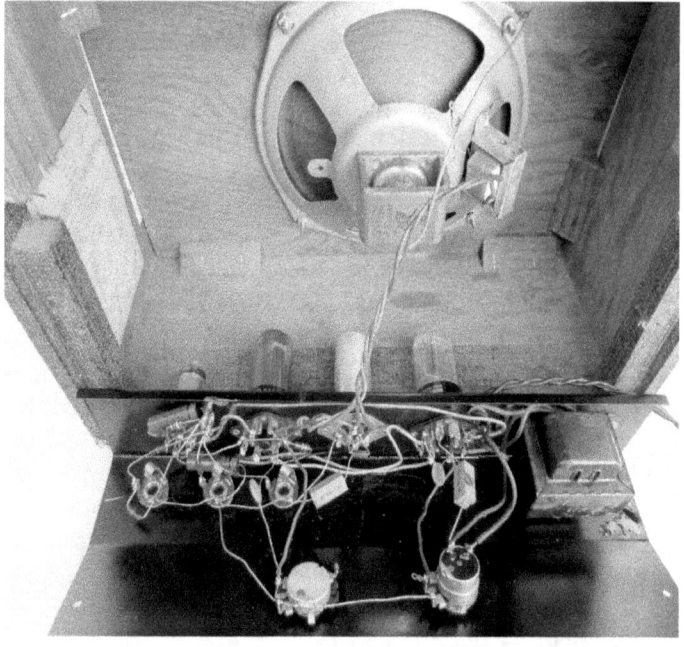

ABOVE: The component side of the chassis and the speaker wiring. This is an example fo true point-to-point wiring, there are no terminal strips or boards.

BELOW: The top view of the chassis and the tube sockets

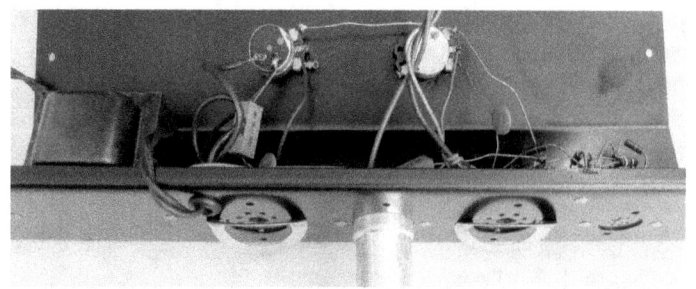

Since its anode-to-cathode voltage was 120V-8V=112V, the quiescent power dissipation of the screen and anode together was P=112*0.053 = 6 Watts, which was very conservative for a tube rated at 10+1.25 = 11.25 Watts!

The anode and screen currents of the first stage can also be calculated from the voltage drops across their resistors: I_A= (118-50)/630 = 68/630 = 1.1 mA and I_S= (118-19)/2,200 = 99/2,200 = 0.05mA. 630k was the measured value of the anode resistor, nominally 690k on the drawing.

At low signal levels, the sound was clean and sparkly. At higher levels, the breakup was excellen. However, the hum on the speaker, measured at 12 mV_{AC} was too high and annoyingly loud.

After: The improved circuit

An isolation transformer T3 was installed (1). In 120V mains voltage countries use a 1:1 ratio (120V primary and 120V secondary). In our case, the transformer is of step-down type, 240V down to 120V. A fuse (2) was added on the primary and neon "Power on" indicator on the secondary (120V side) of the existing transformer T1. That way, the neon indicates the health of both power transformers (3).

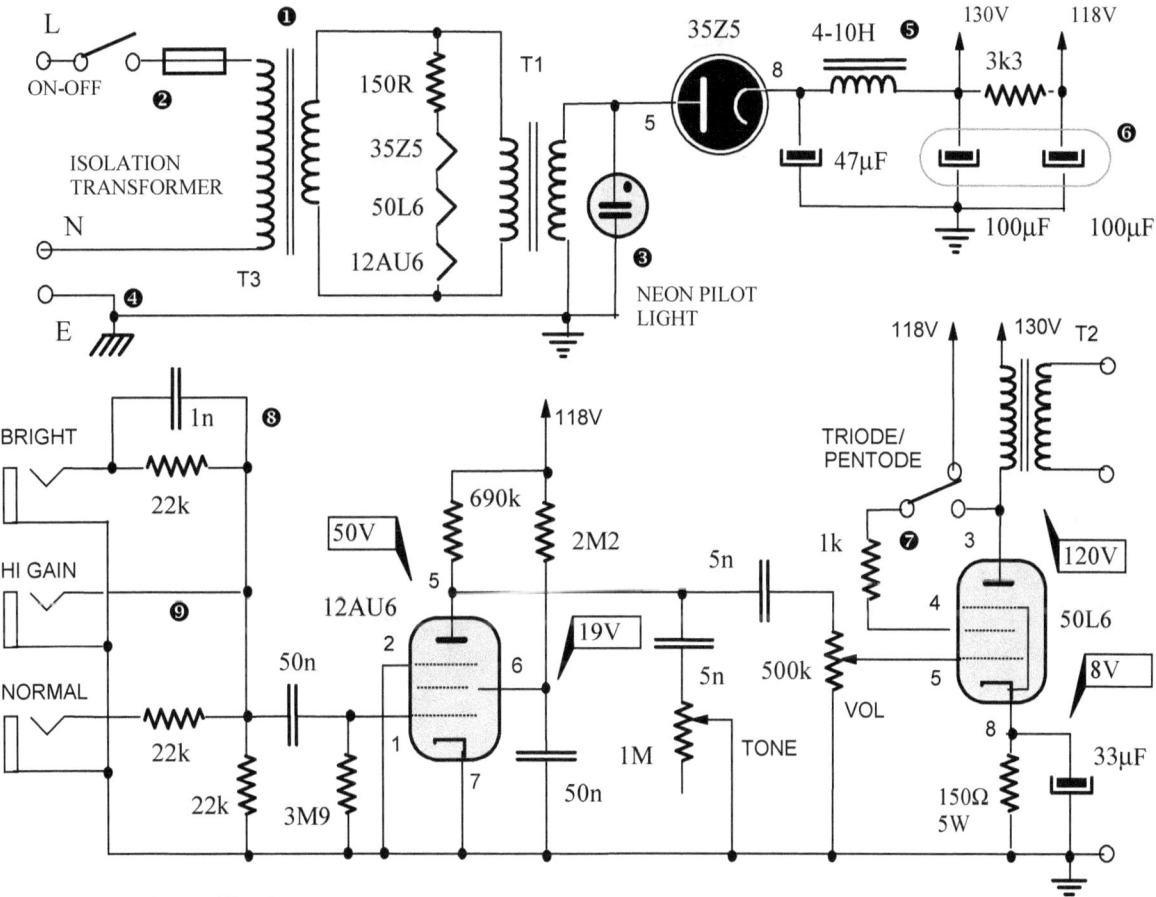

ABOVE: Kay K503A after modifications

A modern 3-core power cable grounded or "earthed" the chassis (4). Finally, to reduce hum, a filtering stage was added after the rectifier, comprising of a 47mF elco and a choke. Choke's inductance isn't critical; anything above 4 or 5H would do. Alternatively, use a 100-390Ω resistor.

After a while, the original multisectional elco started crackling and smoking, so it had to go. We replaced its two 40mF 150V sections with the only multi-section elco we had in our stash, a 2x100μF 500V JJ elco, which was a bit of an overkill, but it saved us $30-50 for one of those overpriced multisectional elcos sold online.

The amp is surprisingly loud for its size, so if you want lower power levels, the easiest option is to install a triode-pentode switch (7). Since the circuit works well, there is no need to change anything except perhaps customizing the three inputs. One option is to make one input a "bright" channel by bypassing its series 22k resistor with a 1-5n capacitor (8).

Since all original inputs feed a 22k-22k voltage divider, meaning the input voltage is reduced to half of its level coming out of a guitar, another possibility is to remove the series 22k resistor on one channel and make it into a "high gain" channel (9).

50L6GT beam power tube

12L6, 25L6, and 50L6 have the same parameters but different heater voltages. The name and the octal socket (even identical pinout) confuse some who proclaim that 50L6 is a 6L6 with a 50V heater. That is not the case. 50L6 has less than half of the 6L6's anode and screen dissipation (10 Watts and 1.25 Watts) and much lower maximum anode and screen voltage limits. The maximum power output of around 4 Watts (single-ended pentode stage) is obtained with the load between 3 and 4.5 kΩ, while 3.8kΩ load results in the minimum distortion.

Perplexingly, the July 1954 GE datasheet calls these tubes "beam pentodes," but the symbol used shows the third grid instead of the beam deflection plates.

TUBE PROFILE: 50L6
- Indirectly-heated beam power tube
- Octal socket
- Heater: 50V, 150mA
- V_{AMAX}=200V, P_{AMAX}=10W
- V_{SMAX}=125V, P_{SMAX}=1.25W
- gm=8 mA/V, μ=200, r_I=25kΩ
- SE pentode P_{OUT}= 3.7 W @ 10% THD

Other "widowmakers"

Budget vintage combo guitar amps without a power transformer, such as Kay 503, were named "widowmakers" since there was no galvanic isolation from the mains, and electrical shocks were likely and often fatal. Harmony H400 used a 12AU6 and 50C5 in the audio section together with a 35W4 rectifier, the same tube complement used in Laurel 900 and Alamo Capri 2560 by Alamo Amps. If you are tempted to buy one of those and use it as a learning or conversion platform, think again. Here is one cautionary tale, a true lemon.

The lemon: Winston

Despite its cute looks in gray cloth and marble-gray cabinet, this is genuinely a "budget" amp: very thin cabinet walls, a minuscule speaker, and a primitive printed circuit board that holds all components. Some were not even soldered properly.

All the tubes were made in Japan, but the front panel says "Made in USA," most likely to instill confidence in American buyers in the days when "Made in Japan" was a fact to be concealed. No matter who made it, this is a true abomination of an amplifier!

The On-Off switch is on the Volume pot, which holds the whole PCB onto the front panel, made of "craft wood." The bracket the volume pot is mounted on has six larger holes drilled, so additional pots can be installed. Should you wish to add a tone control potentiometer, an external speaker jack, or both, there is space on the front panel (2) and the bracket.

Model 461 has an added tone control and a larger speaker, and model 462 has tremolo speed and strength controls as well. There is no nameplate or model printed anywhere, but by some logic, could this be model 460 then?

Since there are so few parts, it is not difficult to follow the signal flow even without a circuit diagram. Once you start tracing, you'll soon notice a striking similarity with the Kay K503A circuit.

The input tube (3), the power tube (4), tiny output transformer (5), triple elco 20-60-40µF (6), half-wave rectifier tube (7), series power resistor that kills the remaining voltage in the 150mA heater chain (8) and the speaker (9).

There is no pilot light, but adding one is easy. The first option is to add a 115V neon light in parallel with the heater string. Interestingly, the 35W4 rectifier tube's heater has a tap (pin 6) designed for that purpose, so a type 40 or type 47 panel lamp can be added, as per the diagram on the right. These are 6.3V, 150mA, 1W incandescent miniature bulbs designed for indication purposes and can still be bought.

POWER SUPPLY MODIFICATIONS AND IMPROVEMENTS

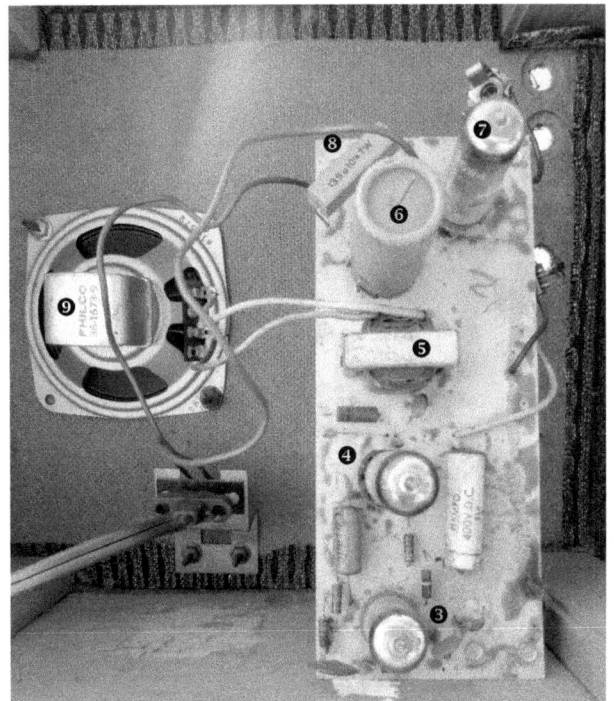

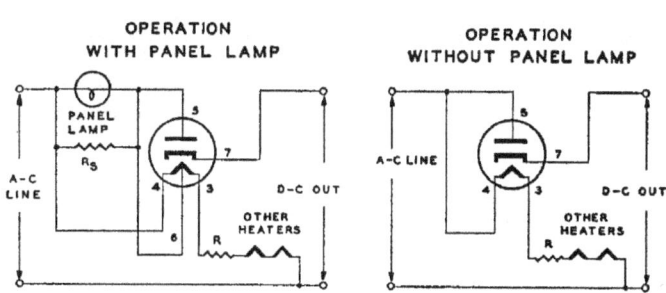

The 35W4 half-wave rectifier tube was designed so a pilot lap could be connected to its heater tap. Winston's designer chose to connect the mains input to that tap (pin 6) instead, in a fashion totally diffeernt from these two topologies suggested in the tube's data sheet.

TUBE PROFILE: 50C5
- Indirectly-heated beam power tube
- Miniature 7-pin socket, 50V/150mA heater
- V_{AMAX}/V_{SMAX}: 135/117V, P_{AMAX}/P_{SMAX}=5.5/1.25W
- gm= 7.5 mA/V, μ=75, r_I=10kΩ
- SE pentode with 2k5 load: P_{OUT}= 1.9 W @ 9% THD

Circuit diagram

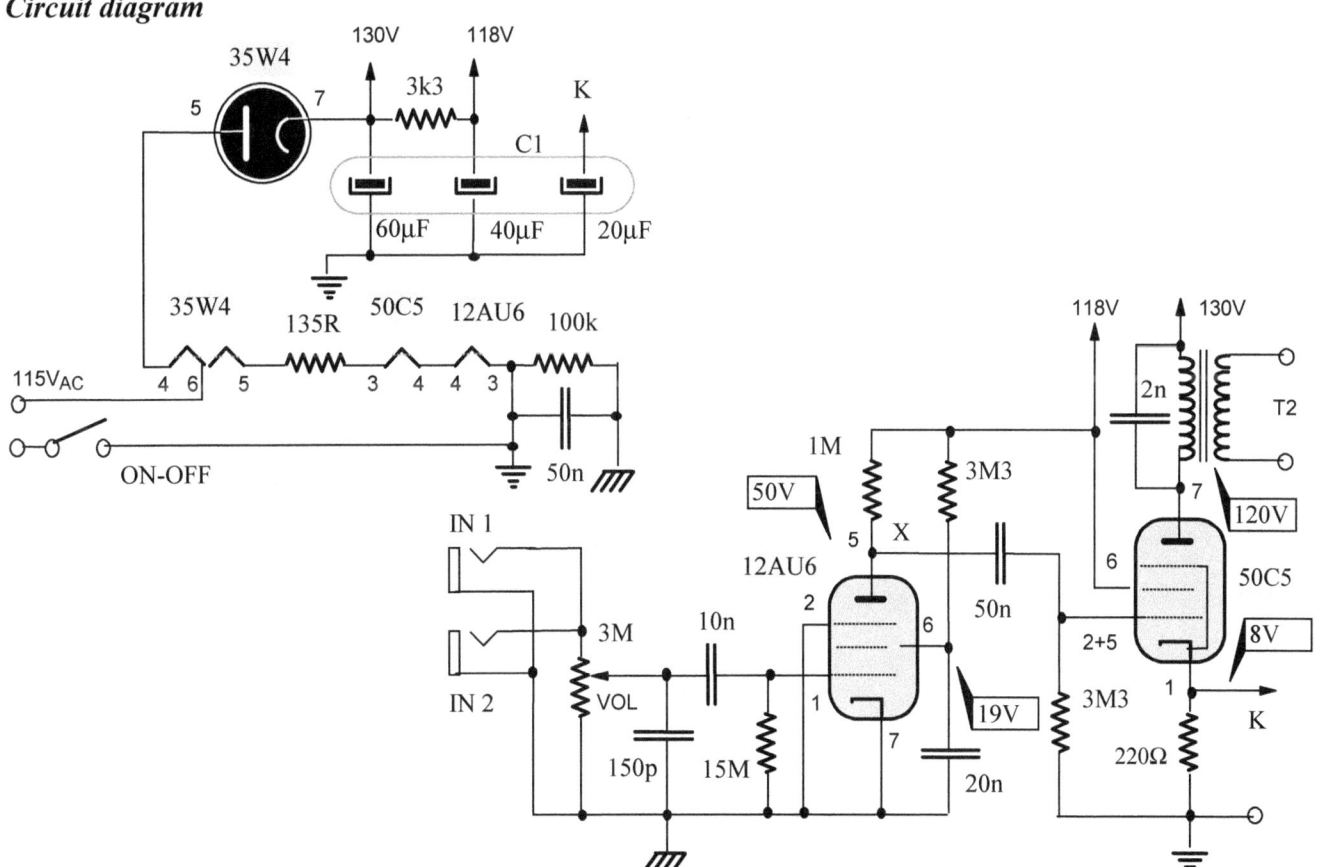

Half-wave rectification is provided by 35W4, a single vacuum diode, just as in Kay K503A. Notice that the input pentode's heater is the first in the chain, with one end grounded (pin 3), presumably done to minimize hum.

Once we cleaned and resoldered all components, the amp worked but with a very loud hum. When we disconnected the signal flow in point "X" (the output of the 1st stage taken from its anode), the hum stopped, meaning it originated in the first stage.

Despite the installation of an isolation transformer, adding a metal shield to the input tube, and other modifications, the hum persisted. We even replaced the power supply with a fully rectified and better-filtered one, to no avail. Finally, we replaced the whole amp with the PCB taken from Epiphone Valve Junior (which was rebuilt in a point-to-point fashion). Everything worked like a charm.

HOW TO FIX DC VOLTAGES: JOYO "SWEET BABY"

This China-made combo amp retailed for AU$229, but this unit, a customer return, was reduced to AU$169 (approx. US$118), including courier from Melbourne to us in Perth.

Considering its budget nature, the amp is very well made, input jacks are metal, not plastic, the chassis is fully chrome-plated, and the cabinet is sturdy. A tube rectifier is a nice bonus, unusual for such a budget amp. Sure, there is Volume control only, but tone control, external speaker jack, and triode-pentode switch can easily be added.

The output transformer is smallish, rated at only ten or so Watts, so don't expect organ thumping deep bass or lots of clean headroom.

The small chassis is completely filled with two PCBs and lots of unnecessarily thick cabling. The mains voltage enters the power PCB (1) where a varistor and a film capacitor (2) act as a simple filter, exits the PCB, and passes the mains fuse and on-off switch to reach the mains transformer's primary winding. All that unnecessary routing takes more than 1/3 of the volume inside the chassis!

The rectifier heater supply is fused (3), as is the heater supply for audio tubes (4). The HV supply enters the power PCB at (5) and has a fuse in each leg.

The power supply HV filtering uses three elcos, (6), (7) and (8), next to which is the cathode resistor of the output stage and its bypass capacitor. At (9) are the speaker connections.

The circuit is plain vanilla. Since there is no gain-sapping tone control, the designer left cathode resistors un-bypassed in both preamp stages. This reduces the cascaded gain considerably, approximately -3dB in each stage or -6dB overall.

The power supply (next page) shows two DC voltage figures in each point, standard voltages with stock 5AR4 rectifier tube and voltages with 5Y3 rectifier in oval frames.

JOYO "SWEET BABY" JTA-05
- Tubes: 1x12AX7, 1x6V6GT, 1x5AR4
- Output power: 5 Watts
- Volume control
- 8" Celestion speaker
- Dimensions: 360(W) x 190(D) x 360(H) mm
- Weight: 9 kg

The mains voltage and anode (plate) power dissipation issue

As with many Chinese-made amps, the power transformer was designed for $230V_{AC}$, so on our 247V mains, the heater voltage was 6.8V instead of 6.3V. Since 247/230=1.074, all voltages would be 7.4% higher; here, they were 7.9% higher.

Apart from the elevated heater voltage, which shortens the life of all tubes in the amp, the most critical issue is the anode dissipation of the output tube. 6V6 is rated at 12 Watts anode dissipation.

POWER SUPPLY MODIFICATIONS AND IMPROVEMENTS

JOYO "SWEET BABY" JTA-05 OUTPUT TRANSFORMER:
- EI48, a=16 mm, S=20mm
- $A = aS = 3.2 cm^2$, $P = A^2 = 10.2W$
- Primary DCR: $R_P = 382\Omega$
- $L_P = 6.4H$, $L_L = 32mH$
- Quality factor: $QF = 6.1/0.032 = 190$
- Primary impedance (8Ω load): 10.5kΩ

Joyo Sweet Baby output transformer (1) is tiny and unremarkable, with a very low quality factor.

The anode DC voltage is 371V, and since the cathode is at 22.1V, the anode-to-cathode voltage across the 6V6 tube is $V_{AK}=349V$. The cathode current is the cathode voltage divided by the cathode resistance or $I_K=22.1/470=47mA$. However, we don't know how that current is divided between the screen grid and the anode.

The cathode current of the first stage is 1.6V/1k5 = 1.07mA, say 1.1mA. The anode current of the 2nd stage is exactly 1mA, so there are 2.1mA flowing through the 22k filtering resistor. The voltage drop across it should be 2.1*22=46.2V, we measured 355-305=50V, that is close enough.

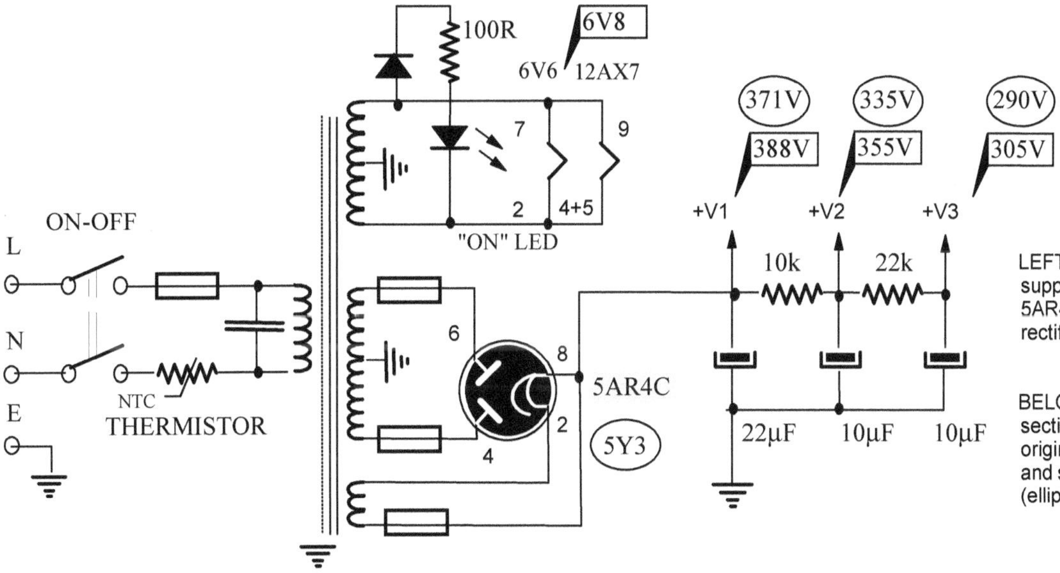

LEFT: Joyo Sweet Baby power supply voltages with the original 5AR4C and substituted 5Y3 rectifier tube

BELOW: Joyo Sweet Baby audio section, DC voltages with the original 5AR4C in square frames and substituted 5Y3 rectifier tube (elliptical frames)

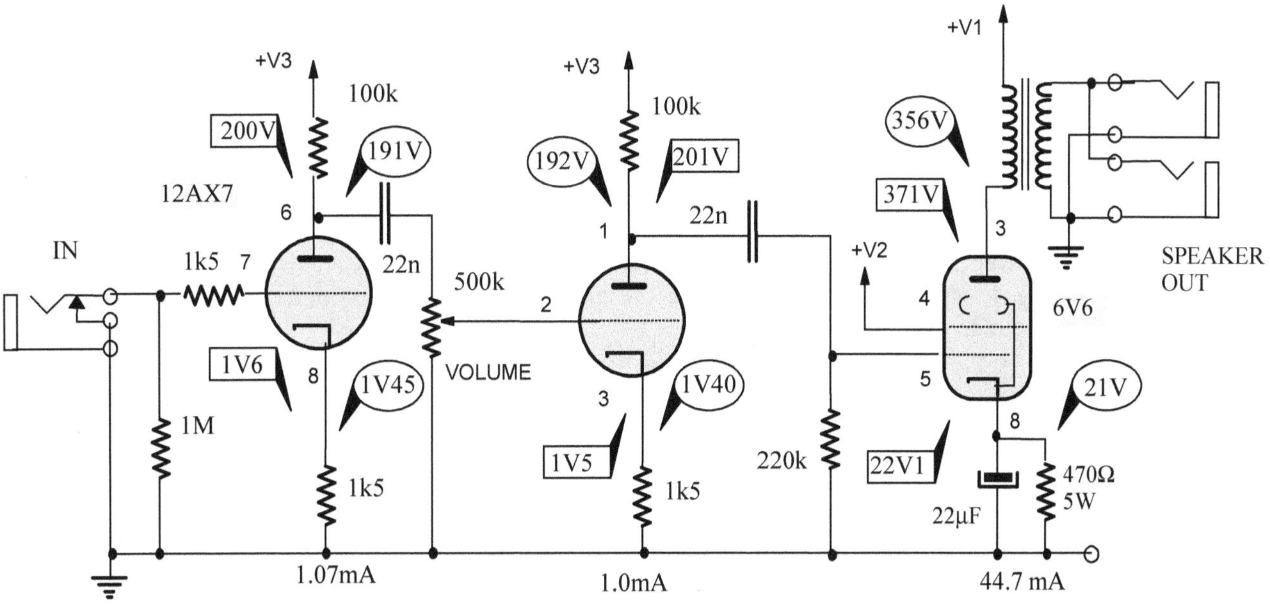

The voltage drop through the 10k resistor supplying the screen grid and the preamp stages is 388-355=33V, so the current through it is 33/10=3.3mA. We know that the preamp stages are drawing 2.1mA, so the screen is only drawing 3.3-2.1=1.2mA. Since the screen grid current is only 1.2mA at idle, the anode current is IA=47-1.2 = 45.8mA. The power dissipated on this poor 6V6 tube is $P_A = V_{AK}*I_A$=16 Watts! Since 16/12= 1.33, that means the tube's power rating was exceeded by a wide margin of 33%! Since our mains voltage is only 7.4% higher, this amp was not appropriately designed even for 230V mains voltage! The output tubes in this amp would have a short and stressful life in any case.

Modification #1: 5Y3 rectifier tube instead of 5AR4C

Another strange design choice is the 5AR4 rectifier tube. That tube can be found in stereo amplifiers such as Dynaco ST70, where it powers four EL34 tubes in push-pull. Its maximum DC current is 225 mA, and this amp's high voltage current draw is only around 1mA+1mA+47mA=49 mA!

An oversized rectifier tube is not a problem in itself. The high anode voltage problem is made worse by the fact that 5AR4 is a modern rectifier with a very low internal voltage drop. 5Y3 rectifier has the same pinout, but it's a directly heated tube of a lower maximum current (125mA). After 5Y3 (a vintage rectifier with one of the highest internal voltage drops) was plugged in, V1 dropped from 388V to 371V, meaning a 17V higher internal voltage drop on 5Y3 than 5AR4.

The anode DC voltage dropped 14V to 356V, and since cathode is now at 21V, the anode-to-cathode voltage is V_{AK}=335V. The cathode current is I_K=21/470=44.7 mA, so the anode current is around I_A=42.7 mA. The power dissipated on the anode is now $P_A = P_{AK}*I_A$=15.0 Watts, which is still way too high.

Mod #2: Reducing screen voltage only

You may remember the fact that anode and screen currents in pentodes are determined primarily by the screen voltage and not the anode voltage, which is much further from the cathode than the screen grid. In other words, assuming the anode voltage does not exceed the maximum allowed for the particular tube we can leave the anode voltage as it is and reduce the screen voltage instead.

First, without removing the PCB from the amp, we unsoldered one side of the 10k resistor in the power supply filtering chain, the one between the anode supply voltage V1 and screen supply voltage V2 and added an 8k resistor in series, for a total of 18k. All DC voltages dropped slightly. The cathode current of the first stage was now 1.4V/1k5 = 0.93 mA, and that of the 2nd stage is 1.3/1.5= 0.87 mA, a total of 1.8mA.

The voltage drop on the 18k series resistor was 375-313= 62V, the total current through it was 62/18 = 3.44mA, of which 3.44-1.8 = 1.64 mA was flowing through the screen grid.

The power dissipated on the power tube was now its anode current of $I_A=I_K-I_s$= 41.7-1.64 = 40.06 mA multiplied by its anode-to-cathode voltage of 340.4 V, so P_A=340.4*0.0406 = 13.6 Watts.

This was better than the original 16 Watts or 15W with 5Y3 rectifier, but still above the 12 Watts maximum, so our intuitive estimation that an 80% increase in the screen filtering resistor would be enough to bring the dissipation below the maximum level was in the ballpark, but not quite enough. So, a 27k resistor may be needed.

Mod #3: Reducing the anode voltage

In all that excitement, we forgot to check if maximum anode and screen voltage limits were respected or not. Remember, the 6V6 tube in the original amp had an anode-to-cathode voltage of 349V, while the maximum allowed is 315V. The maximum screen voltage is 285V, but Joyo had 335-21V=314V!

It seems Joyo's designers were either blissfully ignorant of those limits or didn't care how long the output tube would last in their amp. So, we had to go back and reduce both anode and screen voltages. With a 37mA anode current (estimated figure), 12 Watts/0.037A = 324V, the voltage across the 6V6 tube can be up to 320V (a small margin under the maximum allowed). The bias would then be around -17V, so 320V on the anode and +17V on the cathode means the supply voltage must be around 337V. That is our goal for DC voltage at pin 8 of the rectifier tube. There are a few different ways to achieve that.

Using 5Y3 instead of 5AR4 drops that voltage down from 388V to 371. From now on, we will assume that the 5Y3 rectifier will be kept in operation. In (b), a Zener diode is inserted in the CT of the HV secondary. Since we have 327 V_{AC} on each half of the winding (instead of the nominal 305V), the conversion ratio is 371/327=1.13. Thus, for 320 V_{DC} we need 320/1.17 = 274V_{AC}. Finally, 327-274 = 53V, so a standard 50V Zener diode is needed.

Since the amp works in Class A, the power draw of around 50 mA is constant, which also makes it possible to reduce the voltage V_1 as per diagram at (c), *either* by inserting a series resistor R_S between pin 8 and the first elco *or* a resistor R_{CT} instead of Zener diode between CT and GND.

POWER SUPPLY MODIFICATIONS AND IMPROVEMENTS

2x305V_{AC} nominal (marked on the amp), 327V_{AC} measured

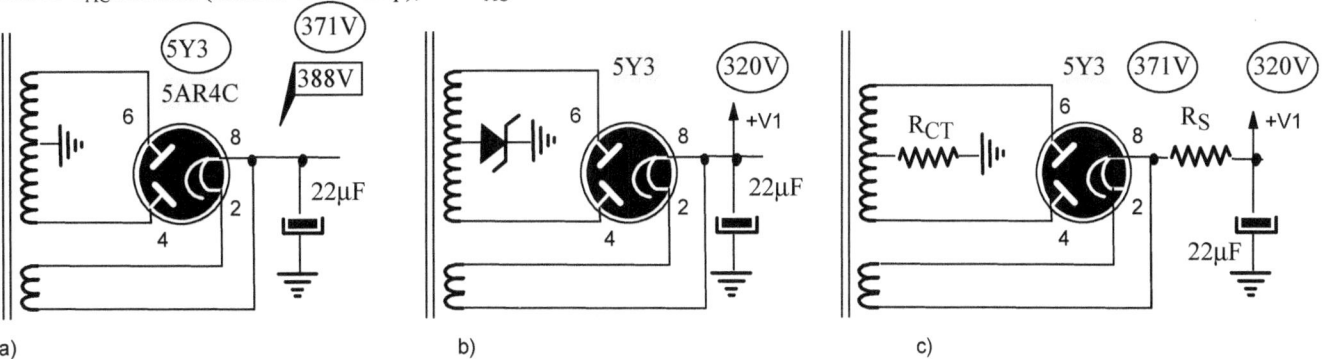

ABOVE: Three ways or reducing anode voltage. a) replacing a low voltage drop rectifier tube (5AR4) with one with a much higher voltage drop (5Y3) b) inserting a Zener diode in the CT leg of the high voltage winding c) inserting a voltage drop resistor R_{CT} in the HV filtering circuit

Due to Joyo's wiring, the easiest way to reduce anode and screen voltages was to add a series resistor between the center tap (CT) of the HV secondary winding and ground (GND). The photo explains why.

From pin 8 of the rectifier tube, below the fuse (1) on the upper PCB, the copper track goes leftwards to the ribbon cable, the rightmost conductor (2). It enters the bottom PCB at (3) and then it goes to the first elco (4) and the rest of the filtering chain. That rightmost strand of the ribbon cable could be cut and extended to reach the pins vacated by removing the AC in and out from the upper PCB (5), where R_{CT} (6) is soldered, but that would be messy.

It is easier and tidier to plug the CT wire from the secondary to one of the bottom two lugs (cannot be seen) and then reuse one of the surplus links (used to connect AC input to the upper PCB) to join the other bottom lug to the GND lug on the PCB (6). That way, no cable needs to be cut or extended, and the whole process is fast and straightforward.

Approx. 50V voltage drop with 50mA current means its resistance must be 50/0.05 = 1,000Ω! The power wasted into heat on R_{CT} will be P=V*I= 50*0.05 = 2.5 Watts, so a 5 Watt rated resistor is needed; a 7 Watter would be better.

Three sets of DC voltages are marked on the diagram (next page). The first (1) was measured while keeping the increased screen resistor of 18k and with the added 1k resistor between CT and GND.

ABOVE: The upper PCB and the power supply section on the lower PCB after modification.

The 6V6 cathode current was 15.7V/470 = 33.4mA.

The power dissipated on its anode was its anode current of approx. 32 mA multiplied by its anode-to-cathode voltage of 292-15.7=276.3V or 8.8 Watts. That was way too low for a 12 Watt tube.

So, the next step was to return the 18k screen filtering resistor to its original value of 10k. The power supply filtering chain voltages increased from 306-256-223V to 300-271-236V, or a 15V increase of the screen voltage. The cathode was now at 16.6V and the anode at 287V, so V_{AK}=270.4V!

The cathode current was 16.6/470 = 35.3mA (33mA of anode current), and the power dissipated on the anode was P=0.033*270.4=8.9V, meaning it hardly changed at all!

Finally, we reduced R_{CT} from 1k to 650Ω (using two 1k5 3 Watt resistors in parallel). The power supply filtering chain voltages increased to 317-286-240V, 17V more on the anode and a 15V increase of the screen voltage.

The cathode was now 1V higher, at 17.6V and the anode at 302V for V_{AK}=284.4V The cathode current was 17.6/470 = 37.5 mA (35mA of anode current), and the anode dissipation increased to P=0.035*284.4=10 Watts.

We could have reduced the value of R_{CT} further, to say 500Ω which would bring the anode dissipation up to 11.0-11.5 Watts, but 10 Watts seemed a nice safe value, ensuring a long and stress-free tube life. Why such a discrepancy between our calculated value for R_{CT} of around 1kΩ and the actual value of 500-650Ω?

Circuit diagram after fixing DC voltage levels

BELOW: Three attempts to fix DC voltages and reduce the anode dissipation of the 6V6 tube to below the rated maximum of 12 Watts.
1) keeping the increased screen resistor of 18k, plus a 1k resistor between CT and GND
2) returning the screen resistor to the original value of 10k, plus a 1k resistor between CT and GND
3) the original 10k screen resistor plus a 650R resistor between CT and GND (two 1k5 3W resistors in parallel)

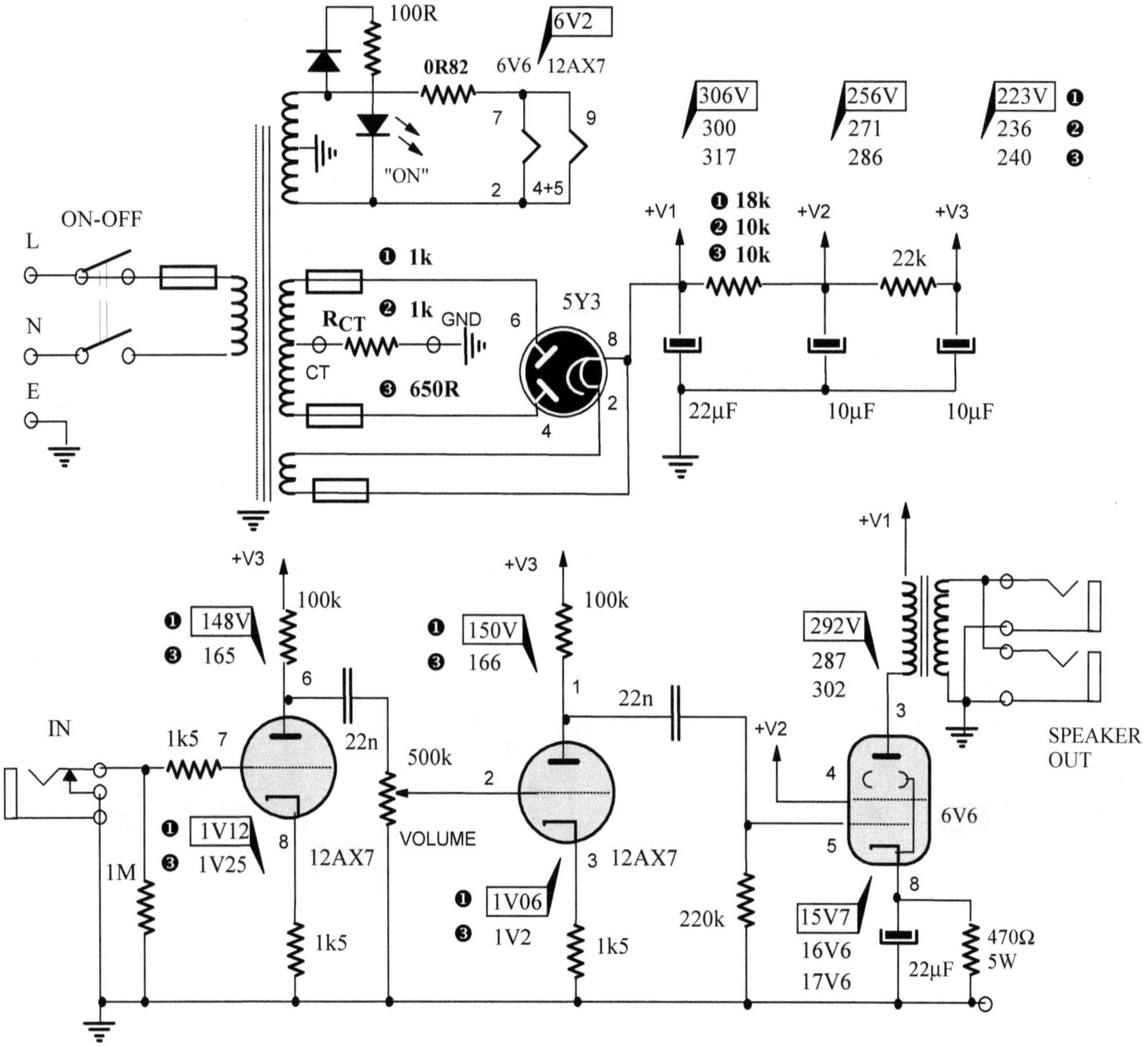

First, look at the original circuit diagram and the mains side where the NTC and the film capacitor are connected. Then, draw the diagram of that part of the PCB (after we've disconnected the AC input and output wires and soldered R_{CT} between two of the four lugs), and you will get a simple diagram (right).

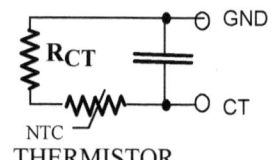
THERMISTOR

Since we hadn't removed the thermistor and the RF filtering capacitor, the thermistor remained in series with RCT. Hence, its resistance in the hot state (current flowing from GND to CT terminal) had to be subtracted from the calculated estimated value of 1kΩ for R_{CT}!

Mod #4: Reducing the heater voltage

Of course, this only fixed the high voltage supply to various gain stages; the heater voltage was still on the high side. To drop the heater voltage from 6.8 to 6.2V (it's better to underheat the preamp tubes slightly) or 0.6V, we need to know the total current draw of the two heaters. 6V6 draws 0.45A and 12AX7's heater at 6.3V nominal voltage draws 300mA. So the series resistance that needs to be added is $R_S = 0.6V/0.75A = 0.8Ω$, which is fortunate, since 0.82Ω is one of the standard values in the most common E12 (10% tolerance) series.

IMPROVING POWER SUPPLY FILTERING: PANAMA CONQUEROR

The design and construction issues

While the list price of this made-in-Panama head is US$349, the combo version sells for only US$50 more, and since it includes a 12" speaker, it seems a better value-for-money. Anyway, this used head was on sale for only US$139, so we got it from the USA.

The pretty two-tone cabinet with a high gloss cedar front panel is quite deep since the chassis and tubes are mounted horizontally.

Conceptually, the amp is OK, but a few issues let it down, the first of which is very sloppy soldering. Instead of feeding wires and component legs through the holes (eyelets) on tube sockets and terminal board, they were simply tacked on.

It seems whoever did this soldering job (most likely an assembly worker without proper training) had no clue about wiring practices and reliability issues.

The power cord is very short. With the head sitting on top of even the smallest speaker cabinet, it would not reach the ground, requiring the use of an extension cord.

To get the chassis out of the cabinet, the four bolts holding the rear cover in place, the four bolts on top (holding the chassis), and the bolts of the handle must all be undone. But, that is not all.

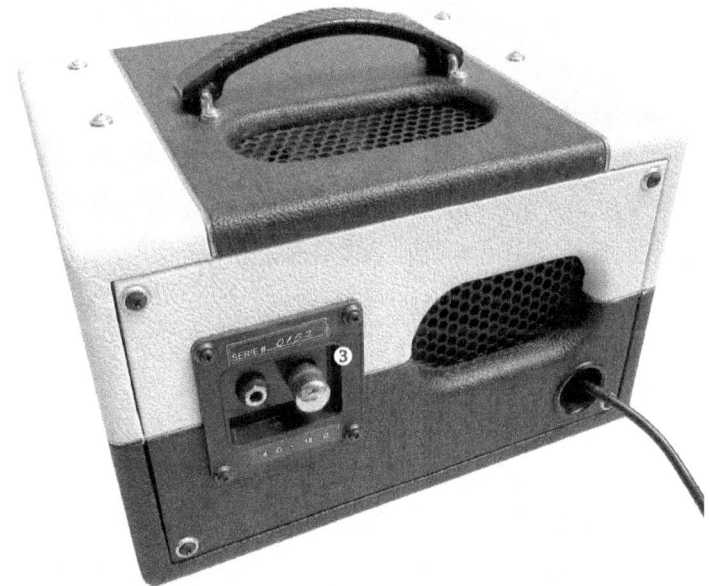

Since the chassis slides out backward, the side timber mounting brackets that hold the rear panel to the frame (1) are in the way and must also be removed.

The power cable exits the chassis on the opposite side from where it exits the rear panel, so it crosses the whole width of the cabinet and is held in place by a bolted-down timber clamp (2), which also needs to be removed.

The front panel is simple, one input jack, power switch, pilot light, volume, and tone control knobs. The L-pad output attenuator is mounted at the rear (3), next to the speaker output jack.

Accessing the fuse holder (4) requires the removal of the rear panel.

Despite all these irritating design and construction aspects, this amp still represents a great value-for-money. The hand-wired version of a similar VOX amp with EL84 power tube sells for more than twice the Conqueror's list price.

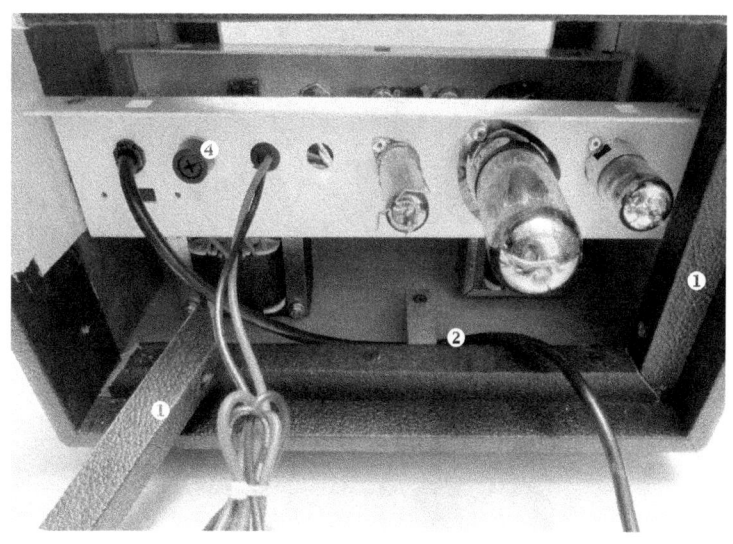

The chassis, transformers and internal wiring

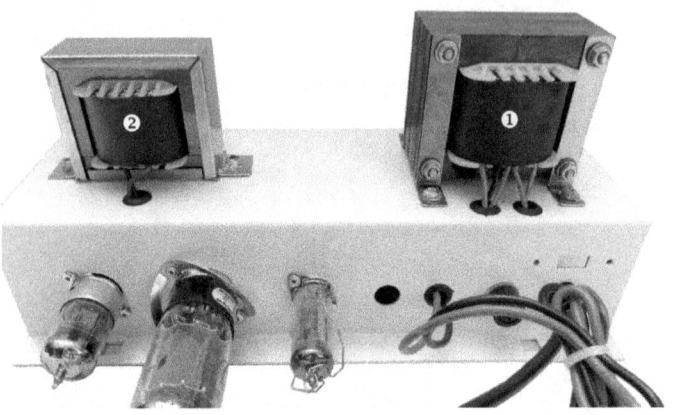

PANAMA CONQUEROR
OUTPUT TRANSFORMER:

- EI66 laminations
- a=22 mm, stack thickness S=28mm
- Center leg cross section $A=aS = 6.1 cm^2$
- Power rating: $P=A^2= 37W$
- Primary DCR: $R_P=397\Omega$
- Primary inductance @120Hz $L_P= 8.75H$
- Primary inductance @1kHz $L_P= 8.9H$
- Leakage inductance @1kHz $L_L=39mH$
- Quality factor: $QF = 8.9/0.039 = 228$
- Primary impedance with an 8Ω load:
  $Z_P=5.3k\Omega$ (attenuator fully CW) and
  $Z_P=3.1k\Omega$ (attenuator fully CCW)

The power transformer (1) is large for such a low-power amp. With EI76.2 laminations and a 40mm stack, its core is rated at 103VA.

The primary DCR of the output transformer (2) is very high, almost 400Ω, meaning it was wound with a small diameter wire to fit as many turns as possible to increase the primary inductance. The test results are about the average of all similar amps.

Panama markets this amp as their take on Fender Champ 5F1, so its tube layout and terminal board bear some resemblance. The three filtering capacitors (3) and two series resistors are wired in the same fashion, but the rest of the RC components on the board are positioned differently.

Conqueror's circuit diagram (next page) differs from the Champ's in few important aspects. There is only one input which does not attenuate the input signal like Champ's input voltage divider does. Notice that the second 68k resistor is switched in parallel with the 1M resistor when only one input is used on the Champ, creating a 68k-68k (approximately, in reality we have 1M in parallel with 68k) voltage divider, which cuts the input voltage in half!

The cathode resistor of the 1st stage is not bypassed in the Champ, thus reducing the gain of that stage.

RIGHT: The wiring diagram of Fender Champ 5F1 amplifier

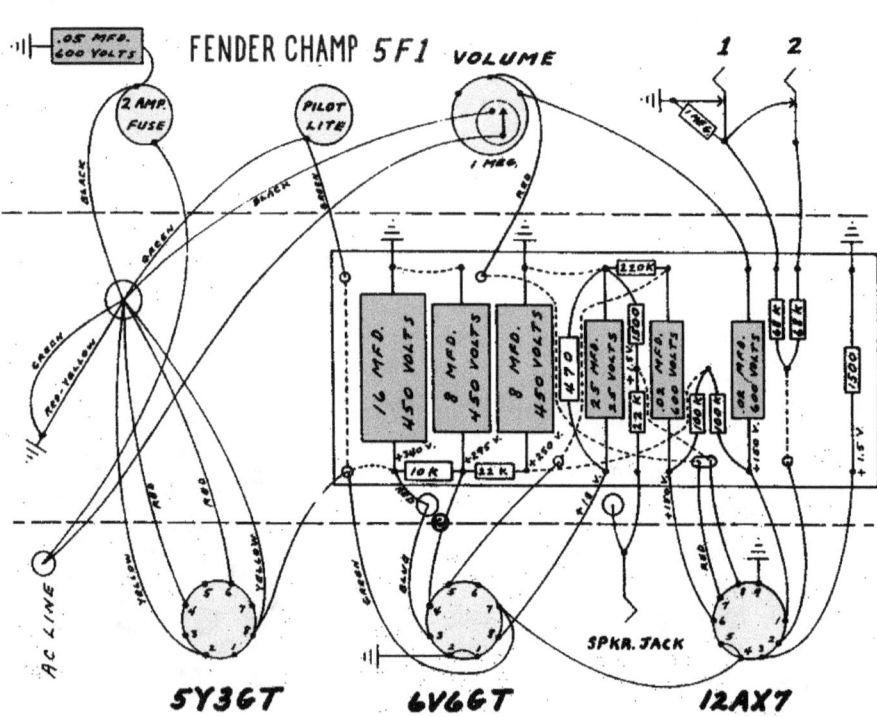

POWER SUPPLY MODIFICATIONS AND IMPROVEMENTS

Thus, Conqueror's input circuitry has much more gain, but that gets eaten away by its tone control circuit, something the Champ did not have. Both amps use negative feedback around the second and output stages, a 22k resistor between the output and the 1k5 cathode resistor of the 2nd stage. Conqueror has higher stored energy in its filtering caps, 22-22-22 µF, compared to Champ's 16-8-8 µF.

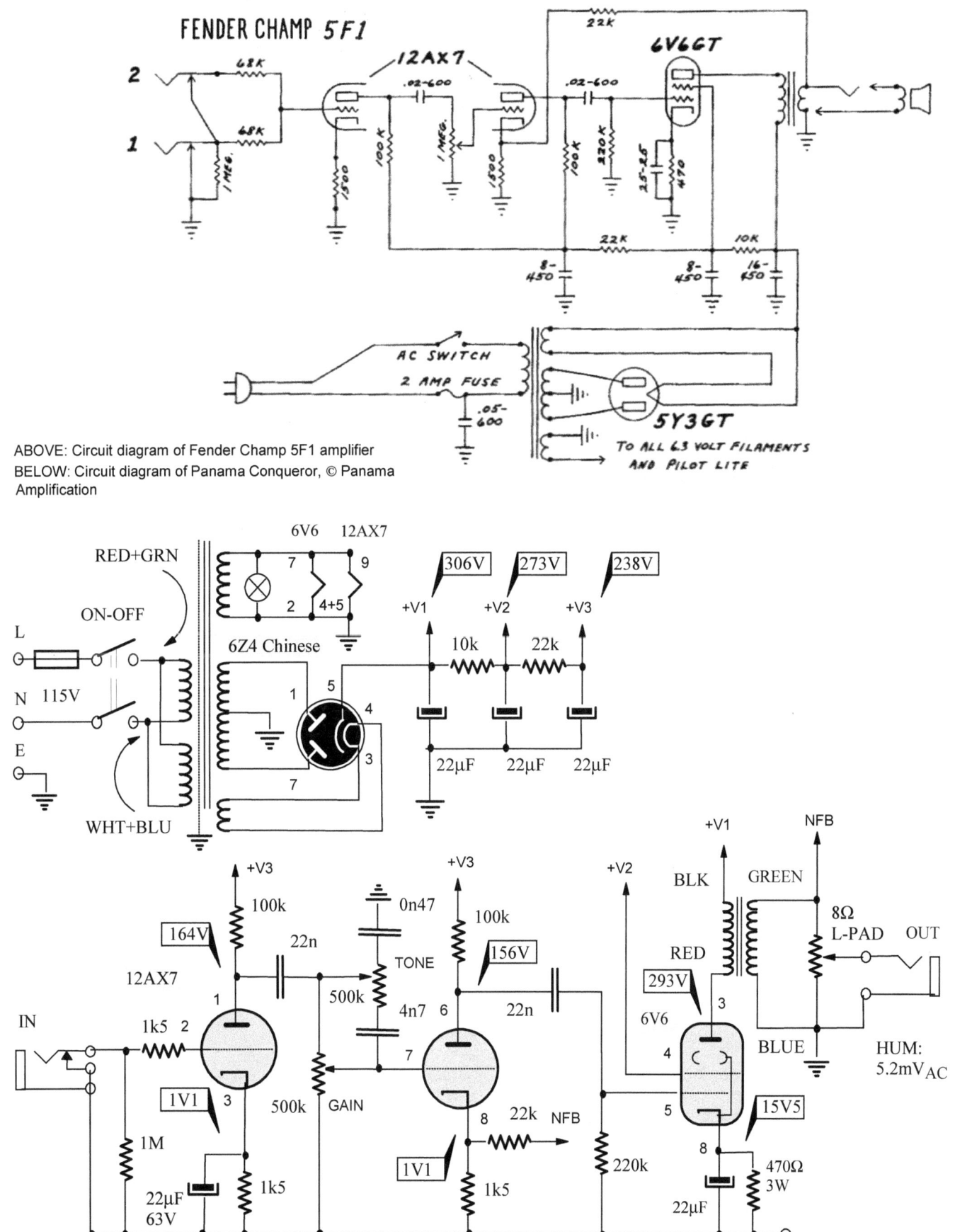

ABOVE: Circuit diagram of Fender Champ 5F1 amplifier
BELOW: Circuit diagram of Panama Conqueror, © Panama Amplification

Primary rewired for 230V mains voltage

The visual inspection of the mains side wiring identified two wires going to each pole of the power switch, RED + GREEN (1) and WHITE + BLUE (2), meaning the power transformer had two identical primary windings connected in parallel. This was excellent news, meaning a simple wiring conversion to a series connection would enable the mains transformer to operate at 230V mains voltage. The only issue was to figure out how to connect the four wires.

We know that wires connected together must carry in-phase voltages, so the voltages at red and green wires are in phase, and voltages at white and blue wires are in phase. But, we don't know which colors are the ends of the two windings. Are the ends of one winding red and white and the other green and blue or the other way around? The only way to ascertain that is to unsolder the four wires from the switch and test them with an ohmmeter.

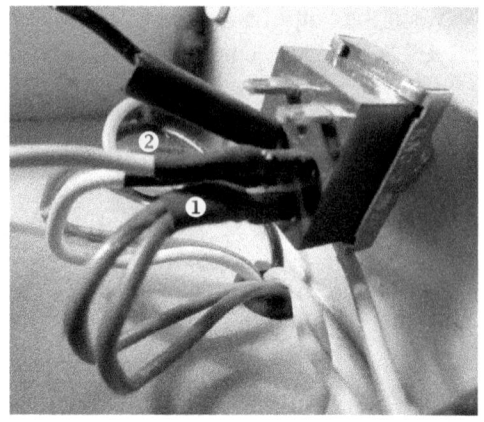

ABOVE: The power switch wiring

The ohmmeter showed a DC resistance of 20.1Ω between red and white and 21.1Ω between green and blue. So the voltage at green is of the opposite phase to the voltage at white (since they were not soldered together before), but now, in a series connection of the two primaries, they should be joined together, and red and blue should go to the switch, as illustrated on the modified circuit diagram.

Operating conditions and test results

With our mains voltage of 242V, the heater voltage was 6.6V (6.6-6.3)/6.3*100= 4.8% above the nominal. Since 242 is 5.2% higher than 230V, this means the power transformer primary windings were designed for just under 115V each or 230V in series.

The idle DC currents through the two input stages are identical, around 1.1/1.5= 0.73 mA. The cathode current of the output tube was 15.5V/470R = 33mA. Allowing, say, 3mA of screen current, the anode current of 30mA with the anode-cathode voltage of 293-15.5=277.5 would result in anode power dissipation of 277.5*0.03= 8.3 Watts, which is way below the 12 Watt maximum for 6V6.

The AC ripple or hum at the output (on the speaker) was 5.2mV$_{AC}$, a level that is too loud and irritating on speakers of average sensitivity. The amp will be way too noisy for recording and too annoying for practice purposes with high-sensitivity speakers.

The culprit was the anode voltage for the output stage, which was taken from the first filtering elco, a point where the ripple is always quite high. To quieten the amp down, another filtering elco and a resistor (or, even better, a choke) need to be added before the high voltage is taken out to the output transformer.

With 80 mV$_{RMS}$ sine signal at the input (1 kHz frequency), the maximum power into an 8Ω resistive load was 4.35 Watts, so Panama was justified in declaring this to be a 4 Watt amp.

The frequency range was very wide, from 44 Hz to 55 kHz, with tone control (TC) in the mid position. Moving TC fully CCW cuts the treble, so the upper-frequency limit is only 18 kHz. Moving TC fully clockwise boosts the treble to an incredible 88 kHz upper limit.

MEASURED RESULTS (before modifications, output attenuator fully clockwise, maximum power output):

- -3dB BW: 44 Hz - 55 kHz (tone control in the mid position)
- -3dB BW: 78 Hz - 18 kHz (tone control fully CCW)
- -3dB BW: 72 Hz - 88 kHz (tone control fully CW)
- P$_{MAX}$= 4.35 W (8Ω resistive dummy load, 1kHz signal, 80mV$_{RMS}$ input)
- Output hum: 5.2 mV$_{RMS}$

Rewiring tube rectifier socket for western 6Z4 or 6X4 tubes

Strangely, the rectifier tube supplied with this amp had no markings at all. So, we had to identify it by its pinout. It turned out that it was the Chinese version of 6Z4 tube, which has a different pinout to the USA and other western-made 5Z4 tubes! Thus, no direct substitutions are possible!

6X4 tube is very similar, with a slightly higher maximum current output (70mA versus 60mA for 6Z4) and a higher maximum allowed filter capacitance (60μF versus 40μF for 6Z4). However, 6X4 has a different pinout again, so the three types of tubes are NOT interchangeable without rewiring the tube socket.

POWER SUPPLY MODIFICATIONS AND IMPROVEMENTS

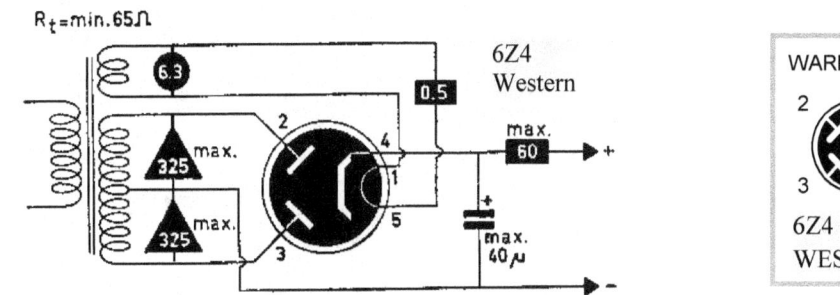
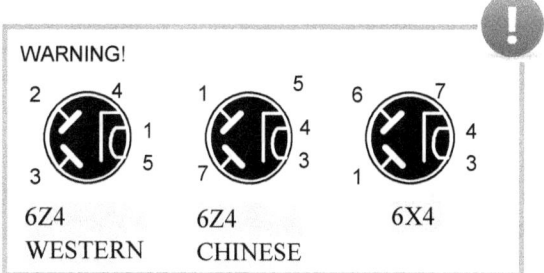

Power supply and audio circuit modifications

Two SPDT (single pole double throw) toggle switches were added on each side of the power switch (1 and 2 on the next page). One switches the output stage into a pentode or triode mode while the other turns negative feedback on and off. A screen grid resistor of 220Ω was added to limit the screen current.

The master volume control would be very easy to implement, but in this instance we chose not to.

The circuit diagram shows the output stage without the L-pad attenuator and with a 6L6 output tube, but 6V6 can be retained, both octal tubes can be used without any changes, in addition to 6F6 and 6K6 lovelies for even more different tonal voicings.

While powering a pentode output stage from the first elco may be a bit noisy but just acceptable, a significant hum of 60mV appeared in triode mode, so we had to add a 47mF filtering elco (3) and a 2H choke (4). That removed all traces of hum and lowered DC voltages by only about 5V (depending on the choke's DC resistance).

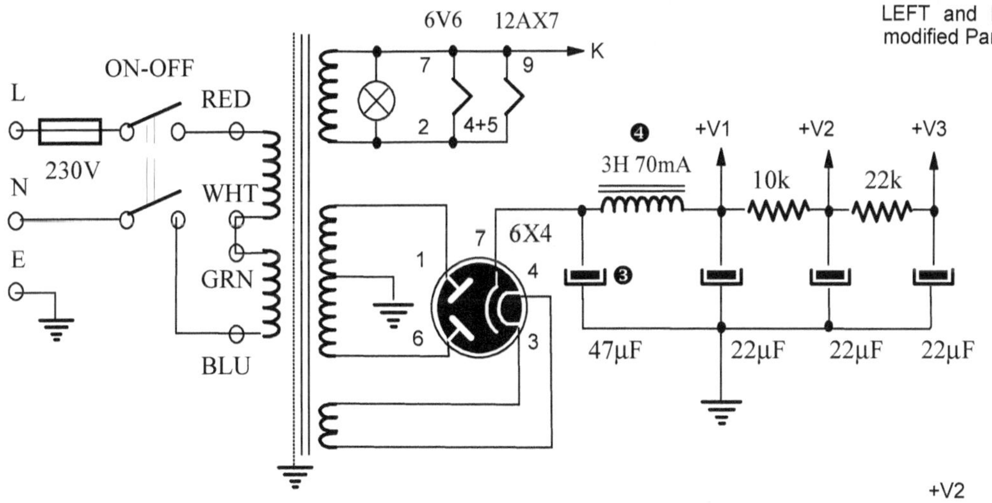

LEFT and BELOW: Circuit diagram of the modified Panama Conqueror amplifier

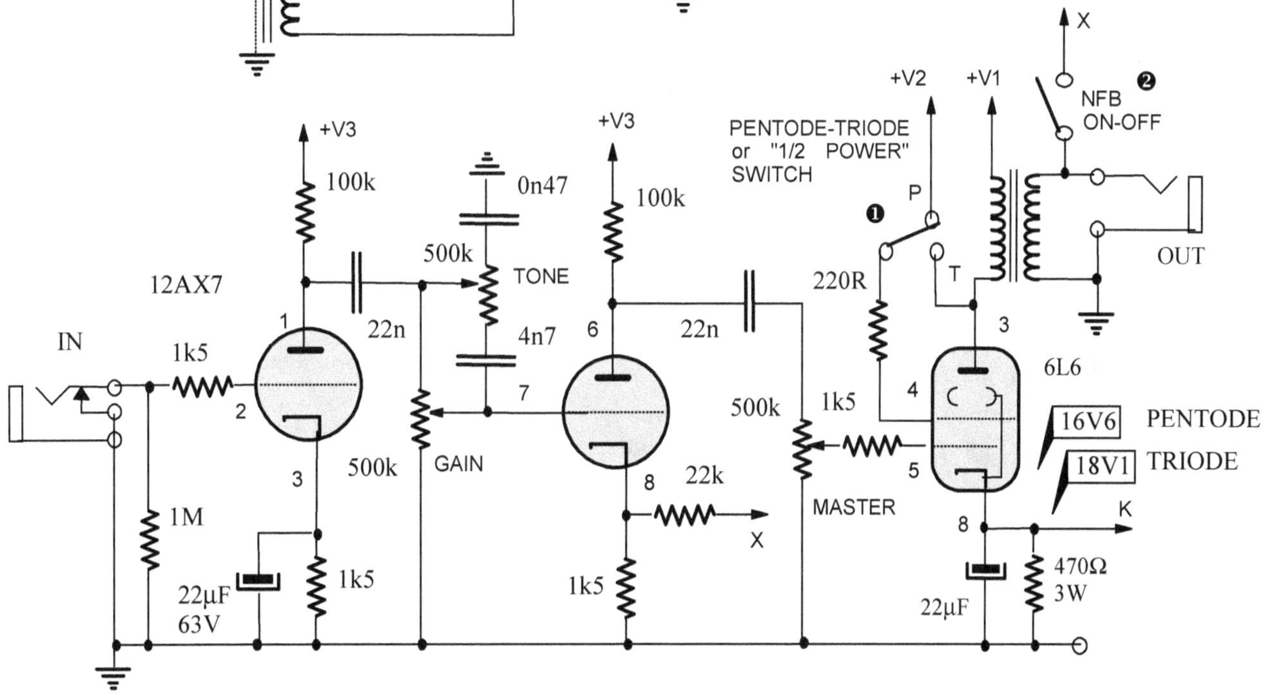

After modifications, the under-the-chassis view shows an axial 47μF elco sitting on top of the existing three 22μF capacitors (3) and a small 3H choke that just fit under the chassis (4). A larger choke would have to be mounted on top of the chassis.

The 220Ω screen resistor (5) straddles the triode/pentode switch and the added two-lug terminal strip (6). The other lug was used to join together two mains transformer's primary terminals in series.

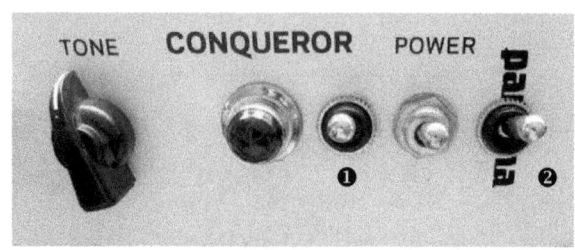

ABOVE: The added triode/pentode (1) and NFB on-off toggle switch (2) were mounted on either side of the power switch. Normally the up position is ON or higher power, so up is " NFB ON" and up is "Pentode or full power" mode.

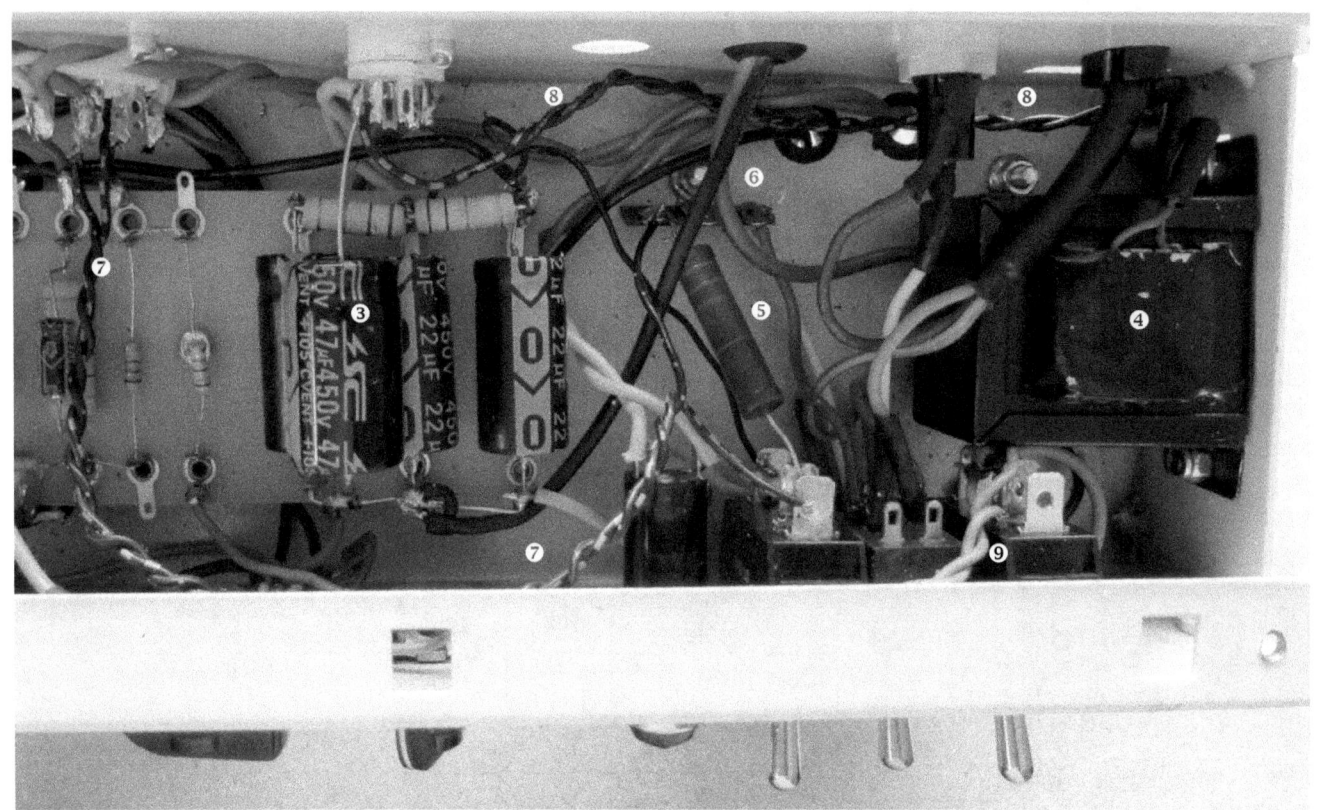

All added cable runs were twisted together to cancel any induced voltages along the way (which could result in increased hum): from the triode/pentode switch back to the output tube socket (7), from the rectifier socket to the choke and back (8) and from the NFB on-off switch to and from the NFB resistor and the output transformer's secondary (9).

Measurements

The triode mode severely narrowed the frequency range, both with and without negative feedback. While the upper limits of 16 kHz and 20 kHz are acceptable for a guitar amp, the lower -3dB (half-power) limits of 270 and 320 Hz mean that bass notes will be noticeably attenuated in that regime.

With 6L6 plugged in instead of 6V6, the output power in pentode mode with NFB jumped from 4.35 W to 6.1 Watts. Again, due to a very mild NFB, the maximum power without feedback was only marginally higher at 6.4 Watts.

In triode mode, the power figures were slightly higher than 1/2 power, 3.2 Watts with NFB and 3.4 Watts without it.

MEASURED RESULTS (AFTER MODS):

- 6V6 TRIODE MODE, NFB ON:
- -3dB BW: 270 Hz - 20 kHz (tone control in the mid position)
- P_{MAX}= 2.7 W
- 6V6 TRIODE MODE, NFB OFF:
- -3dB BW: 320 Hz - 16 kHz (tone control in the mid position)
- P_{MAX}= 2.9 W
- 6L6 TRIODE MODE, NFB ON and OFF:
- P_{MAX}= 3.2 W / 3.4 W
- 6L6 PENTODE MODE, NFB ON and OFF:
- P_{MAX}= 6.1 W / 6.4 W

THE OUTPUT TUBE OPERATED WAY ABOVE ITS MAXIMUM DISSIPATION LEVEL: EPIPHONE ELECTAR TUBE 10

This small combo amp hails from the late 1990s. Made in South Korea, it features a very sturdy cabinet with a closed back and switch-selectable (1) dual voltage mains transformer (2), a very handy feature for those of us living in 230-240V countries.

There is a preamp-out jack (3) but no headphone jack or any power control adjustments for that matter. Since solid-state rectification is used (four diodes on the PCB), a standby switch is a welcome inclusion (4).

Components are easily accessible on the relatively spaced-out circuit board (5), with quite a few jumper wires on the component side, so simple modifications are possible even without removing the PCB and exposing the copper side.

Although built with the lowest possible price in mind, the quality of this amp is acceptable, except for the tiny output transformer (6) and low energy storage.

All three filtering capacitors are identical, 22μF, so the first cap could be upgraded to 47μF. Upgrading the OT (output transformer) is tricky, since a bigger unit cannot fit inside the shallow chassis. There is plenty of room on the other (top) side of the chassis, but the speaker's magnet is in the middle, directly behind it, so a larger OT would need to be moved to the other side of the speaker (7).

SPECS

Epiphone Electar Tube 10

- 10W Single-ended, class A combo amplifier
- Construction: PCB, closed-back cabinet
- Rectifier: solid state diodes
- Reverb: No, Tremolo: No
- 3-band EQ (Bass-Middle-Treble)
- Gain and volume controls, preamp out
- 8" 4Ω speaker
- Ext. speaker output: 4Ω
- Tubes: 1 x 6L6, 1 x 12AX7
- Weight: 17.6 lb.
- Dimensions: 14.75" x 8" x 12"

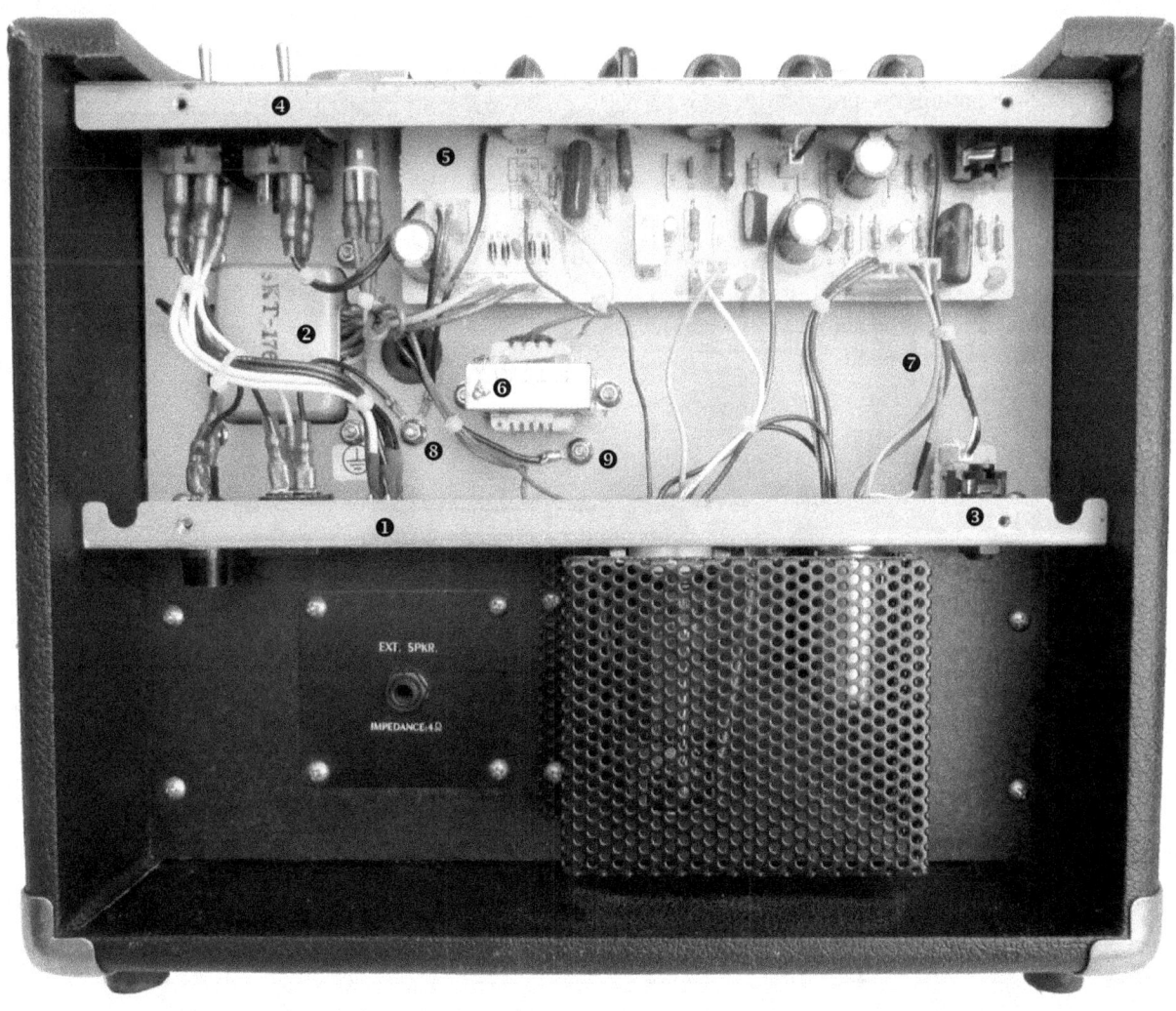

ABOVE: The internal view of Electar Tube 10 amp with rear top panel removed. A very neat and easy to follow wiring, easy to modify.

If you decide to go to all that trouble, you may as well install a small filtering choke in the original location of the output transformer.

The incoming mains earth terminal and the earth wire from the PCB are bonded together to the chassis in point (8), but the CT (center tap) of the HV secondary winding has a separate connection to the chassis in point (9).

The audio circuit

0.9mA though the first and 1mA of anode current in the 2nd stage, a fairly standard circuit needs no further elaboration.

The negative feedback signal through the 15k resistor arrives at the cathode of the second 12AX7 triode and would normally work across its un-bypassed 1k5 cathode resistor. However, notice that the 22µF capacitor bypasses the 1k5 cathode resistor, so the whole feedback signal voltage is shunted to the ground, making the feedback inoperative at most frequencies.

You don't have to perform circuit analysis or do any silly SPICE simulation, simply choose a few signal frequencies and see what happens.

At 1kHz, the impedance of the bypass cap is $Z_C=1/(2\pi fC) = 7.3\Omega$, 200+ times lower than 1k5 cathode resistance, meaning it's practically a short circuit for the NFB signal. At 100 Hz, its reactance is ten times higher, but still only 73Ω. Since at that frequency, the NFB voltage divider feeds back roughly $73/(1,500+73)*100\% = 4.6\%$ of the signal at the speaker, which is considered very mild feedback. We checked its effect at 100Hz and even down to 50Hz - there was no change in the output voltage level with or without feedback connected.

ABOVE: The dominating big & bold Epiphone front plastic logo livens the otherwise BBB combo looks (black, bland & boring).

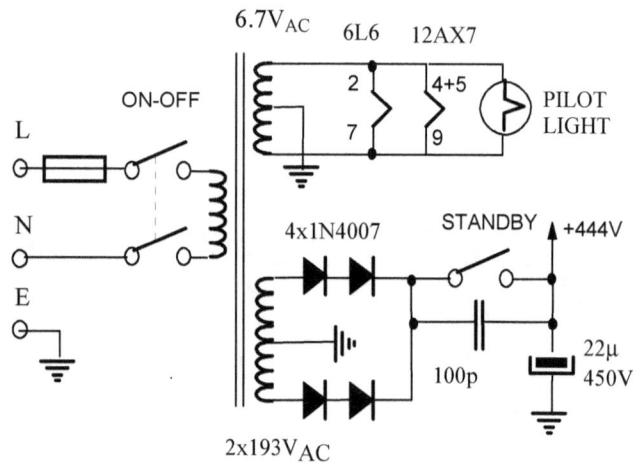

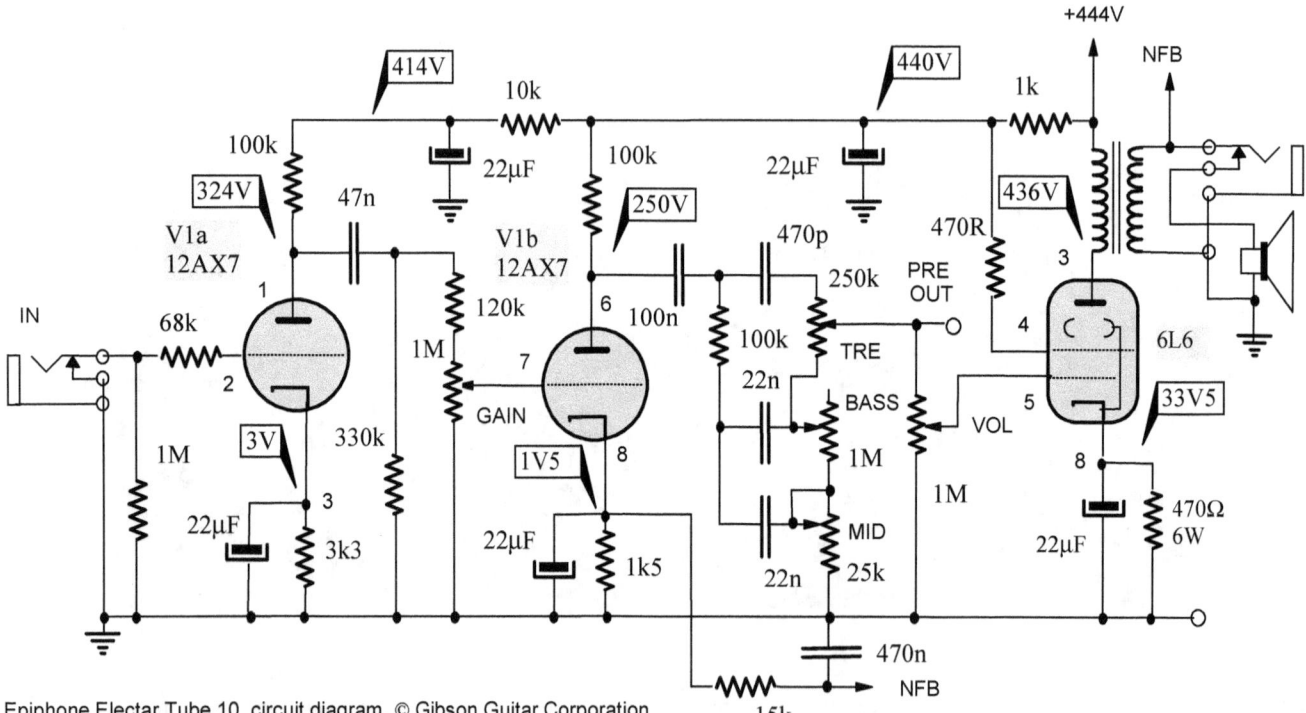

Epiphone Electar Tube 10 circuit diagram, © Gibson Guitar Corporation

The output stage

The cathode current through the output tube is 33.5V/470Ω = 71.2 mA. The screen current is only 2mA, meaning the anode current is just under 70mA. With the anode-to-cathode voltage of 436-33.5 = 402.5V, the anode power draw is 402.5*0.07 = 29.2 Watts!

The Sovtek tube marked 5881/6L6WGC that came with the amp (not sure if that was a later replacement or if the amp left the factory with it) has an anode rated at only 23 Watts, much lower than western 6L6GC varieties rated at 30 Watts, and should not be treated as a direct replacement! Even a 30 Watt-rated tube dissipating 29.2 Watts will not last very long.

Circuit diagram after power supply and other modifications

The heater voltage was reduced from 6.7V_{AC} to 6.4V_{AC} by placing a 0.47Ω resistor in series with each half of the secondary winding (1).

The anode voltage of the output tube was left unchanged but the screen voltage was reduced by replacing the 1kΩ series resistor with an 18kΩ resistor (2). The anode supply voltages for two preamp stages were also proportionally reduced. The screen voltage in pentodes is much more critical than the anode voltage, so by fine-tuning the screen voltage, you will change the operating regime of the tube, not by changing the anode voltage!

The DC cathode current of the output stage dropped from 70mA to 61.7mA, of which about 60mA was anode current. With the new anode-to-cathode DC voltage of 430-29 = 401V, the power dissipation on the tube was lowered to P=I_0*V_0 = 0.06*401 = 24 Watts, way under the 30 Watt rating of beefier 6L6 versions and only slightly above the Sovtek tube's rating. If it bothers you, increase the 18k series resistance to 22k, and all should be well again.

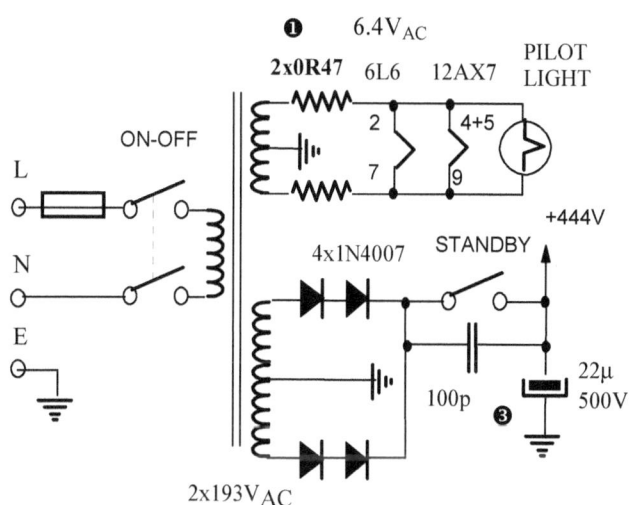

The first filtering elco was rated at 450V_{DC}, but in standby mode, that voltage was exceeded. Even in operation, the 444V was too close for comfort. In such situations, use a 500V-rated capacitor (3) or two 400V caps in series (paralleled by equalizing resistors).

Since it wasn't doing much, the negative feedback was removed. You can play around with cathode bypass capacitor values of the preamp stages to voice the amp. For a fatter sound, increase the value; for a brighter, less boomy sound, decrease the cap's value.

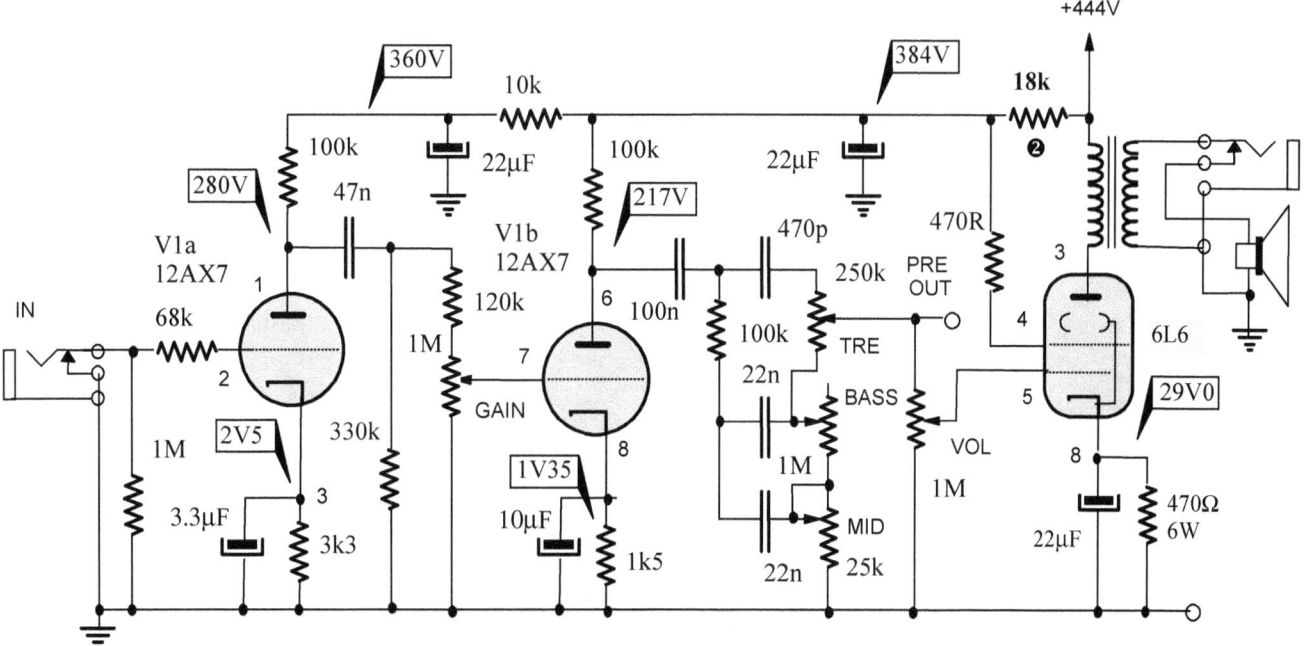

ABOVE: Epiphone Electar Tube 10 after a few basic modifications

VARIABLE POWER CONTROL ADDED: BLACKHEART LITTLE GIANT

The original range of controls of Blackheart BH5-112 combo amp was limited. There was no master volume, no overdrive or standby switch. There was a 5W/3W (Pentode/Triode) switch (1), although the difference in volume between five and three Watts was not significant. A 5W/1W switch would be more useful.

This particular amp was modified by RAT Amplification in the UK and imported by its owner into Australia.

Andy Dokken is the chap behind "RAT Amplification." Visit his website www.ratvalveamps.com/ and check out the examples of various mods he has done over the years; you will definitely get an idea or two.

BLACKHEART BH5-112
- Single-ended combo amplifier
- PCB construction, solid state rectifiers
- Reverb: No, Tremolo: No
- 3-band EQ (Bass-Middle-Treble)
- Pentode (5W)/Triode (3W) Switch
- Pentode: 5 Watts at 22% THD
- Triode: 3 Watts at 9% THD
- Speaker outputs: 4, 8 and 16Ω
- Tubes: 1 x 12AX7, 1 x 6BQ5/EL84
- Speaker: Eminence model 1216A, 12"

The fact that we are discussing some of his mods here does not mean that we agree with them (we may agree with the goal but not the means used to attain those ends). There are many ways to achieve power control in a tube amp, some elegant, others overcomplicated. Once you study this section, make up your own mind.

We've seen BH5's audio circuit (or something very similar) many times in this book, so there is no need to repeat it here; we can go straight to the modifications. Since this is a section on modifying power supplies, we are primarily interested in how the power control or "scaling" was achieved.

The chassis is quite spacious; there is plenty of room for a choke next to the power transformer (2) or other additions.

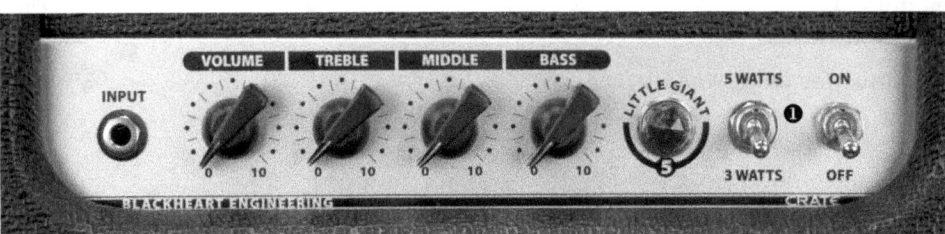

ABOVE: The original faceplate and controls of Blackheart BH5
BELOW: The top view of the modified chassis.

The two switches next to the input jack were added (3), as were the two pots on the other side of the pilot light (4), one marked "MASTER," the other "POWER SCALE."

The On-off switch was moved to the back, and a Master volume control replaced the 5W/3W (Pentode/Triode) switch.

A 3-position switch, shown on the next page, functionally replaces the 5W/3W switch (5). The middle position is standby, the upper position is "PS1" and the bottom "PS2", referring to two different power scaling levels.

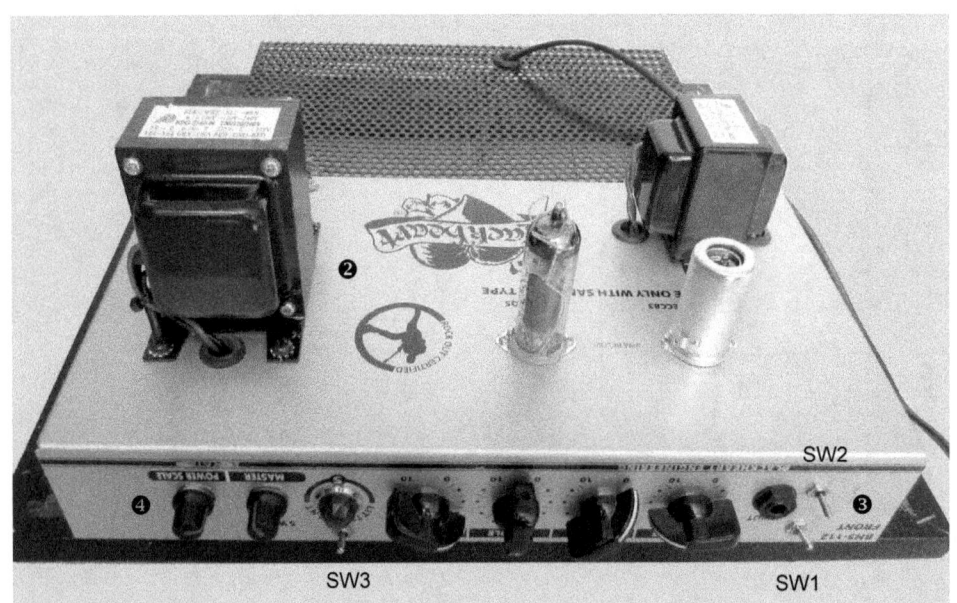

POWER SUPPLY MODIFICATIONS AND IMPROVEMENTS

ABOVE: The modified rear panel BELOW: The original rear panel

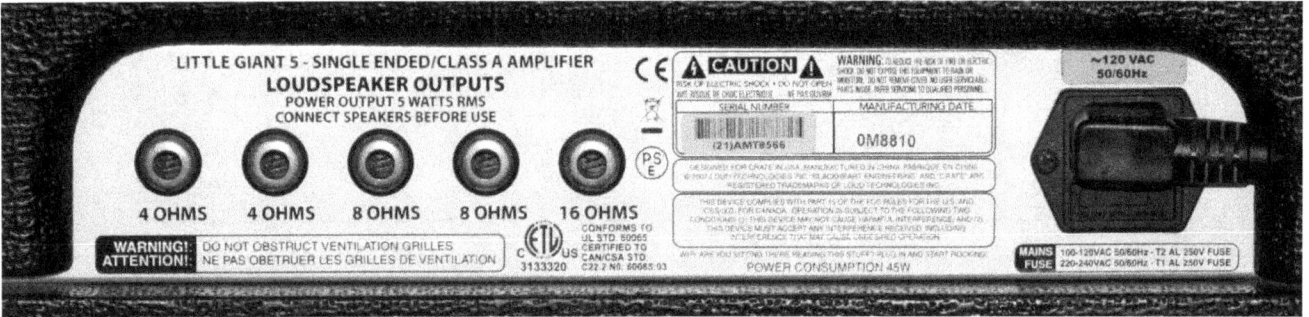

The rear panel had no less than five speaker jacks (!) and no other features, so RAT Amplification converted one to a line output and another to the headphones output. The middle hole was used to install a 3-position speaker impedance switch (4-8-16 Ω). A new faceplate was made covering everything except the original IEC input/fuse holder assembly (6).

There are two braided wires running from the mains transformer to the PCB, marked "REC-0Vac" and "REC-5Vac". This 5V heater supply for a rectifier tube isn't used but will come very handy if you decide to install a tube rectifier. This mod will be discussed soon.

The original circuit

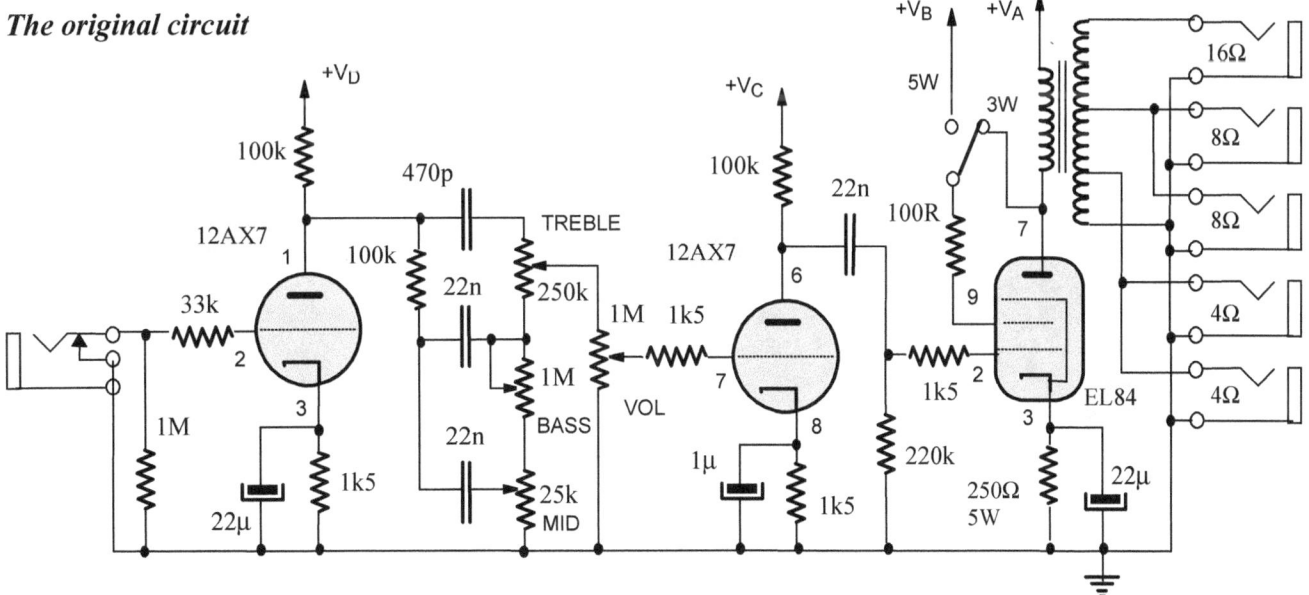

ABOVE: The original audio section, © LOUD Technologies

The circuit after RAT modifications

Switch #1 has two contacts (next page). One switches a 22mF cathode bypass capacitor of the 1st stage in and out (1), the other switches local negative feedback in and out (2). Switch #2 either closes the bottom of the MID potentiometer down to the ground (stock circuit) or connects it to GND via a 540k resistor, which effectively switches off the tone stack for higher gain (3).

Likewise, switch #3 switches the cathode bypass capacitor of the power stage in and out (4). Finally, a master volume control was installed, an obvious choice (5).

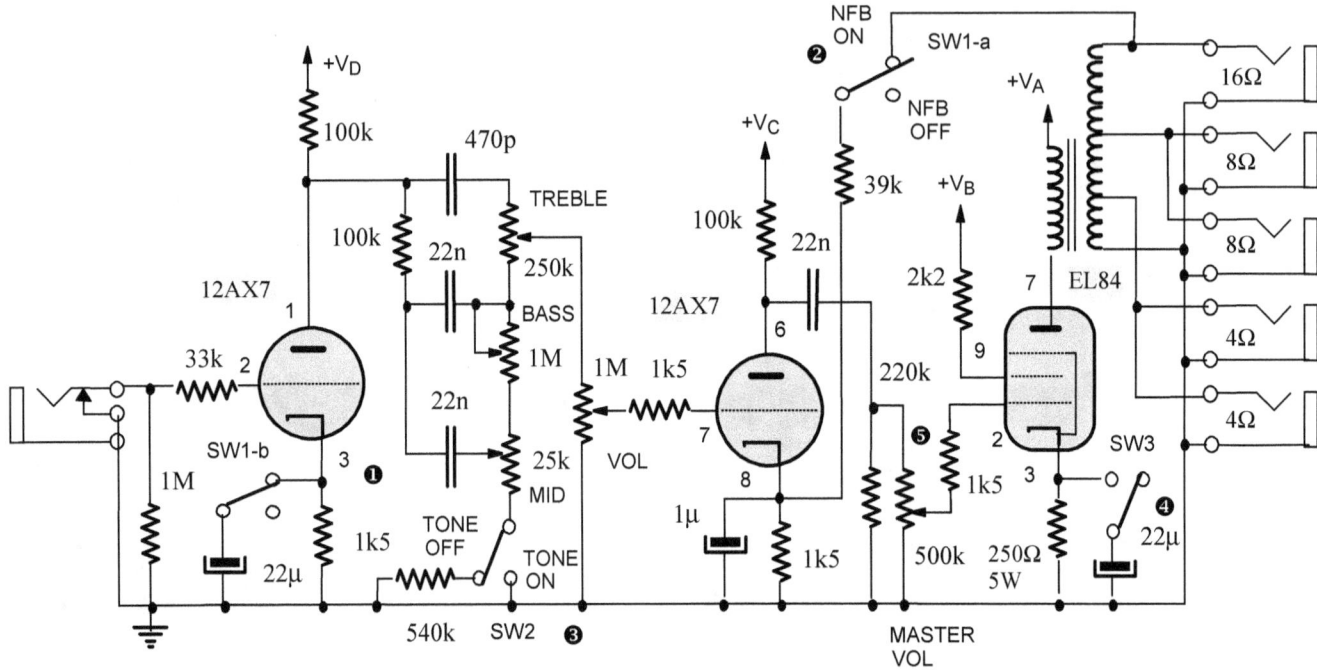

ABOVE: The RAT modified audio section BELOW: The inside view of the modified amplifier

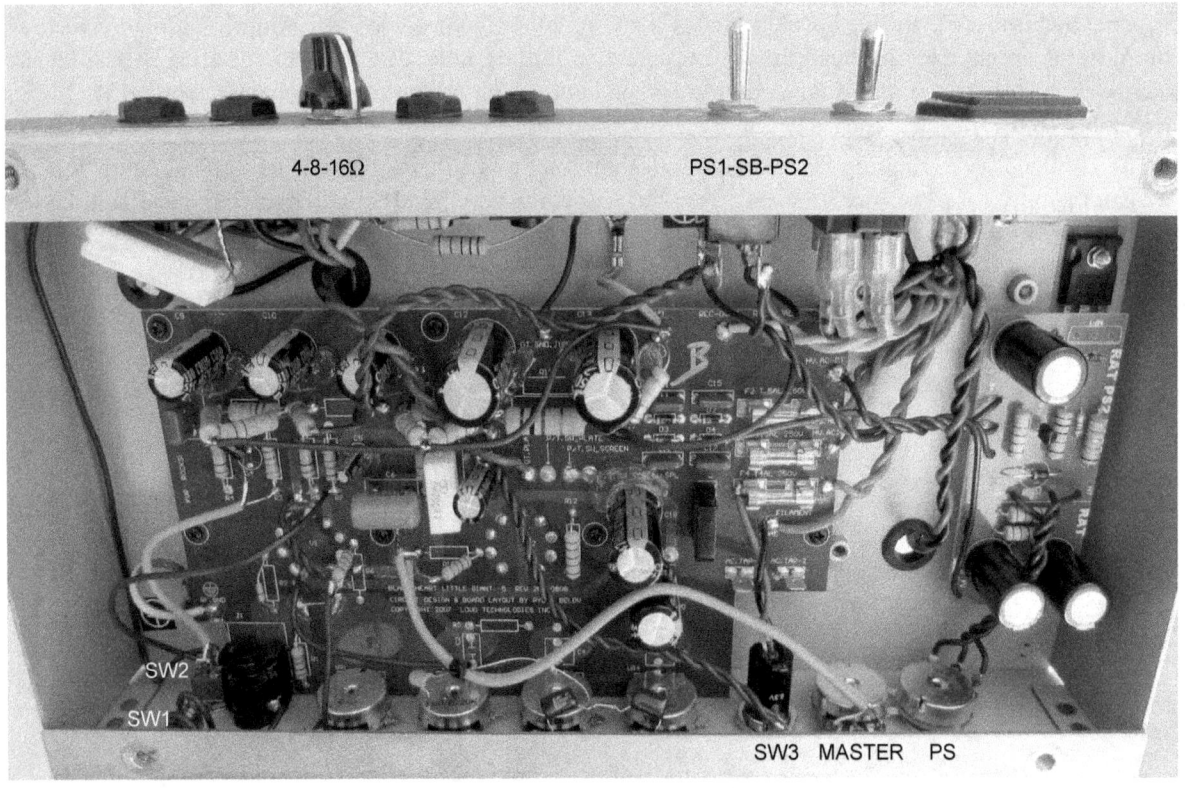

Power supply modifications

Despite solid-state rectification, there was no standby switch. Notice two mains (power) transformer features, the center-tapped (CT) high voltage winding and the unused $5V_{AC}$ heater winding, obviously intended for a rectifier tube. Strangely, RAT modifications made no use of either.

Instead, a three-position switch was inserted between the rectifier and the filtering stage. In the middle (standby) position of the switch, there is no connection between the two stages and no high voltage anywhere. In both PS1 and PS2 positions, the output of the rectifier enters the RAT PS2 module at "B+ IN" terminal. Also, in both positions, the output of the module supplies the first two stages of the filtering chain. Voltages $+V_A$ and $+V_B$ are variable, and so is the output power of the amplifier.

In the PS2 position, as illustrated, the variable output of the 2nd filtering stage (point "X") is switched to the input of the 3rd stage (point "Y"), making voltages $+V_C$ and $+V_V$ also variable.

POWER SUPPLY MODIFICATIONS AND IMPROVEMENTS

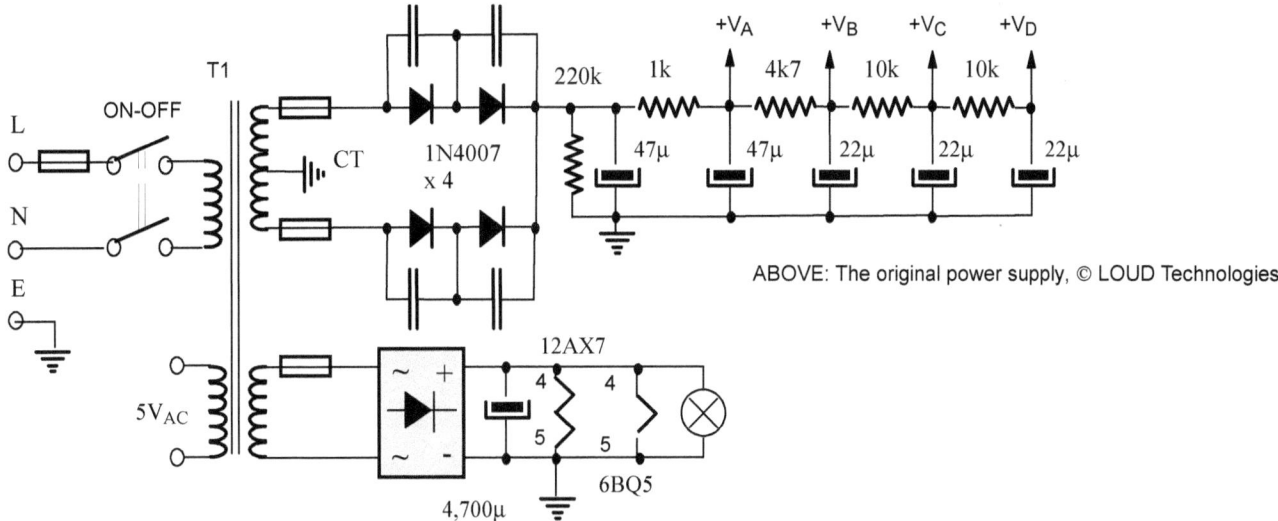

ABOVE: The original power supply, © LOUD Technologies

How does the variable section of RAT PS2 module work? In PS1 position, the input of the filtering chain for preamp stages (point "Y") is switched to the FV output of the PS2 module, which is the output of its fixed voltage section.

IRFP460 is an N-channel power MOSFET in TO-247 housing. It is used here as a series pass element, basically working as a variable resistor, changing the voltage drop across itself and thus making the output voltage variable. The more voltage is dropped across DS (drain-source) terminals, the lower the output voltage. The detailed operation of this voltage regulator will not be explained here; its principles are the same as high voltage tube regulators extensively covered in Vol. 2 of my book "Audiophile Vacuum Tube Amplifiers."

The 1M potentiometer sets the referent voltage that biases the gate of the MOSFET and determines its degree of conduction. BC547 is a general-purpose NPN-type silicon transistor, which senses the voltage drop across the 0.33Ω current sensing resistor and thus acts as a current limiter.

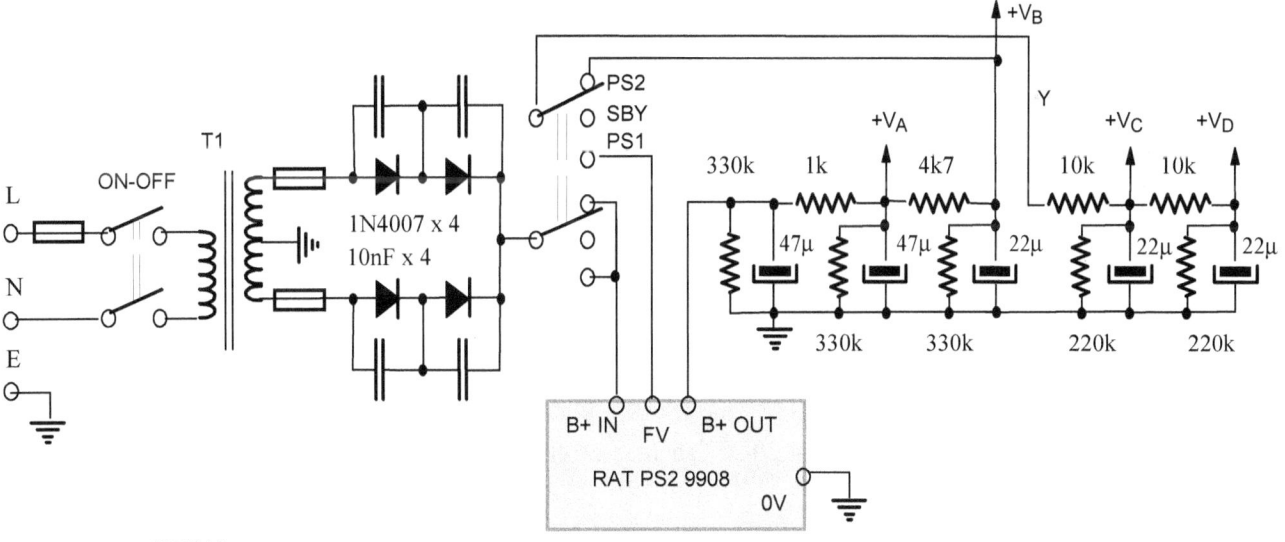

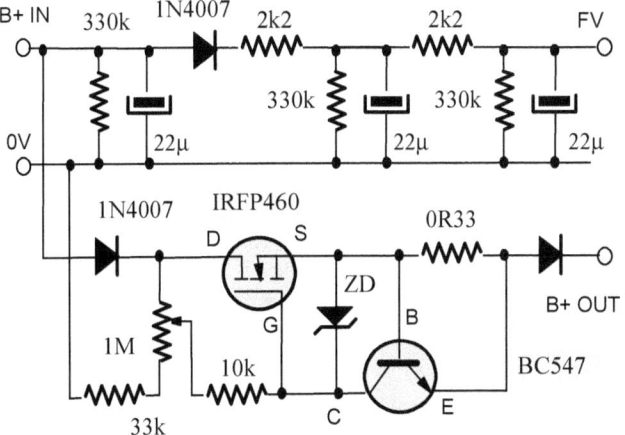

The modified power supply and The PS2 module © RAT Amplification (the unchanged heater supply not shown)

If the load current draw increases for some reason (the output tube loses its negative bias, for instance, and fully opens), the voltage drop across the resistor increases, increasing the base-emitter voltage.

The Zener diode keeps the source-gate and base-collector voltage constant and thus protects the transistor from overvoltage and possible damage or destruction. The V_{GS} limit for IRFP460 is 20 Volts.

The dual rectifier mod

Adding any more features to The Little Giant amp already so extensively modified would make things too complex and controls too confusing - as if they aren't already! However, adding a tube rectifier would be an obvious choice if your amp is in its original condition. You could even make a switch-selectable dual rectifier power supply (tube-solid state).

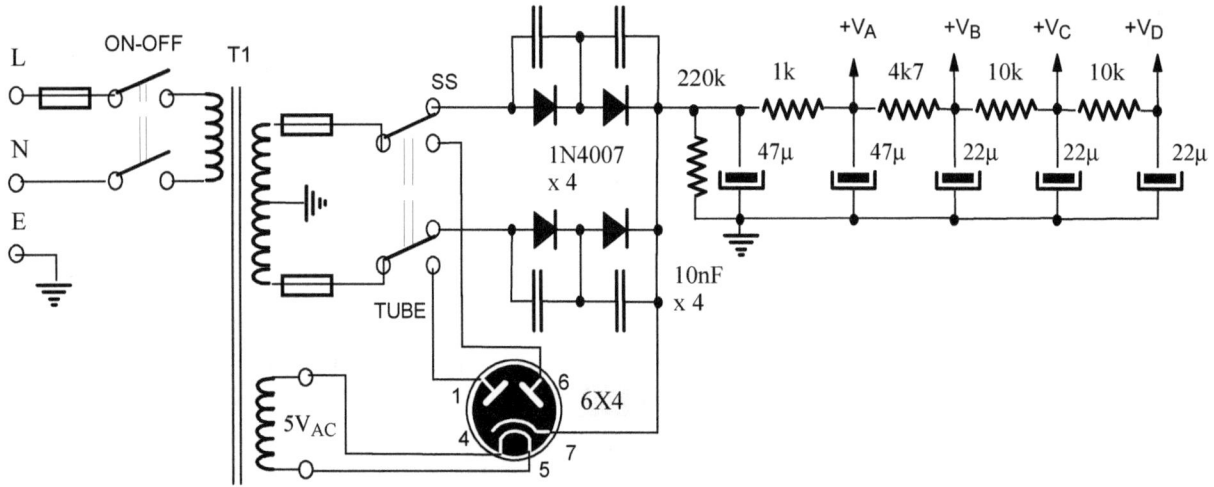

ABOVE: The dual rectifier power supply mod - option 1 - full switching
BELOW: The dual rectifier power supply mod - option 2 - bypassing

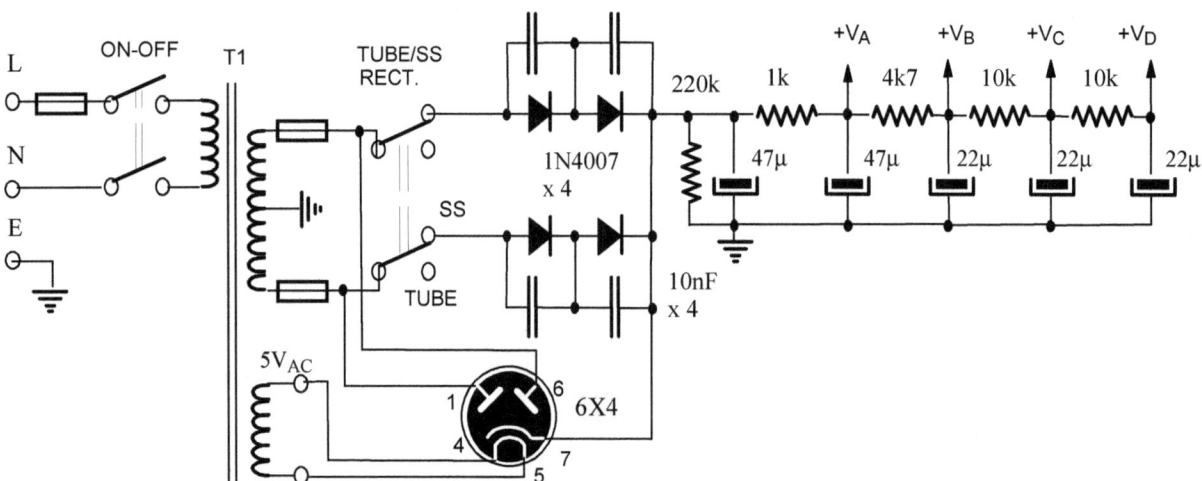

The voltage drop on each silicon diode is around 0.6-0.7V (a total of 1.2-1.4V for two in series, to double the PIV value), while a 6X4 tube would drop around 20-25V across its electrodes. Thus, in the SS mode, all DC voltages would be approximately 20 Volts higher. A DPDT (double-pole-double-throw) switch toggles between the two rectifier modes.

With a full switching option, once the switch is in the SS position (as shown), the anodes of the rectifier tube are fully disconnected from the circuit. Since the internal resistance of silicon diodes is many times lower than that of a tube rectifier, the simpler "bypassing" version would leave the rectifier tube's anodes always connected to the HV secondary, and the solid-state diodes would be "added" or switched in parallel with its anodes.

Because current always takes the path of least resistance, almost all of the current would flow through solid-state diodes and none through the tube rectifier, which would idle along in its warm and glowing state.

TONE TWEAKS

- CHANGING PREAMP TUBES
- CHANGING POWER TUBES
- CHANGING RECTIFIERS
- OTHER MODIFICATIONS

Every design choice and every component inside a tube guitar amp impacts its sound and playability. However, not all affect the amp's voicing equally; some contributions are significant and immediately noticeable, while others are more subtle.

Quite a few profitable businesses sell replacement transformers and speakers to guitar players who want to improve the sound of their amps. While these may be justified in some cases, similar improvements in tone can often be achieved without such drastic "surgery" and rewiring by straightforward and cheap means.

A different brand of the same tube type or a different tube plugged in (say 12AT7 instead of 12AX7) may bring about just the tonal tweak you've been looking for.

CHANGING PREAMP TUBES

While there are some tech-savvy guitar players, most don't have the knowledge or time for modifications to their amps (replacing output transformers, resistors, or capacitors and rewiring circuits). Tube rolling is thus the simplest and often the cheapest gain- and tone-changing modification one could make to a tube amp.

The 9-pin family of pin-compatible duo-triodes, most commonly used in guitar amps (although some vintage amps used octal tubes such as 6SN7 and 6SL7), can be subdivided according to their m or voltage amplification factor ("gain") into four distinct groups. Since the most commonly used tube, 12AX7, has an extremely high gain, substitution of any other tube will reduce the gain of that stage, which can often improve the sound.

LOW μ (16-20)	MEDIUM μ (30-44)	HIGH μ (60-70)	EXTRA HIGH μ (100)
12AU7, 6189, 7730, 5814, 6913 (17)	12AY7 (44)	12AT7 (60)	12AX7, 12BZ7, 12DF7, 12DT7, 12DM7, 6681, 7025, 7058, 7494, 7729
6955, 7318 (16)	12AV7 (41)	12AZ7 (60)	
7489 (20)	6072 (44)	6201 (60)	
5963 (21)	6829 (47)	6679 (60)	
	7490 (32)	5751 (70)	

Four groups of pin-compatible Noval duo-triodes. While most tubes in the table have heaters that draw 300mA at 6.3V and 150mA for 12.6V, some draw more and some draw less heater current.

The plug & play contenders

Apart from 12AU7 and similar low μ duo-triodes, half a dozen triodes could be plugged in instead of the most common 12AX7 tube, without any changes to the anode voltages or anode and cathode resistors. All have lower amplification factor m, ranging from approx. μ=70 for 5751 to around 50 for 12AT7 and 40 for 12AY7 and 12AV7. The table below summarizes the main parameters of these duo triodes. A range of anode currents I_A is given for each tube, and the range of the internal resistance r_I and μ that corresponds to those I_A ranges.

The amplification factor μ is most constant for 12AX7; it hardly changes as the anode current varies from 0.2 to 1.7 mA. For other triodes, μ varies widely; it is always the lowest at low anode current levels and rises as I_A goes up.

Of the four main replacement contenders, two are more similar to 12AX7 in terms of anode dissipation and low anode current levels, namely 5751 and 12AY7. Both have slightly higher anode power ratings and also higher anode currents, up to 2.5mA for 5751 and up to 4mA for 12Y7.

The other two, 12AT7 and 12AV7, are higher power tubes with 2.5-2.7 Watts anode dissipation.

SIDE-BY-SIDE	12AX7	5751	12AT7	12AY7	12AV7
Heater	6.3 V, 0.3A	6.3 V, 0.35A	6.3 V, 0.3A	6.3 V, 0.3 A	6.3 V, 0.3 A
	12.6 V, 0.15A	12.6 V, 0.175A	12.6 V, 0.15A	12.6 V, 0.15A	12.6 V, 0.15A
Equivalents	12AD7, 6681, 7025, 7729		CV4024, 6201	6072	5965
P_A max. [W]	1.0	1.1	2.5	1.5	2.7
V_{AK} max. [V]	300	330	330	300	300
I_A [mA]	0.2 - 1.7	0.2 - 2.5	0.5 - 14	1 - 4	3 - 26
μ	100	60 - 75	30 - 60	40 - 44	30 - 45
r_I [kΩ]	140 - 58	150 - 40	60 - 28	40 - 30	17 - 4k2

The impact of tube's internal resistance

Tone controls significantly attenuate the guitar signal. This attenuation is primarily the result of their subtractive nature. These passive tone controls cannot boost or amplify the signal; they can only reduce or attenuate it.

So, when we talk about the "bass boost," this isn't so at all; the "bass boost" is actually "treble cut," just as the "treble boost" is a misnomer, it works by reducing the bass rather than increasing the treble frequencies!

However, another reason for such significant signal attenuation is the internal resistance of the previous tube stage, the one driving the tone stack.

The graph (next page) shows how the internal resistance of the three most commonly used Noval (9-pin) preamp tubes varies with anode current. Of course, the anode-to-cathode voltage (the DC voltage across the tube) must be identical; in this case, that is $100V_{DC}$, one of the published curves. Graphs are published in steps, such as 50-100-150-$200V_{DC}$. For any other V_{AK} voltage the graphs can be graphically interpolated.

Two things are immediately obvious. Firstly, triodes' resistance increases significantly at very low anode currents. Thus, designers should avoid operating tubes in that "starvation" region.

TONE TWEAKS

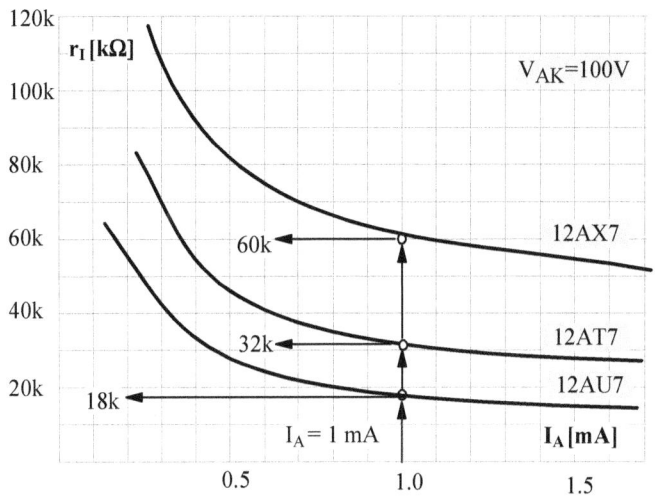

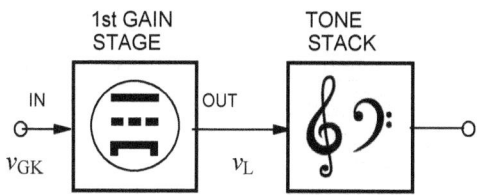

ABOVE: A typical input topology of a tube guitar amp, 12AX7 common cathode stage driving a tone stack

LEFT: The internal resistance of three common Noval duo triodes as a function of anode current

The μ-versus IA graph, also published by tube makers, shows that gain drops in that region, making things worse by reducing the overall gain of the common cathode stage even further. Secondly, the higher tube's gain (μ or "mju"), the higher its internal resistance.

Let's look at what happens when we plug a 12AT7 tube instead of a 12AX7 input tube in an amp. The current through the stage will change depending on various circuit parameters (the bias resistor, the anode resistor, and the anode supply voltage). For this discussion, just to illustrate the principle, let's assume that the 1mA current of the 1st gain stage remains constant. At that current level the internal resistance r_I of 12AX7 triode is around 60kΩ, while 12AT7 has a much lower r_I of around 32kΩ, so almost half.

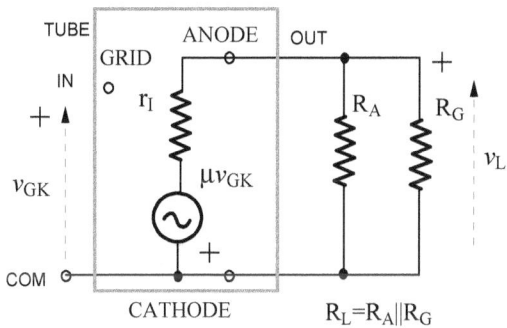

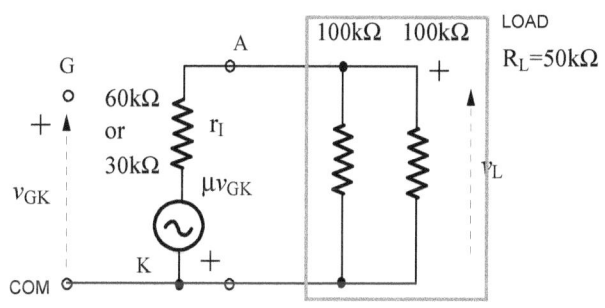

ABOVE: The voltage source model of a common cathode triode stage. Full bypass for all signal frequencies of the cathode resistor is assumed, so cathode K is always at the COM or zero signal level.

ABOVE: With a typical 100k anode resistor and a 100k tone stack contour resistor, the 50k load on the driver stage is the same order-of-magnitude as the internal resistance of the tube!

A typical value of anode resistor R_A is 100kΩ for a 12AX7 stage, and the typical input resistance of tone stack is also 100kΩ. Again, all models are simplifications, but their results are surprisingly close to real-life results!

With 12AX7 r_I =60kΩ, the signal voltage on the load is v_L=50/(50+60)μv_{GK}= 50/110 = 45.4% of the μv_{GK} value. With 12AT7 r_I =30kΩ, the input voltage to the tone stack is 50/(50+30)= 50/80 = 62.5% of the μv_{GK} value, which is 62.5/45.4 = 1.375 or 37.5% increase compared to the 12AX7 situation.

Now, assuming a 10mV input signal from a guitar pickup on the grid (v_{GK}), the μv_{GK} value for 12AX7 is 0.01*100=1V, while with 12AT7 μv_{GK} = 70*0.01=0.7V. Finally, the load voltage for 12AX7 is 1V*0.454 = 0.454V and the load voltage for 12AT7 is 0.7V*0.625 = 0.4375V, which is almost the same as for 12AX7!

So, despite its higher voltage gain, the output voltage of the high internal resistance 12AX7 stage is the same as the output voltage of a lower gain stage with 12AT7 due to the lower internal resistance of the 12AT tube and the voltage divider (attenuation) effect those internal resistances cause.

The real life plug & play example

Now that we understand the internal impedance issue, let's see what happens when a 12AT7 is plugged into an amplifier designed for 12AX7. In this example, it's the input stage of the Silvertone 10XL amplifier.

Anode current can be determined from the cathode circuit. With 0.9V on the cathode we get I_A=0.9/2.2=0.41mA and since the voltage drop on the 220k anode resistor is 34, we get I_A=34/100 = 0.34mA. This discrepancy is due to resistor tolerances; the cathode resistor has drifted to 2k6, so the anode current should be 0.9/2.6= 34.6mA.

The DC voltages in major points enable us to draw the load line on 12AX7 curves. The curve for 0V9 bias ($V_G = -0.9V$) is not given, but we can approximate it. The intersection of that curve and the 100k load line gives us the operating point or quiescent point Q.

The V_0 is the anode-to-cathode voltage. The graph shows 90V but in real life we measured $102 - 0.9 = 101$ V. The graph shows an anode current of 0.45 mA, while in reality, it is around 0.35 mA.

This discrepancy will depend on the actual tube used and is relatively small here. The difference between graphic analysis and actual measurements is usually much bigger. The published curves are for rough estimates only, or as they say, for "orientation purposes"!

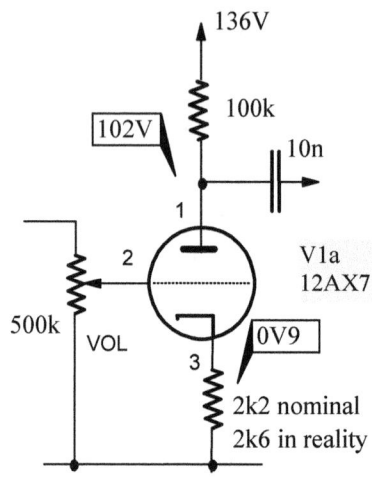

The input stage of Silvertone 10XL amplifier

ANODE VOLTAGE GAIN OF COMMON CATHODE STAGE WITH UN-BYPASSED CATHODE RESISTOR

$$A_A = -\mu R_L / [r_I + R_L + (1+\mu) R_K]$$

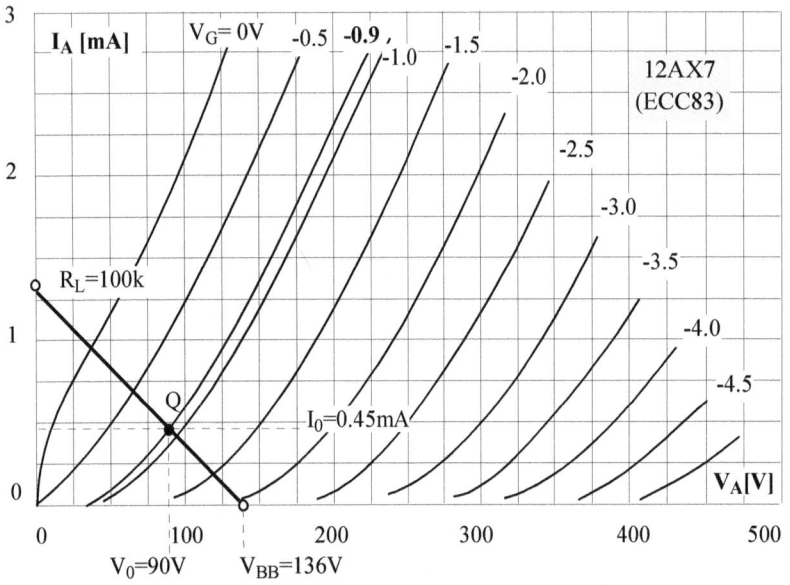

Using the small-signal model for triodes, the voltage gain of the 1st stage is $A_A = -\mu R_L / [r_I + R_L + (1+\mu) R_K]$. For 12AX7 in its operating point, as discussed, the parameters are approx. $\mu = 100$ and $r_I = 100 k\Omega$. These can be read from graphs that show the three static parameters (μ, gm, and r_I) as a function of anode current I_A.

In our case we get $A_A = -100 \cdot 100 / [100 + 100 + (1+100) \cdot 2.6] = -100 \cdot 0.216 = 21.6$!

This is very low gain from a 12AX7 stage. The reasons are:

1. The relatively low value of the anode resistor (only 100k).

2. The low value of anode current (only 0.35mA), which means the internal impedance of the tube is increased, and the gain is lowered because a higher proportion of the signal is lost on the internal resistance of the tube (the tube and the load form a voltage divider).

ABOVE: The DC loadline and Q-point for the input stage of Silvertone 10XL amplifier
BELOW: The same gain stage once 12AT7 triode is substituted

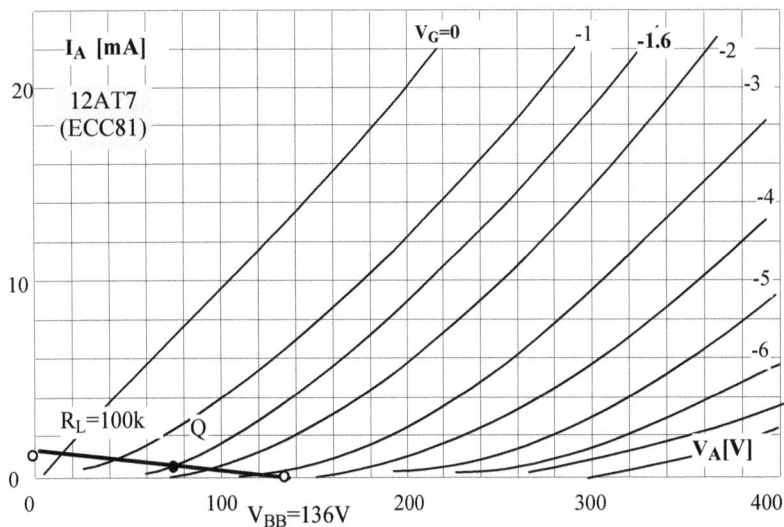

3. The local negative feedback caused by the un-bypassed 2k6 cathode resistor.

If we bypass it with a capacitor, the $(1+100) \cdot 2.6$ factor in the formula will disappear and the gain will increase to $A_A = -100 \cdot 100 / [100 + 100] = -100 \cdot 0.5 = -50$! The gain would be increased $50 / 21.6 = 231\%$, more than doubled.

The graph on the next page is interesting. It shows you that the tube's internal resistance is higher at higher anode voltages (in this case, 250V compared to 100V). It also says that both gm and μ are higher at low anode-to-cathode voltages for the same anode current levels.

For instance, at 6 mA of current 12AT7 has µ=70 at V_{AK}=100V but only µ=55 at V_{AK}=250V. For example, that could significantly differentiate between a clean gain stage and one breaking up into distortion.

So, the conclusion is that for any serious design, it's never enough to look at the basic short-form data for a particular tube. Always download its long-form data, including various graphs, and determine its parameters for each specific circuit!

If we draw V_{BB}=136V and the 100k loadline on the curves for 12AT7, we notice an immediate problem. Since 12AT7 is a high anode current triode, the vertical scale on the published graphs is imprecise at such low anode currents (less than 1mA), so we cannot possibly analyze that way.

As I write this we have the Silvertone amp on the bench and can simply plug in a 12AT7 and measure DC voltages. The anode voltage dropped from 102V to 99.5V and the cathode was 1.6V above ground, meaning the anode current was approx 0.615 mA.

The anode voltage of the second triode dropped from 168V to 112V while its bias increased from -1.9V (+1.9V on the cathode) to - 3.2V!

For 12AT7 tube in its operating point of 0.615mA we read (the graph above) that its internal resistance is around 40kΩ and its µ is around 36 times, so:

$A_A = -\mu R_L / [r_I + R_L + (1+\mu)R_K] =$
$= -36*100/[40+100+(1+36)*2.6] = 15$!

Due to a very low anode current and local NFB through an un-bypassed cathode resistor, the gain only dropped from -21.6 with 12AX7 to -15 with 12AT7, which wasn't a huge change.

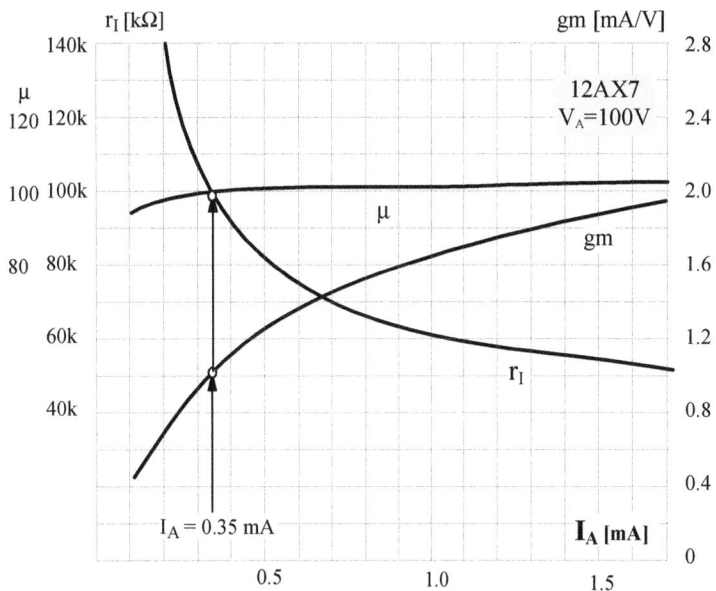

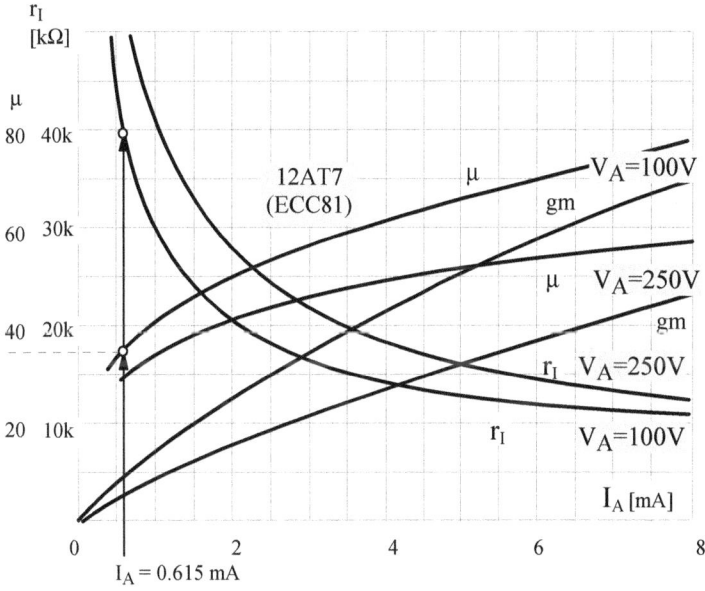

ABOVE: The three main static parameters of 12AX7 triode as a function of anode current I_A. Note: Valid only for V_{AK}=100VDC!
BELOW: The three main static parameters of 12AT7 triode as a function of anode current IA. Note: Valid only for VAK=100VDC and VAK=250VDC .

Gain reduction or redistribution without rewiring

In many cases, replacing a very high gain (also a very high internal resistance) duo-triode such as 12AX7 with a high gain 12AT7 or low gain 12AU7 results in a sub-optimal gain distribution.

For instance, we may still want the first stage to remain a high gain stage while aiming to reduce the gain of the second stage and use a lower gain triode there. If that 2nd stage also drives a tube stack, that stage would benefit from tubes with progressively lower internal impedance such as 12AT7 (10-45kΩ, depending on the anode current) and 12AU7 (6-15kΩ). This would reduce the gain loss due to the driving tube's lower internal resistance.

One solution is to use a duo triode with two dissimilar triodes, for instance, 12DW7, a.k.a. 7247. Its triode #1 has the same mju as 12AX7 (100), while triode #2 has the same gain as 12AU7 (17). However, with duo triodes having two identical systems (such as 12AX7), it does not matter which system is used for the 1st (pins 1,2,3 or pins 5,6,7) and for the 2nd stage.

You may find that in some amps, the input stage uses pins 1, 2, and 3, and in 12DW7, that is the low m triode. If you want to keep that stage as the high gain stage, the triodes need to be swapped over. If a PCB is used, that is too messy and too difficult, but point-to-point wiring makes it possible. So, this option may or may not work in your particular amp.

JJ currently manufactures ECC832, which is 12DW7 equivalent. They also identified the pinout problem with most amps wired, so tube 1 is pins 3,2 &1, so a "reverse" 12DW7 was released, called ECC823, whose high gain triode is on pins 3,2 &1.

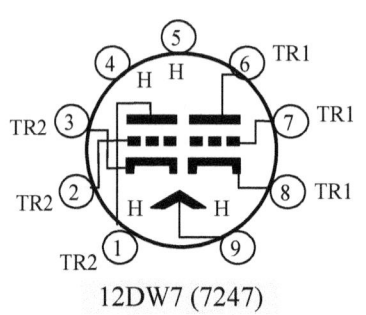

12DW7 (7247)

TUBE PROFILE: 12DW7 (7247)
- Indirectly-heated dissimilar double-triode
- Noval socket
- Heater: 6.3V, 0.3 A - 12.6V, 0.15A
- Maximum anode voltage 330 V_{DC}
- $V_{HKMAX} = 200\ V_{DC}$
- Triode 1 (pins 6-7-8): µ=100 (P_{AMAX}: 1.2W)
- Triode 2 (pins 1-2-3): µ=17 (P_{AMAX}: 3.3W)

CHANGING POWER TUBES

Providing preamp tubes have the same pinout and reasonably similar electrical parameters, tube rolling is usually safe - very little damage can be inflicted on an amp. Power tube changes are much more dangerous. Even if a tube with an identical pinout is substituted, for instance, 6L6 instead of 6V6 (or the other way around), certain maximum limits could be exceeded (max. anode power dissipation or max. anode voltage) or the amp's high voltage or heater power supply can be overloaded, resulting in a damaged or destroyed tube or burned-out power transformer.

Extreme caution is required. If unsure, consult a specialist. In the meantime, let's get a better idea about the factors involved in such changes through a few case studies.

6973 replacements

Since EH (Electro-Harmonix) in Russia started making 6973 tubes again, the issue of a suitable replacement is not critical anymore. However, most guitar players don't like being limited to one brand only. Plus, as an educational exercise, this discussion will teach you the methodology that can be applied to any substitution of power tubes with their close-but-not-identical stand-ins.

The first step is to do a comparative study of the original tube's characteristics, especially the limiting (maximum) values. Then, determine the operating voltages and conditions in the amplifier you are retubing. Increased heater power can be an issue in budget amplifiers whose transformers have no spare capacity.

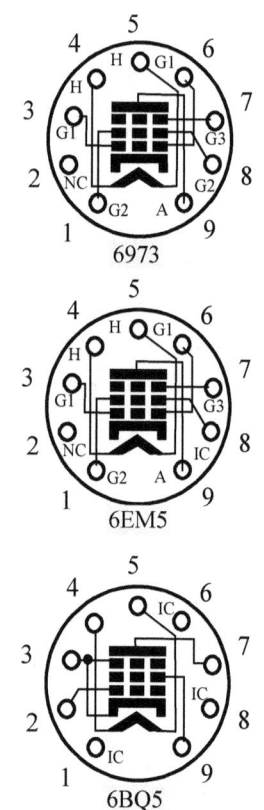

SIDE-BY-SIDE	6973	6EM5	EL84 (6BQ5)
TYPE	Beam tetrode for audio service	Beam tetrode for vertical deflection	Power pentode for audio service
Heater	6.3 V, 0.45A	6.3 V, 0.8A	6.3 V, 0.76 A
P_A - P_{G2} max.. [W]	12 - 2	10 - 1.5	12 - 2
V_A - V_{G2} max.. [V]	440 - 330	315 -285	300 -300
AVERAGE CHARACTERISTICS SE PENTODE:			
V_{G1} [V]	-15	-18	-7.3V
I_A / I_{G2} [mA]	46/3.5	35/3	48/5.5
rI / R_L [kΩ]	73 / 5.5		40 / 5.5
gm [mA/V]	4.8	5.1	11.3

Replacing 6973 with 6BQ5 (EL84)

This substitution requires a major rewiring on pins 3, 7, 8, and 9. Also, EL84 needs 0.3A more heating current per tube, so a push-pull amp with four tubes would draw a 1.2A higher heater current! The existing power transformer may not have that much spare capacity; an add-on 6.3V heater transformer may be required.

Finally, in push-pull circuits, 6973 needs 20 V more grid drive than 6BQ5, and the bias of EL84 is almost half that of 6973, so the input and driver stages of a 6973 amp will need to have their gain lowered for EL84 tubes which require a smaller grid signal.

Replacing 6973 with 6EM5

Issue #1: Pin 8 on 6EM5 is internally connected, while on 6973, it is connected to G2. Any component wired up to that pin in the actual amplifier needs to be moved to pin 1, which is also G2 on 6973. Nothing should be connected to that pin if 6EM5 is to be used. A simple jumper to pin1 is not a solution.

Issue #2: 6EM5 seems to be much less efficient (just like EL84), needing 0.35A more heating current per tube, so a PP amp with four tubes will need 1.4A higher heater current! Most existing power transformers will not have that much spare capacity, so an add-on 6.3V heater transformer will be required.

Replacing 8417 tubes

A few guitar amps (CMI SG100) and some bass amps such as Guild Quantum from the early 1970s used the then newly-developed 8417 beam power tubes at their outputs. 8417 was one of the last power tubes to be developed but was plagued with problems. 8417 tubes are rare, not robust at all, and thus unreliable and undesirable in guitar amps.

There are a few candidates for their replacement, although all require some changes to operating conditions (bias changes), socket rewiring, or both.

If 8417 are replaced by KT88, KT90, KT100, or even KT120, no socket rewiring is needed, but higher bias voltage and a higher gain are required from preamp stages. With 12AX7 at the front, this should not be a problem. Reducing NFB will also increase gain.

7027A is the best replacement, albeit at a reduced output power level. The heater load on the power transformer will reduce by 1.4A, and the anode load on the high voltage power supply will also be lower, so the mains transformer will run cooler and quieter. The distortion will go down, and the amp will sound cleaner.

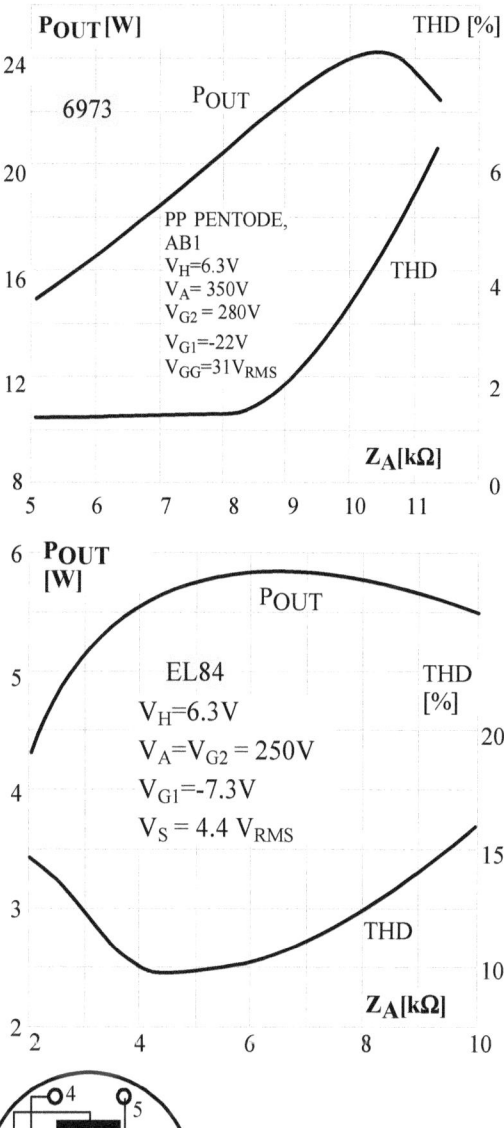

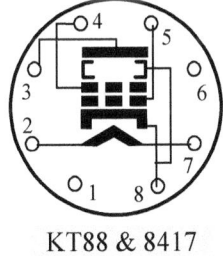
KT88 & 8417

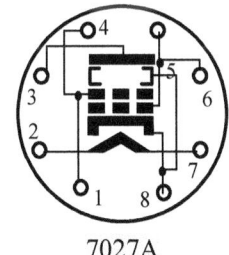
7027A

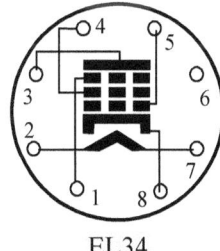
EL34

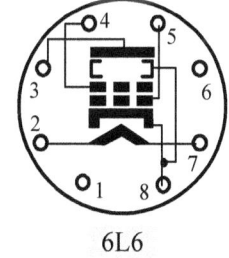
6L6

SIDE-BY-SIDE	8417	7027A	EL34	KT88
Heater	6.3 V, 1.6A	6.3 V, 0.9A	6.3 V, 1.5A	6.3 V, 1.6 A
P_A - P_{G2} max. [W]	35 - 5	35 - 5	25 - 8	35 - 6
V_A - V_{G2} max. [V]	660 - 500	600 - 500	800 - 800	800 - 600
R_{GMAX} [kΩ]	250	500	500	100 - 220
TYP. U/L PP OPERATION				
V_A	445	450	450	460
V_{G1} [V]	-25	-30	-32.5V	-59V
I_{KO} / I_{KMAX} [mA]	146 / 314	95 / 190	65 / 85	50/140
Z_{AA} [kΩ]	3.5	6.0	7.0	4.0
P_{OUT} [W]	70	50	40	70
THD [%]	2.5	1.5	4.5	2.0

HIGHER THD, LOWER OUTPUT POWER → (7027A, EL34)
LOWER OUTPUT POWER → (7027A)
HIGHER GRID DRIVE & CIRCUIT CHANGES REQUIRED ← (KT88)

Replacing 7355 with 7591 or 6L6

JJ (Slovakia) and ElectroHarmonix (Russia) produce 7591A tubes. There are still vintage used or NOS USA-made tubes available, so the availability is much better than that of 7355, which is not in current production and wasn't widely used even while it was in production.

Remove any components soldered to pin 4 of 7355. Pin 4 is often used as a tie-in point since it is not connected to any of 7355 internal pins, but it is connected to G2 of 7591!

Notice that 7591 requires a lower bias voltage (-15.5V versus -21V for 7355). This should be an easy adjustment in fixed bias amps, assuming the trimmer pot has a wide enough range. In cathode-biased amps, the cathode resistor would need to be replaced.

SIDE-BY-SIDE	7355	6L6	7591
Heater volts	6.3	6.3	6.3
Heater amps	0.8	0.9	0.8
Max. anode voltage [V]	500	500	550
Max. screen voltage [V]	400	450	440
Plate dissipation [Watts]	18	30	19
Screen dissipation [Watts]	3.5	5.0	3.3
Typical class AB1 PP amplifier with fixed bias	7355	6L6	7591
Anode voltage [V]	300	360	350
Screen voltage [V]	250	270	350
Control grid bias [V]	-21	-22.5	-15.5
Zero signal plate current [mA]	100	88	92
Max. signal plate current [mA]	185	132	130
PP load resistance [Ω]	4,000	6,600	6,600
THD [%]	2.0	2.0	2.0
Power out [Watts]	28.5	26.5	30

Replacing 6L6 with 7027A

While the distortion of the ubiquitous 6L6 in SE triode mode is up to 10%, 7027A is a much more linear and better-sounding tube. So, if you have a bass amp with 6L6 or a guitar amp that would benefit from an overall cleaner tone, especially at higher power levels, 7027A is the tube to choose.

For the same heater power, 7027A has a 35Watt anode power rating, higher than 30Watts of 6L6 varieties, meaning it is also a more efficient tube. Thus, not only you'll get less distortion, you may also get higher output power (depending on the amp's design).

If 6L6 tubes are run at the unusually high anode/screen voltages (making some currently produced 6L6 tubes struggle and suffer high failure rates), 7027A babes will take up to 100V higher voltages, so the reliability (or the lack of) problem will be solved as well.

SIDE-BY-SIDE	6L6	7027A
Heater volts / amps	6.3/0.9	6.3/0.9
Max anode voltage [V]	500	600
Max screen voltage [V]	450	500
Anode dissipation [Watts]	30	35
Screen dissipation [Watts]	5.0	5.0

The two tubes are pin-compatible, providing that in amplifiers designed for 6L6, nothing is connected to pins 1 and 6! Pin 1 is a screen grid in 7027A, and pin six is connected to pin 5, the control grid. Sometimes designers of point-to-point wired equipment use such unused lugs on tube sockets as tie-points to save on terminal lugs. This is a risky practice and should be avoided.

So, if the designer/maker of your 6L6 amp used those pins, the first mod you'll need to make would be adding a small terminal strip with a few lugs to carry the components currently soldered onto pins 1 and 6.

Usually, those are screen grid resistors connected between pins 4 and 1 or 4 and 6, or control grid stopper resistors connected between pins 5 and 6 or pins 5 and 1.

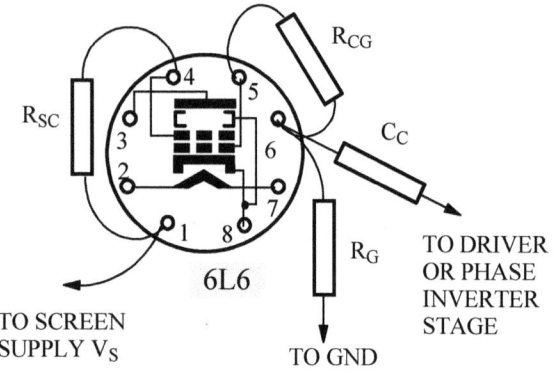

BEFORE: Unused 6L6 tube socket's pins used as component support lugs

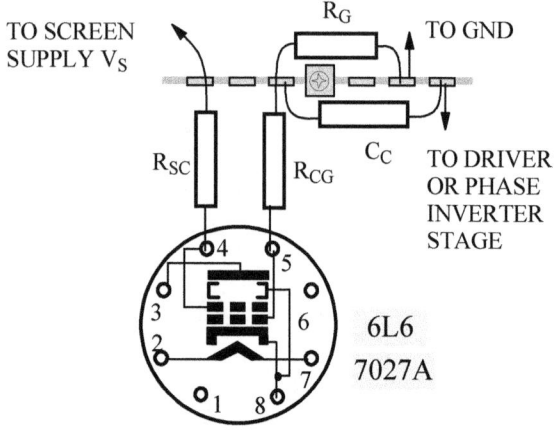

AFTER: 7-lug terminal strip (single center support) added so no components are connected to pins 1 and 6, which must be left free if 7027A is plugged in instead of 6L6!

EL84 substitutes: 6EM5

6EM5 was designed for vertical deflection service in TV sets. The electrical characteristics are similar to those of the European Magnoval socket EL508, although EL508 has higher anode and screen dissipation and is generally a superior tube. However, 6EM5 is a Noval tube, and NOS are widely available and thus cheap.

Electrically, the tube could be substituted for EL84, but unfortunately, the pinout is totally different, so the only option is to use an adapter. The ever entrepreneurial Chinese offer all kinds of adapters for sale online; however, 6EM5-to-EL84 isn't one of them, so you'd have to make your own.

TUBE PROFILE: EL84 (6BQ5)
- Indirectly-heated power pentode
- Heater: 6.3V, 0.75 A
- $V_{AMAX}=V_{G2MAX}=300$ V
- $P_{AMAX}=12$ W, $P_{SMAX}=2$ W
- TYPICAL SE PENTODE OPERATION:
- $V_A = V_S = 250V$, $V_G = -7.3V$
- Signal (grid voltage): $6.1 V_P$, $4.3 V_{RMS}$
- $I_{A0} = 48$ mA, $I_{S0} = 5.5$ mA
- Load resistance: 4.5 kΩ
- gm=11.3mA/V, μ=430, r_I = 38kΩ
- gm_{G2}=1.8 mA/V, μ_{G2G1}=19 (μ triode)
- P_{OUT} = 5.7 W @ THD = 10%

TUBE PROFILE: 6EM5
- Indirectly-heated miniature beam power tube
- Noval socket, heater: 6.3V, 0.8A
- V_{AMAX}=315V, V_{G2MAX}=285V
- I_{KMAX}=60mA, P_{AMAX}=10W, P_{G2MAX}=1.5W
- As a triode: gm=5.8 mA/V, μ=8.7, r_I = 1.5kΩ
- TYPICAL OPERATION PENTODE:
- V_A=250V, V_S=250V, V_G=-18V
- I_{A0}=40 mA, I_S=3 mA
- gm =5.1 mA/V, μ=255, r_I=50kΩ

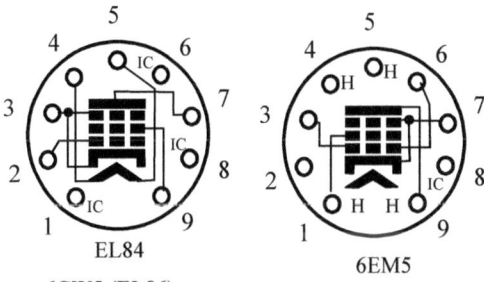

EL84
6CW5 (EL86)

6EM5

TUBE PROFILE: 6CW5 (EL86)
- Miniature Noval beam power tube
- Heater: 6.3V, 0.76 A
- V_{AMAX}= 270V_{DC}, V_{SMAX}= 200V_{DC}
- V_{HKMAX}=100V_{DC}
- P_{AMAX}=12W, P_{SMAX}=1.75W

EL84 substitutes: 6CW5 (EL86)

The American 6CW5 (European EL86) pentode was developed for use as an audio amplifier in TV sets; two would be connected in a single-ended push-pull output stage, driving a high impedance loudspeaker directly, without an output transformer. The heater draw is almost identical to that of EL84, and the pinout is the same, as is the maximum anode power dissipation. The only caveat in substituting EL86 for EL84 are its lower maximum anode and screen voltages, 270 & 200V_{DC} compared to 300V_{DC} for EL84.

EL84 substitutes: 6M5 (EL80)

6M5 was quite a common power tube in Australian radios and radiograms such as Astor, HMV, and Kriesler, probably because it was made locally by Philips, Mullard, and AWA, and thus cheaper. A few USA tube makers also made it for a while, but it was not common or popular either there or in Europe.

With use, 6M5s go dark inside due to the buildup of a silverish layer, so used 6M5s look ugly. Good looks aren't important in guitar amps since tubes aren't usually seen from the outside, but aging 6M5s also start to draw grid current, which changes the bias and increases anode current significantly. Ultimately, if left in such a state long enough, it damages or destroys the tube, or, even worse, the output transformer's primary winding burns out.

If you compare their tube profiles, 6M5 and EL84 are very similar but not identical. 6M5 has a lower anode power dissipation, and at a similar bias of -7V and 250V on anode and screen, it draws a lower anode current (36mA) than EL84 (48mA). Also, it requires a higher load impedance, typ. 7kΩ in a SE pentode stage.

The usual 5kΩ primary output transformers used for EL84 would work with 6M5, of course, but the maximum output power will be reduced, so instead of 5-6 Watts with EL84, you can expect around 3 Watts with 6M5.

6M5 has only one pin different from EL84; pin #1 is the screen grid, EL84's screen is pin #9. Since pin nine is not used, we could strap pins 9 and 1 together on the amp's power tube socket, and 6M5 would work as a substitute for EL84. However, in EL84, pin #1 has an internal connection, so plugging EL84 into such a socket would bring the screen DC voltage onto pin#1, and the tube could burn out. Thus, a switch is required to connect pin#1 to pin#9 for CM5 and disconnect it for EL84.

OTHER MODIFICATIONS

Converting tube into solid-state rectification

This is one of the easiest modifications of all. Unplug the rectifier tube and solder two silicon diodes to the tube socket, as illustrated. Leave the rectifier heater wiring in place so the mod is reversible. Even if a rectifier tube is plugged back in later, the solid-state diodes will still work, and the rectifier tube in parallel will not change anything.

Alternatively, there are commercial solid-state plug-in modules available. They are moderately priced and since diodes (and sometimes series resistors to provide voltage sag) are soldered internally, they require no soldering - plug one in instead of the rectifier tube.

What do we gain by this conversion:

1. Less load on the power transformer. Rectifier heating is typically 5V@2A or 5V@3A, so 10-15 watts less load on the mains transformer, which will run cooler and may become quieter. Mains transformers on most vintage amps are undersized, and any load reduction is welcome!

2. The voltage drop on SS diodes is only around 0.6V, while on tube rectifiers, it ranges from 20 to 50 Volts. So, all the anode and screen voltages will rise proportionally, resulting in higher output power or more headroom in the preamp stages. This voltage rise can also create problems. If the increased voltage exceeds filtering capacitors' voltage rating, they must be replaced with caps of a higher voltage rating. In amplifiers designed to push the output tubes to their limits, their screen or anode voltages may be too high, exceeding the maximum power dissipation.

3. Solid-state rectifiers allow for larger filtering capacitors; instead of a 22-47 μF elcos, we can use 100 μF or even 330 or 470 μF units! This would significantly increase the stored energy in these power supply capacitors, which is $E=CV^2/2$ [Joule]. With a tube rectifier and two 22μF elcos in a CRC filter, that energy is approx. $E=CV^2/2 = 2*22*10^{-6}*400^2/2 = 3.5$ Joule.

By using SS diodes, we would get 30V higher plate voltage, and assuming two 330μF elcos are used in a π filter, the energy would now be $E=CV^2/2 = 2*330*10^{-6}*430^2/2 = 61$ J!

This increase in available energy would manifest itself in the situations when the output stage demands high power flows from the power supply. The bass would improve significantly.

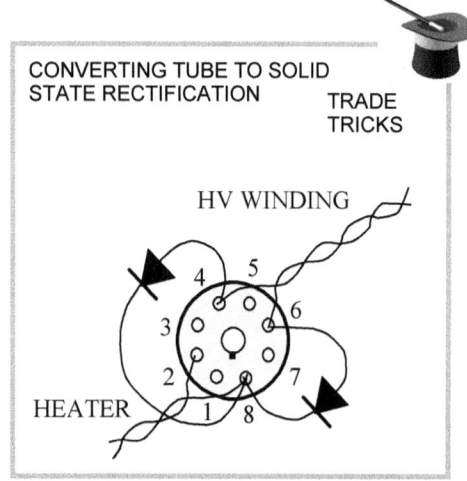

TUBE PROFILE: EL80 (6M5)
- Indirectly-heated power pentode
- Heater: 6.3V, 0.71 A
- $V_{AMAX}=V_{G2MAX}=300$ V
- $P_{AMAX}=9$ W, $P_{SMAX}=3.3$ W
- TYPICAL SE PENTODE OPERATION:
- $V_A = V_S = 250$V, $V_G = -7.0$V
- Signal (grid voltage): 3.8 V_{RMS}
- $I_{A0} = 36$ mA, $I_{S0} = 5.2$ mA
- gm=10mA/V, μ=400, $r_I = 40$kΩ
- $P_{OUT} = 3.9$ W @ THD = 10%

CONVERTING TUBE TO SOLID STATE RECTIFICATION — TRADE TRICKS

ABOVE: SS modules such as TubeDepot SSR are simply an octal plug with two 1N5408 silicon diodes soldered inside.

However, some guitar players don't like high-energy ("stiff") power supplies, claiming that the bass becomes too boomy or muddled, so caution is required.

Adding the loudness control using a frequency-compensated attenuator

In audiometry, the unit of loudness is a phon, but in acoustics, a decibel (dB) is used. These curves, called "equal-loudness" or Fletcher-Munson curves, illustrate the relationship between the two, or how human hearing varies with signal frequency and sound pressure levels (SPL).

The curves for various phon levels have a similar shape. Two things are apparent. Our ears are most sensitive in the frequency range between 500 Hz and 3,000 Hz, where most of the human speech content falls.

As the frequency of sound drops towards the bass region, higher SPL levels are needed to achieve the same perception of "loudness" by human test subjects. That is how these curves were derived, by asking people to adjust a volume to achieve the same subjective level of "loudness".

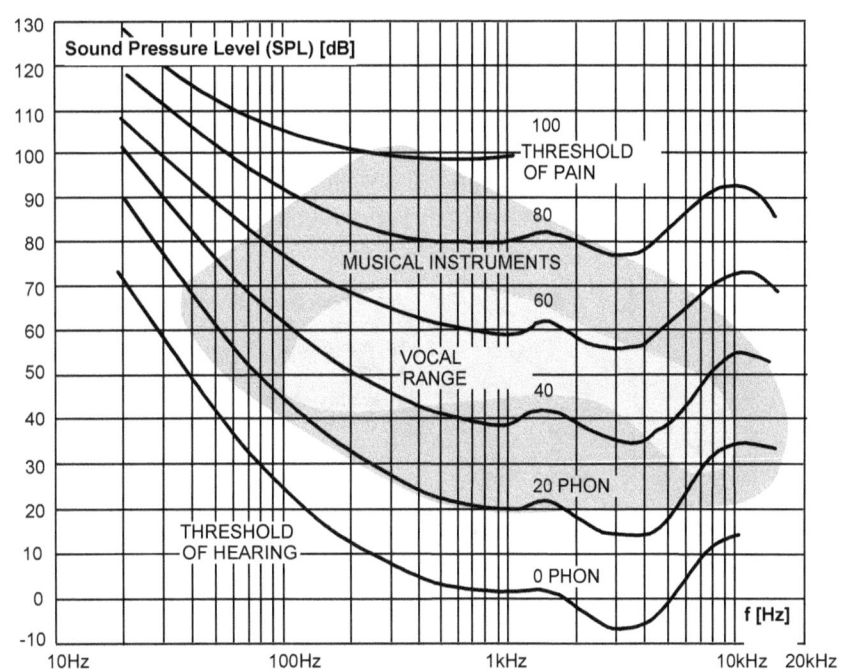

ABOVE: Equal-loudness (based on ISO226:2003 revision) or Fletcher-Munson curves

The same conclusion applies to higher frequencies, although to a lesser extent. As a result, bass frequencies need to be amplified more than the midrange signals to achieve the same perception of loudness, especially at lower loudness levels.

This becomes a factor of great annoyance to audiophiles, whose hi-end tube amplifiers and preamplifiers usually have no tone controls at all, so it is impossible to compensate for this nonlinearity. Thus, even the best hi-fi system sounds "thin" at low levels since bass frequencies seem less loud than midrange band and louder volume control levels.

Frequency-compensated volume control

Both sonically and from the reliability perspective (they become scratchy, lose contact and suffer from tracking errors), volume control potentiometers are considered the weakest link in the signal chain. Audio attenuators are considered superior mainly because the thin carbon or conductive plastic film is replaced by a quality switch and a resistive divider.

The minimum number of switched steps is 12, although most commercial units used in hi-fi amps use switches with at least 23 or 24 positions. Obviously, the higher the number of discrete positions, the finer the resolution of volume control.

The steps or jumps in resistance between the switched positions are usually expressed in dB. They may be uniform but need not be. At lower volumes, finer steps may be used, say 2 dB, and at higher volumes, larger jumps can be tolerated, 3 or 4 dB, for instance.

The simplest type of audio attenuator is the series type, where a string of resistors is connected between the input and common (ground). As the switch is cycled from the lowest (zero) volume (the bottom position on the drawing where V_{OUT}=0), more and more resistors are connected in series between the output and ground and less and less between the input and output.

The input impedance of the series attenuator is constant. One pole is enough for a guitar amp or a mono hi-fi amp, so the switch used would have only one wafer.

Ladder and shunt-type attenuators require two poles or wafers for a mono amp. A frequency compensated volume control adds frequency-dependent RC circuits (filters) at various positions along the way. The mathematical analysis of such a complex circuit is beyond the scope of this book and the math skills of a typical guitar player or amp maker. Suffice to say that the attenuator boosts the bass frequencies as the volume is reduced.

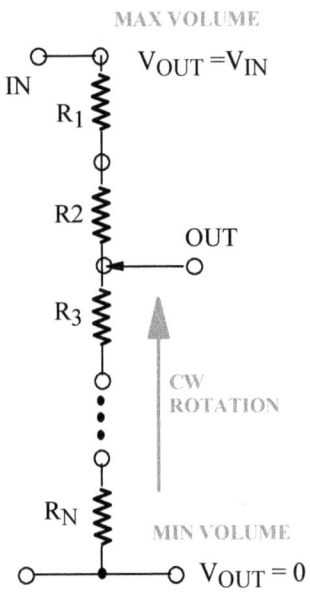

Or, more precisely, it attenuates the bass frequencies less than the high frequencies, in accordance with Fletcher-Munson's curves and the frequency-dependent sensitivity of the human ear.

As the switch moves from the upper position downwards, the volume is reduced, but the bass increases at a rate that offsets the ear's tendency to do the opposite. This eliminates the "thin" sound at low volumes, which lacks "body" and dynamics.

RIGHT: A frequency-compensated volume control (also called loudness control) using an 18-position rotary switch with make-before-break contacts (also known as "shorting type")
LEFT: Series attenuator

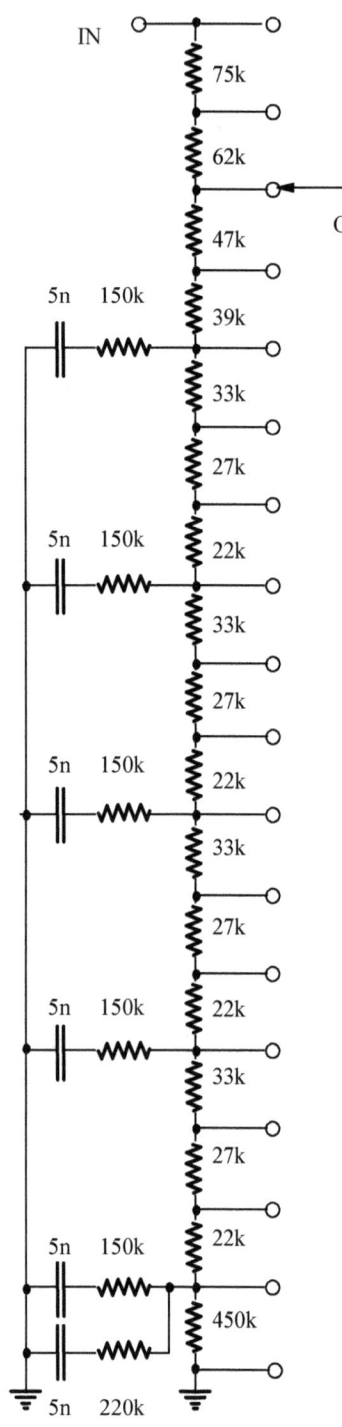

An overview of other possible changes

Most of these changes have already been described more than once in this book's two volumes, so we will not repeat them here. Instead, let's summarize the major lines of thought and common options.

CHANGE	BENEFIT	COST
Install a larger power transformer	Firmer power supply, less sag at higher power levels.	Expensive, wiring changes and chassis modification may also be needed.
Install a filtering choke	Better ripple (hum) filtering, improved amplifier dynamics & response.	Relatively low cost, easy mod.
Replace a tube rectifier with solid state diodes	Higher output power, less sag at higher power levels, different amp feel and tone.	Cheap, easy and reversible. Just unplug the rectifier tube and solder two SS diodes on its socket's pins.
Replace a solid state with tube rectifier	Lower output power, more sag at higher power levels, different amp feel and tone.	Impossible unless the existing power transformer has a CT high voltage winding. Chassis drilling, punching and rewiring needed.
Replace the output transformer	Higher power levels possible, cleaner bass, more headroom, better tone.	High cost of currently produced transformers, use salvaged vintage trafos instead.
Replace anode or cathode resistors	Less hiss and hum if noisy carbon composition resistors are replaced by film type. Different tone and feel.	Cheap and easy mod.
Replace power supply electrolytic caps with HV film type	Cleaner and tighter bass, more transparent sound.	Costly. HV high capacitance film caps are physically large and may not fit under the chassis.
Install a triode-pentode switch	Significant power reduction in triode mode, different tone and feel.	Wiring precautions needed, can introduce hum. Better filtering of the HV power supply is also required.

MODERN PUSH-PULL AMPS

- ORANGE TINY TERROR
- JET CITY JCA20H (with Bitmo™ Sufra™ modifications)
- FENDER EXCELSIOR (PAWNSHOP SERIES)
- LORDEN TL-15R
- EPIPHONE ELECTAR CENTURY

In this chapter, we review and analyze five low-to-medium power push-pull amplifiers. All are modern designs, except Epiphone Electar Century, a reissue of their vintage amp, albeit short-lived and unsuccessful.
In contrast, Orange Tiny Terror has been a very popular and well-received model despite its steep price.
Jet City JCA20H is a decent head which we use here to illustrate commercial modifications by Bitmo™.
Fender Excelsior is a solid yet Spartan combo, making it a perfect platform for modifications and additions.
Lorden TL-15R is a cheap & cheerful push-pull amp with spring reverb and lots of improvement potential.

ORANGE TINY TERROR

This used head still had a sticker with an AU$849 price tag, quite expensive for such a small Chinese-made amplifier. Just as in hi-fi, you are paying for its brand name and the perceived prestige that supposedly comes with it.

The range of controls is very basic, Volume, Tone and Gain, two speaker output impedances, 8Ω and 16Ω, and the 15W-Standby-7W toggle switch. No tremolo, no reverb, no FX loop, no headphone or preamp output.

The two 12AX7 duo-triodes are shielded, while the two EL84 pentodes have spring retainers. Although the tubes are mounted in the upright position, such retainers (usually only used in amps where tubes hang downwards) are a prudent inclusion.

Apart from the small board with three speaker jacks, the whole amp was built on one large PCB. While the resistors can be (un)soldered without removing the PCB, capacitors are only accessible from the other side and require its removal.

The cutout for the mains IEC inlet was cut too large, so the factory used silicon sealant to try to glue the IEC socket to the chassis (without success), the socket can be pulled out by hand!

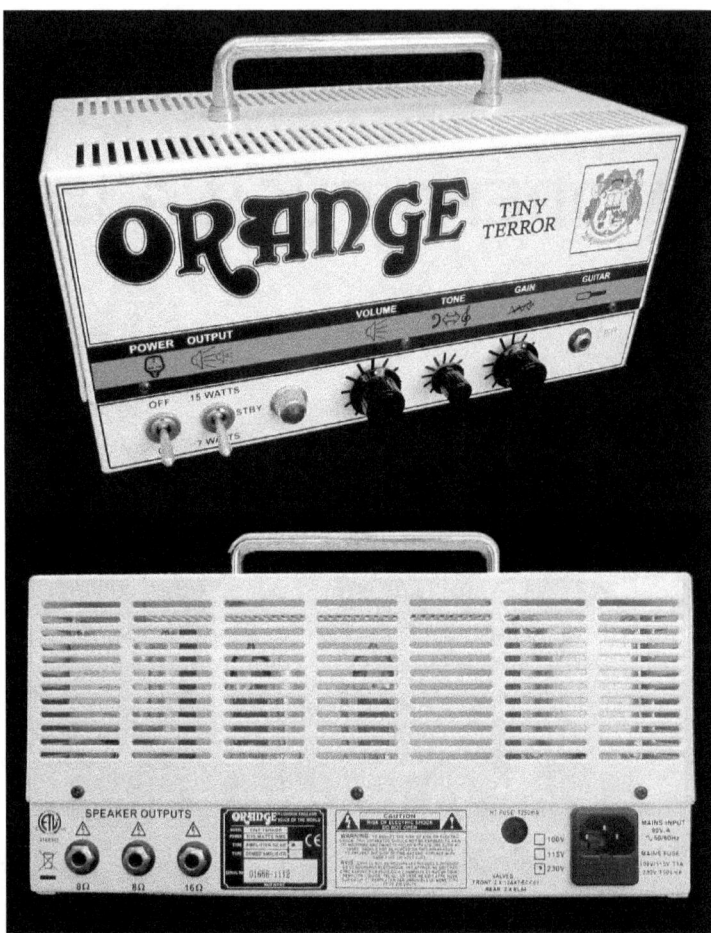

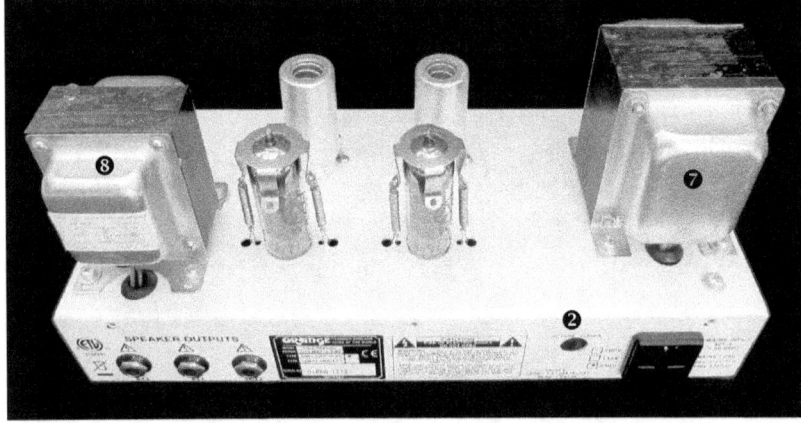

1) Side-mounted mains voltage selector switch (115/230V)
2) HT ("High tension", UK term for high voltage) fuse
3) Heater circuit fuse
4) Hum balance trimmer potentiometer
5) 150Ω cathode resistor of the output stage
6) 470Ω 5W EL84 screen resistors
7) Power transformer
8) Output transformer
9) Unused primary taps for 100V mains

The circuit

Interestingly, both the Volume and Gain potentiometers are dual (stereo) pots, so let's look at the circuit diagram to see why they were used.

The first two stages are identical and their outputs are controlled by a dual 500k *Gain* pot in unison. The *Volume* control pot is positioned after the long tail pair phase splitter, thus the need for a dual pot. The *Tone* control is a simple treble bleed RC circuit between the two grids, 2n2 film cap and 500k potentiometer. The rest is fairly conventional.

Notice identical anode resistors in the phase inverter. To equalize anode outputs one of these resistors must have a lower value, typically 100k and 92k. Orange designers were blissfully unaware of such requirements or deliberately introduced an asymmetrical drive signal to the output stage to increase distortion.

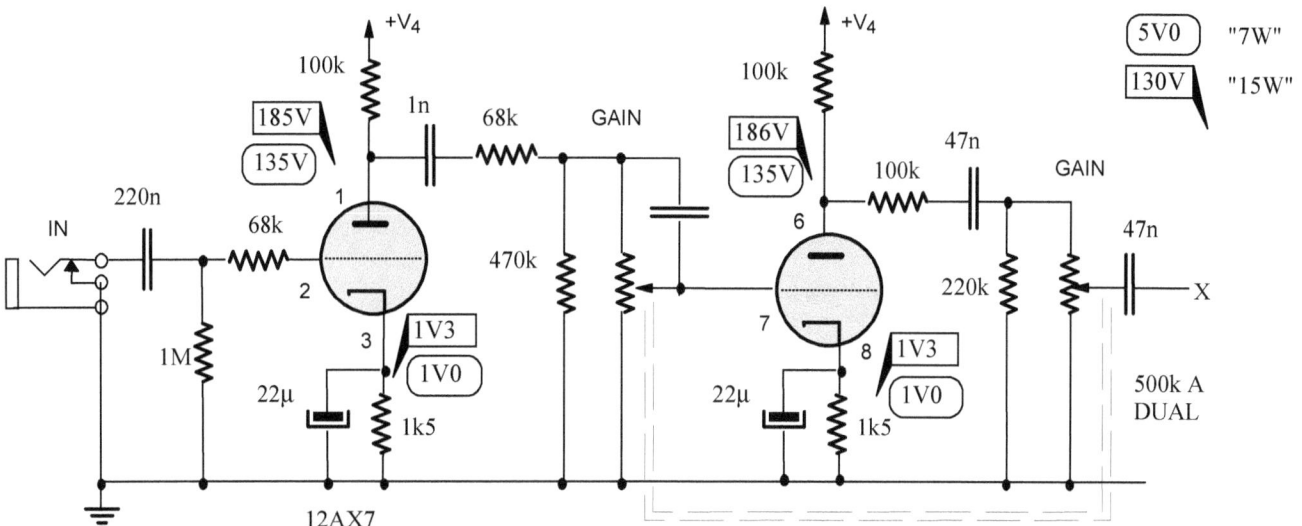

Tiny Terror input section, © Orange Musical Electronic Company, Ltd.

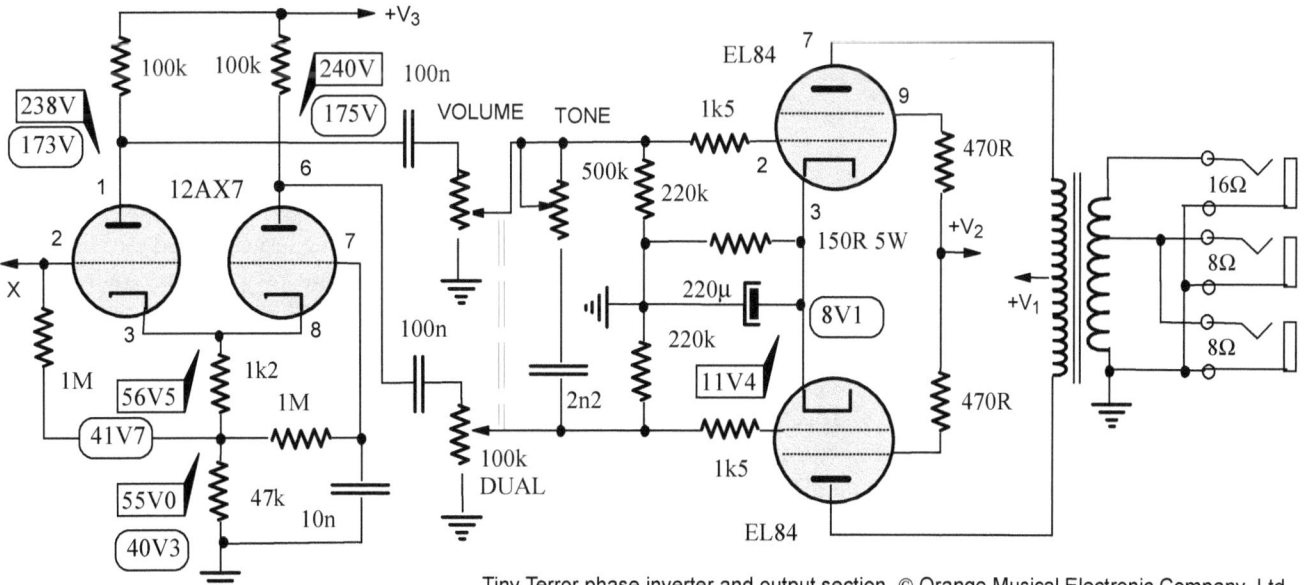

Tiny Terror phase inverter and output section, © Orange Musical Electronic Company, Ltd.

The power supply

$6.6V_{AC}$ heating is used for all tubes. The heater secondary winding has no CT to be earthed, but two 100R resistors create an artificial earth reference to GND. The *Hum Balance* trimmer potentiometer is not seen often in modern tube amps, but the designer felt the need for it here for some reason.

Notice that the 15W-7W switch is not a Pentode-Triode selector; it simply feeds a higher or lower secondary AC voltage to the rectifier-filter chain and thus varies the maximum output power obtainable. In both cases, the output tubes work in a pentode connection.

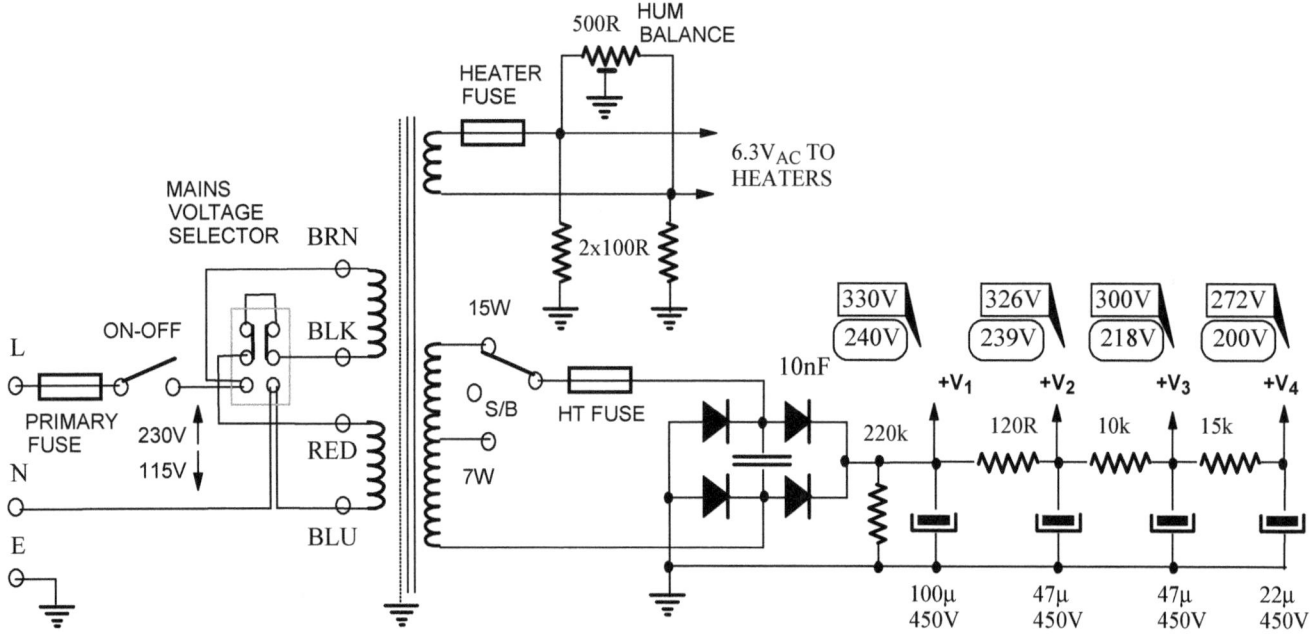

Tiny Terror power supply, © Orange Musical Electronic Company, Ltd.

The output transformer

The output transformer is of average size for a 15 Watt amp. Made of ordinary silicon steel (not grain-oriented), its primary inductance is low, only around 7H, so don't expect a strong bass from this amp.

However, at around 7mH, the leakage inductance is also very low, thus raising the quality factor to the above-average (for a guitar amp, of course) figure of almost 900. This means it's capable of producing plenty of shimmering highs, due to its extended high-frequency range.

ORANGE TINY TERROR
OUTPUT TRANSFORMER:

- Brand: Yuyao Huachi Electronic
- EI66, a=22 mm, stack thickness S=25mm
- Center leg cross section $A=aS = 5.5cm^2$
- Power rating: $P=A^2= 5.5^2= 30W$
- 1/2 Primary DCR (BRN-RED): $R_P=122\Omega$
- 1/2 Primary DCR (BRN-BLK): $R_P=123.2\Omega$
- Primary inductance @120Hz $L_P= 7.2H$
- Primary inductance @1kHz $L_P= 6.6H$
- Leakage inductance @1kHz $L_L=7.4mH$
- Quality factor: $QF = 6.6/0.074 = 892$

JET CITY JCA20H

JCA20H is a popular medium-to-high gain design by Mike Soldano. Although it uses one large printed circuit board (photo on the next page), the layout is neat and easy to follow. Since there is plenty of space around most components, modifications are easier to implement than on most PCB-based amps, although not as easy as on amps built using the point-to-point method.

The amp has two high gain 12AX7 stages (1st and 3rd). The 2nd stage has an un-bypassed cathode resistor which reduces gain by about 30% or around -3dB but the cathode-coupled or "long-tail" phase inverter also has voltage gain. The gain of the output stage has been reduced by the local negative feedback from the 16Ω tap back to the cathode of the phase inverter.

SPECS

JET CITY 20H

- 20W push-pull amplifier head
- Construction: PCB
- Rectifier: solid state diodes
- Bias: Fixed
- 8 and 16Ω speaker outputs
- Tubes: 3 x 12AX7, 2 x 6BQ5/EL84
- Dimensions: 64 x 31 x 32 cm
- Weight: 9.7 kg

MODERN PUSH-PULL AMPS

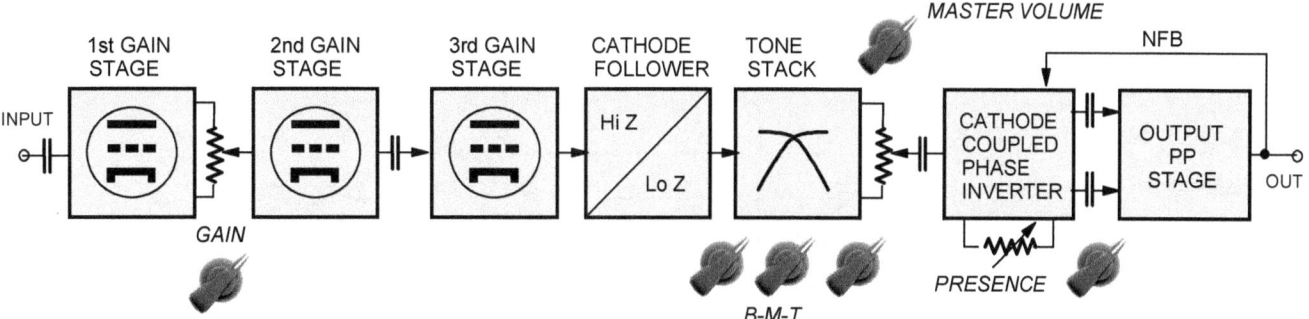

The tone stack is driven by a cathode follower with very low output impedance, always a great idea. However, notice the location of the tone stack after the first three gain stages, just before the master volume and the phase inverter. Perhaps it would work better (have more impact on the sound) if placed between the 2nd and 3rd gain stages.

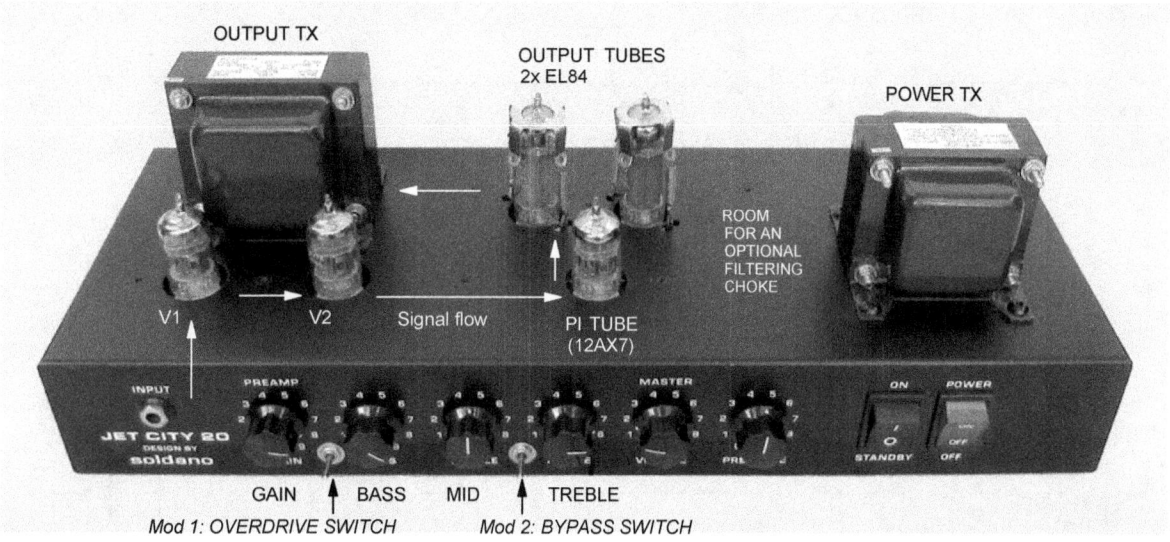

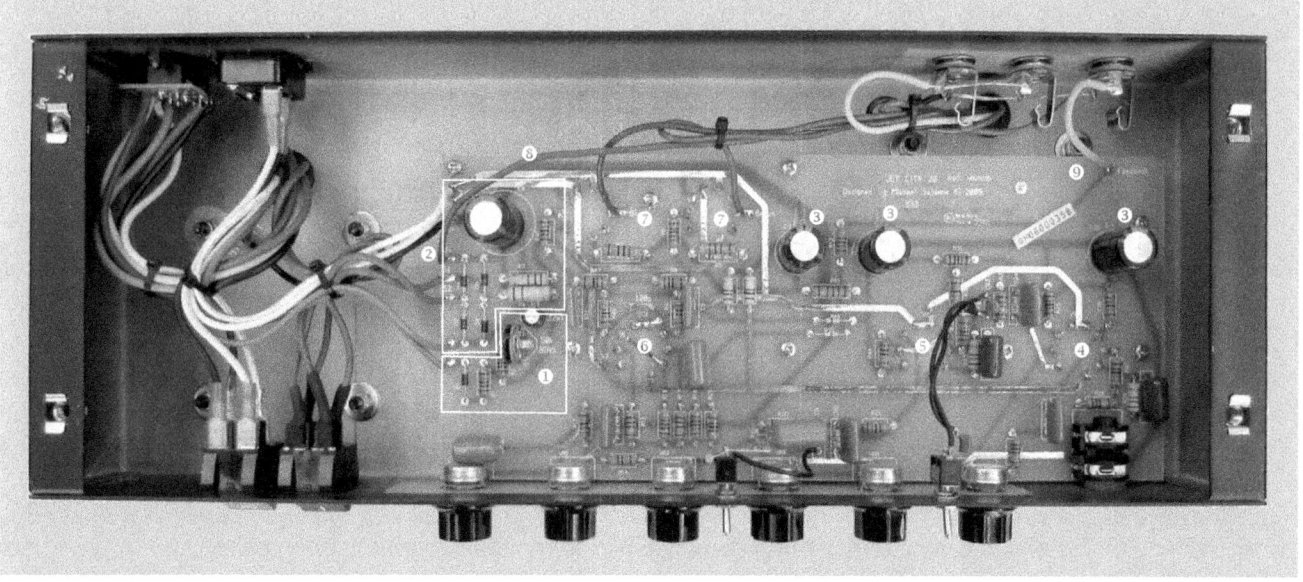

ABOVE: The internal (under-the-chassis) view of JET CITY 20: 1) Bias supply section with the "Bias adjust" trimpot 2) HV rectifier diodes and the first elco 3) HV filtering capacitors (22µF each) 4) input 12AX7 tube 5) Second preamp tube, also 12AX7 6) phase inverter tube, also 12AX7 7) output EL84 tubes 8) B+ connection to the CT of the output transformers, straight from the first elco/rectifier bridge output 9) NFB connection

Circuit diagram

The "long tail" phase inverter and push-pull pentode output section are fairly standard; their operation has been covered elsewhere in this book. $+V_1$, $+V_2$, and $+V_3$ (next page) are DC voltages produced by a solid-state HV power supply (not shown here). V_{BIAS} is the adjustable negative bias voltage (both power tubes together).

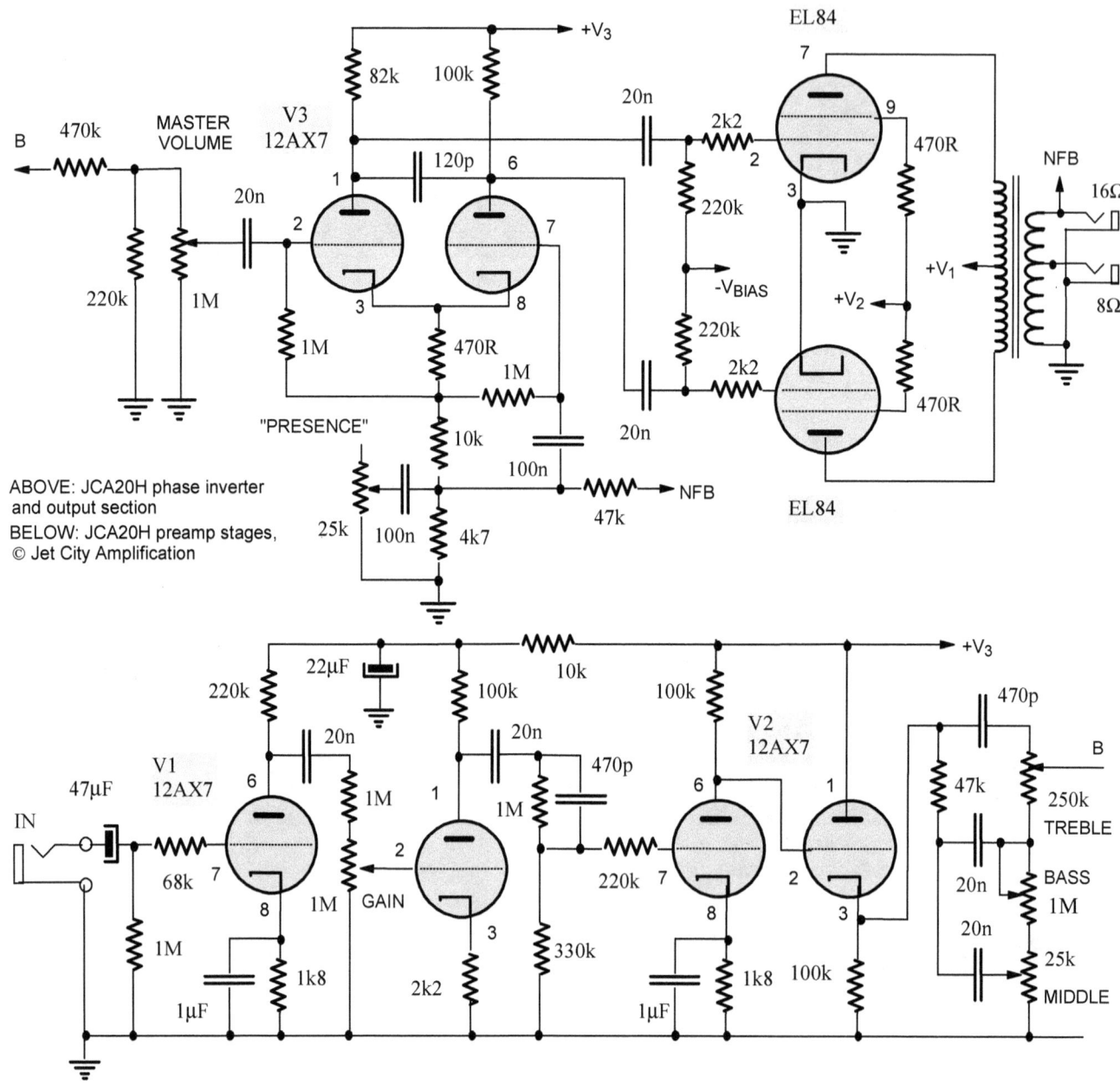

ABOVE: JCA20H phase inverter and output section
BELOW: JCA20H preamp stages,
© Jet City Amplification

The inclusion of master volume is always a good idea. The negative feedback cleans up the power stages but can easily be disconnected and made switchable or variable should you desire more tonal flexibility.

Notice a few interesting design choices in the preamp stages. The input signal passes through a series electrolytic capacitor (47μF) while the output of the high gain 1st stage (220k anode resistor) is attenuated by the voltage divider formed by the fixed 1M resistor and 1M gain potentiometer. So even with the gain at maximum, the attenuation is 2:1, so only half of the signal reaches the grid of the 2nd stage. Even stronger attenuation happens at the output of the 2nd stage, since 330/1,330 = 0.248, only around 25% of its output reaches the grid of the 3rd stage.

Since the attenuating series resistor (1M) is bypassed by a 470p "bright" capacitor, the high frequencies above 338 Hz are not attenuated as much, since a high pass shelving filter is formed between these two stages.

Notice also the identical cathode resistors and bypass caps in stage 1 and 3 cathodes. The high pass filter thus formed attenuates low frequencies because the cathode resistor is bypassed only for HF; for LF there is a local NFB (negative feedback) and attenuation. The corner frequency is $f_L = 1/(2\pi R_K C_K) = 88$ Hz.

Bitmo™ Sufra™ modifications

The preowned amp we bought included two modifications, apparently by BitmoTM. Who installed them is not clear, but ultimately, it does not matter. Of interest to us from an educational perspective is the intent of such mods and how they were achieved.

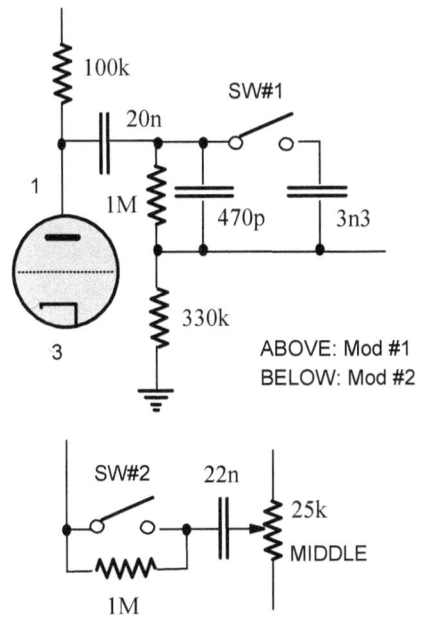

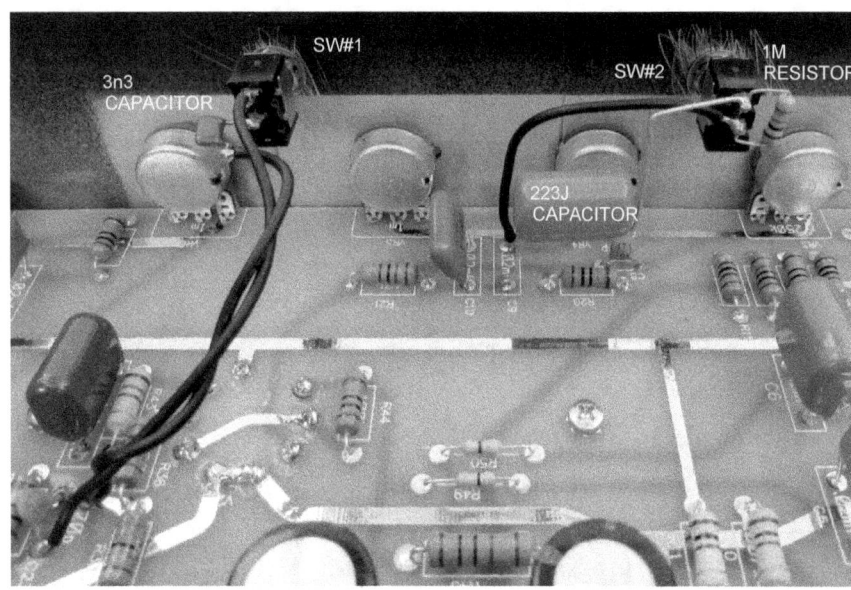

ABOVE: Mod #1
BELOW: Mod #2

ABOVE: The close-up view of the two BitMo™ modifications

When engaged, SW#1 simply adds another 3n3 film capacitor in parallel with the existing 470p "bright" or bypass capacitor. This lowers the original 338Hz boost limit down to 42Hz. Thus, practically all frequencies that a guitar can produce bypass the 1M series resistor and so achieve an overall boost. The switch is called "overdrive", and the instructions that come with the Sufra kit state that the 1st switch "slathers on a dollop of fatback overdrive ..."

The other mod ("Bypass") removes C9, the 20nF capacitor at the wiper of the Middle tone control pot, and adds a 22n capacitor in series with a 1M resistor in parallel with SW#2. When SW#2 is closed, the resistor is shorted, and the circuit operates as before; the Bass and Middle controls are operational.

With an open switch, the high resistance of the 1M resistor practically acts as an open circuit, thus disabling the Bass and Middle controls. The Treble control remains operational. Since the tone stack significantly attenuates the signal, increasing the gain.

The mod does not bypass the tone stack as the name of this mod suggests The instructions from BitMo claim that thus mod "...adds loud clean headroom and punch ..."

The new capacitor has practically the same capacitance as the old one but is much larger physically. Retaining the old capacitor in place and just unsoldering its one leg would have done the job, so perhaps the replacement was done for voicing reasons; perhaps the new capacitor sounds better than the stock one.

Bitmo™ modification kits are a brainchild of Bruce Hutcheon and are available from his website http://www.bitmomusic.com.

The output transformer

The output transformer uses either EI85.8 or EI86 laminations (it was hard to determine the exact dimension using a simple ruler since these two sizes are so close). Regardless of which is the case, its magnetic core is rated at around 65 Watts, not bad for a 20 Watt amp.

The primary inductance is low, under 7H, and the leakage inductance very high (0.18 H), making this transformer's quality factor the lowest we have ever seen, only 39!

JCA20H OUTPUT TRANSFORMER:
- EI85.8, a=28.6 mm, S=28mm
- Center leg cross section $A=aS = 8$ cm^2
- Power rating: $P=A^2= 64W$
- Voltage ratios: 28 - 20
- Primary inductance L_P= 6.9H
- Leakage inductance L_L=178mH
- Quality factor: QF = 39

Gain reduction or re-distribution

If you find the overall gain too high for your taste (there are four gain stages, (there are four gain stages, including the phase inverter!), replace either V1 or V2 (12AX7) with 12DW7, a dissimilar duo-triode. The 1st (or 3rd) stage will remain a high gain stage (pins 6-7-8 is a high gain triode within 12DW7), but the 2nd (or 4th) stage will become a low gain stage (pins 1-2-3 are the low gain triode of 12DW7, 12AU7-like).

If you like what you hear, try replacing both V1 and V2 with 12DW7! Alternatively, replace V2 with 12AU7 (ECC82 or 5963) for an even lower gain.

FENDER EXCELSIOR (THE PAWNSHOP SERIES)

Fender Excelsior combo amp is a push-pull design with a cathode-biased 6V6 output stage, similar to the vintage tweed Princeton. Excelsior is one of the very few amps that use a split chassis design.

The power section is located at the bottom of the enclosure, with only the preamp section at its top. While this separates the hum-sensitive preamp stages from the hum-inducing power transformer and the power supply, this topology is more expensive to manufacture.

Another notable positive feature is the 15" speaker, unusually large for such a relatively compact amp. There is plenty of room on the power chassis to add a standby switch, a triode/pentode (1/2 power switch) or NFB On-Off switch in position (1), and even a choke in position (2). The speaker jack is hard to reach behind the output transformer (3); if it bothers you, move it to location (4).

As for the downside, the lack of controls is hard to comprehend. There is only volume control ("gain" in essence) and no master volume. There is no tone control either, only a bright-dark switch. However, the bright voice is too bright and shrill for many players, and the dark voice is dull, lacking life and dynamics.

Replacing the "dark-bright" switch with a simple one-pot tone control adds more versatility to the amp's tone. If "mic" and "accordion" inputs aren't needed (if you only use the "guitar" input), those two 1/4" sockets can be removed, and their holes used to install a couple of switches to turn various mods on and off!

The tremolo only has a speed (frequency) control, no depth or intensity control. There is no gain boost or footswitch controls of any kind either.

Just as with Epiphone Valve Junior, the other controls were omitted either deliberately to make the amp more appreciated by the DIY and tweaker community and thus in higher demand or as a result of an astonishing lack of thinking by the amp's designers.

The internal inspection

Since the componentry is split between two chassis and three circuit boards (photo on the next page), the Excelsior is easy to comprehend (even without a circuit diagram), easy to work on and modify.

Furthermore, there are no semiconductors, as in the other two Pawnshop series models, Ramparte and Vaporizer, so the Excelsior wins in terms of DIY and modification friendliness.

Apart from the main (primary side of the power transformer) fuse (1), there are two fuses on the boards, the 7A heater fuse (2), soldered onto the board, and the 250mA HV supply fuse (3).

Following the bridge rectifier (four individual diodes), are two 7 Watt 270Ω resistors in series (4).

ABOVE: The chocolate brown version of the amp truly looks vintage, but perhaps a bit too morbid for some. Cream and light blue versions look more optimistic.
BELOW: The back of is open, no cover of any kind, not a good idea..

FENDER EXCELSIOR

- Inputs: guitar, microphone, accordion
- Controls: Volume, Tremolo (speed), Bright/Dark tone
- 2x6V6 power tubes, 2x12AX7 preamp tubes
- Solid state rectification, PCB construction
- Output power: 13 Watts into 8Ω
- 8-ohm 15" speaker
- Dimensions: 9" (22.9 cm) deep x 19.5" (49.5 cm) wide x 19.5" (49.5 cm) tall
- Weight: 33 lb. (15 kg)

They will produce a voltage drop that is proportionate to the total current drawn by the amp, and since such a current raises significantly with the output power level (being a Class AB design), that will produce a voltage "sag" or drop, akin to that produced by tube rectifiers. Should you wish to eliminate such a voltage sag and thus increase the headroom and maximum output power, replace the two sag resistors with wire jumpers.

All four filtering elcos are in line and can easily be upgraded and larger capacity units installed. The third power resistor (5) provides the cathode bias for the pair of 6V6 output tubes. Notice the most common mistake amp designers make, placing the bypass elco next to it, as if they deliberately wanted to make its life short and stressful. That poor elco will get very hot during amp's operation and eventually dry out. Again, demonstrating an incredible lack of foresight and thinking from the user's perspective, the speaker jack is located behind the output transformer, where it is hard to reach (6). If you are using various external speakers at your gigs, move it to (7).

Despite solid-state rectification, there is no standby switch, so you should install one in location (8), next to the on-off switch.

Finally, notice a few turns of wire wrapped around the ferrite ring (9). That is the wire carrying the signal into the 2k2 grid stopper resistor and the first amplification stage. The Fender factory circuit diagram does not show this RF suppression measure; perhaps it was an afterthought due to radio frequency interference. Sometimes amps pick up local radio stations quite well. As a teenage keyboard player, I had an Italian Farfisa solid state amp from the early 1970s that worked better as a radio tuner than a keyboard amp!

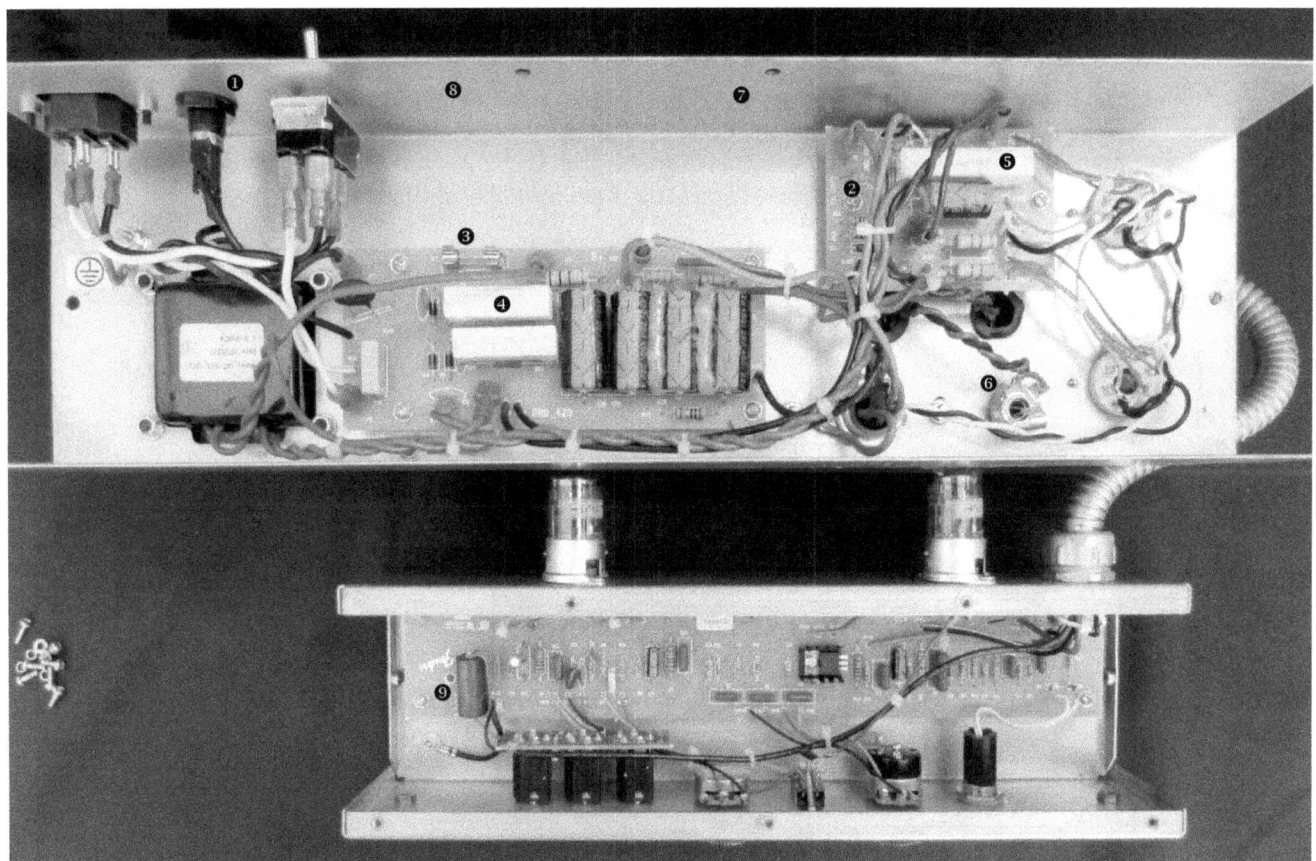

The circuit critique

The promotional blurb touts that the three inputs "each have individually optimized circuitry". Let's redraw the three input circuits separately to see how Fender "optimized" them. We will assume that only one instrument is plugged in at any time.

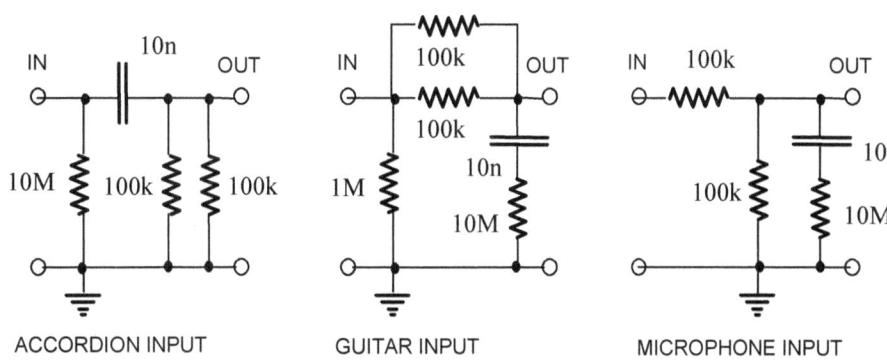

Notice that if two instruments are used simultaneously, the circuit's configuration changes significantly! The trick, as always, is in the type of jacks used and how they are connected. The switchable contact (the small black triangular tip or arrow) on the ACC input is not connected to anything. The switched contact on the GTR input is grounded, while the switched contact of the MIC input is connected to the "hot" contact of the GTR input..

With the ACC input, the two 100k series resistors of the other two inputs are paralleled (equivalent resistance of 50kΩ) and grounded, thus forming a CR or high pass filter with the series capacitor, with -3dB frequency of $f_L = 1/2\pi RC = 318$ Hz. When the GTR or MIC inputs are used, 10n+10M of the ACC input are actually in series. The 1M resistor to GND is shorted with the MIC input, and the two 100k series resistors now form a 2:1 voltage divider. Since their resistances are equal, the output voltage is half of the input voltage. There is no such attenuation with the GTR input; the two series resistors are now strapped in parallel, after the 1M resistor to GND.

The first stage uses a high bias (-2V) and has high gain due to a large 220k anode resistor and a bypassed cathode resistor. The 2nd stage is also a high gain beast with a lower bias (-1.3V), so it will distort before the first stage! It is followed by a split load phase inverter, whose output is amplitude-modulated by the output of the tremolo circuit, a simple phase shift oscillator with a single 12AX7 triode. The tremolo modulation signal goes through a MOSFET source follower, which provides a high input and low output impedance, just like a cathode follower. The tremolo on-off switch is part of the tremolo speed pot (on the same shaft, indicated by the dotted line).

The 29.4V cathode voltage and 470Ω cathode resistor of the output pentode stage (next page) indicate a cathode current of 29.4/470=62.6mA or 31.3mA per output tube. Notice two silicon diodes reversely polarized between each 6V6 anode and ground, protection against induced negative voltage spikes that can damage or destroy the output transformer or the power tubes.

Also, notice that after the first pair of 470k resistors (below), the out-of-phase push-pull signals pass through shielded cables onto another pair of 470k resistors to the ground (next page) This is necessary since those shielded cables run between the top and bottom chassis together with the high voltage supply and the heater wires. As always, only one end of the shield is grounded, this time not directly, but through 10n film capacitors.

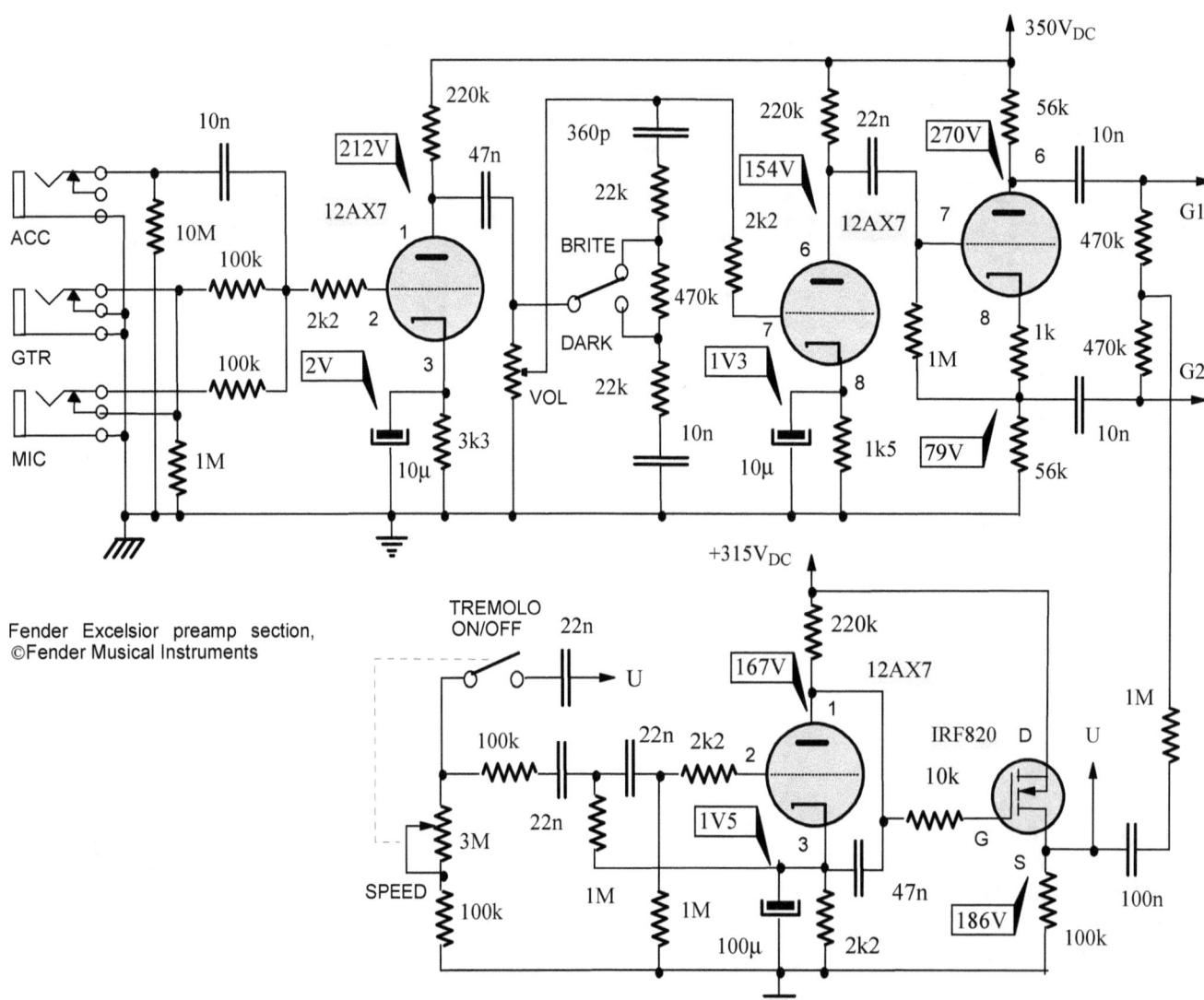

Fender Excelsior preamp section,
©Fender Musical Instruments

MODERN PUSH-PULL AMPS

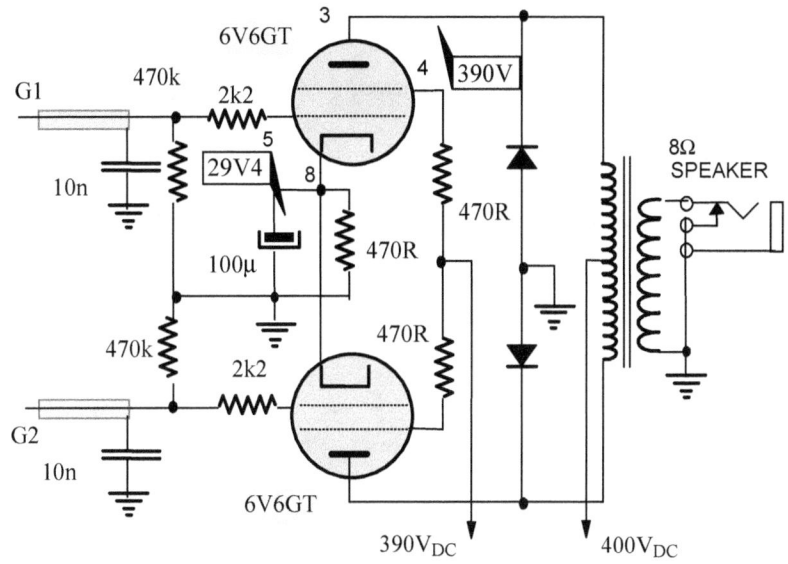

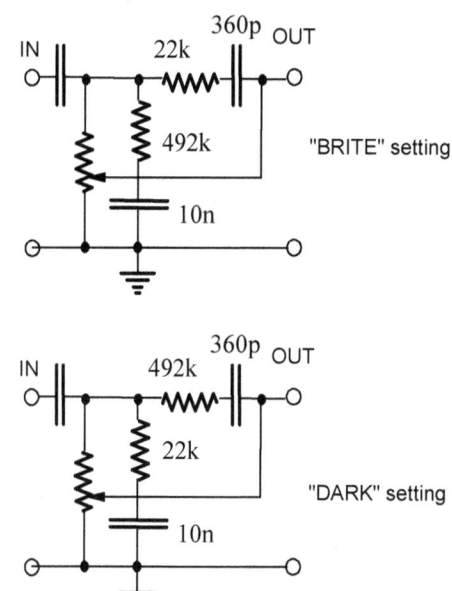

Fender Excelsior output stage, ©Fender Musical Instruments

The BRITE-DARK switch connects the 470kΩ resistor either in series with the bottom shunt capacitor (10n), or in series with the series capacitor (360p). The tone control circuit is parallel with the VOL potentiometer, allowing more (BRITE) or less high frequencies (DARK) to bypass it and thus achieve HF boost or cut. Let's redraw it for educational purposes (above). In "BRITE", 22k+470k (492k) are strapped to the 10n and GND, so less signal is diverted to the ground due to such high resistance. When 22k+470k (492k) are strapped in series with 360p and only 22k go to 10n cap and GND, most high frequencies are diverted to the ground.

The output transformer is a mixed bag. If you want plenty of early distortion and not much bass, then it is okay. However, forget about lots of clean headroom and lower distortion. This can even be deduced just by looking at its tiny size. The high primary inductance helps with the bass somewhat, but its magnetic core is rated at only 14 Watts.

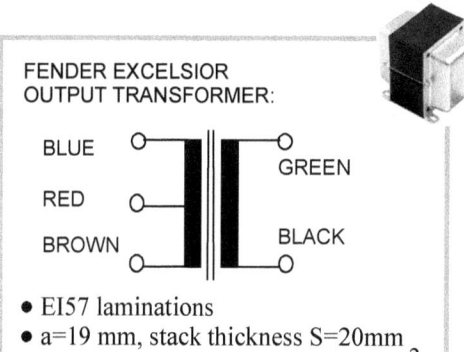

FENDER EXCELSIOR OUTPUT TRANSFORMER:

- EI57 laminations
- a=19 mm, stack thickness S=20mm
- Center leg cross area $A=aS = 3.8 cm^2$
- Power rating: $P=A^2= 3.8^2= 14W$
- Prim. inductance @120Hz L_P= 36.6H
- Prim. inductance @1kHz L_P= 16.7H
- Leakage inductance L_L=6.6mH
- QF = 16.7/0.0066 = 2,530

Making tremolo depth or strength adjustable

While Excelsior's tremolo speed is adjustable, its depth or strength isn't. However, this can easily be implemented. Identify Q1, which is IRF820, a very common power MOSFET transistor. This circuit works as a source follower, an equivalent to cathode follower in tube technology. Remember, just like vacuum tubes, FETs are voltage-controlled devices with high levels of insulation between the control electrode (gate), equivalent to the tube's grid, and the other two electrodes, the source (equiv. to cathode) and the drain (equiv. to anode).

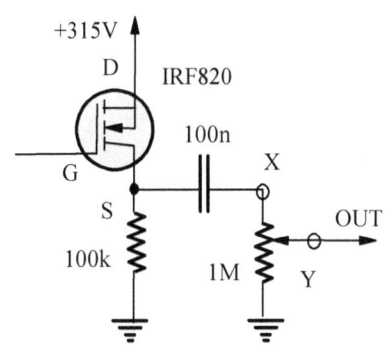

Thus, although their internal workings are different, their outward behavior is analogous to vacuum tubes. The signal exits the source follower, passes through the decoupling series capacitor (100n) and a series 1M resistor, and then enters the common point where the two 470k at the phase splitter's output are joined.

The series 1M resistor needs to be unsoldered and removed from the PCB, and two wires need to be soldered in those two points, X and Y, on the diagram here. One wire will go to one fixed need of a 1M potentiometer, the other wire (Y) to its middle terminal (the wiper). The other fixed end of the pot must be grounded.

Replacing the BRITE-DARK switch with adjustable tone control

Apart from the obvious choice of only two voicings, the main complaint of Excelsior owners is this BRITE-DARK control. The BRITE sound is too shrill, and the DARK setting is too dull and lifeless, lacking sparkle. So, it is impossible to get a decent sound out of this amp as it is! This improvement solves that serious problem.

Remove the preamp PCB and turn its copper side up. Unsolder resistor R8 (470k). The two wires from points X and Y on the PCB go to two changeover terminals on the switch. Unsolder all three wires from the switch and remove the switch.

Drill a hole suitable for the potentiometer you are installing (measure the threaded part, not the shaft) where the rectangular switch cutout is now. Usually the required hole is around 9mm in diameter. Install a 500k LIN (marked "B") potentiometer.

Take the wire that went from point Z (one end of the VOL pot) to the switch, and solder that wire to the middle lug on the newly installed pot (its wiper). Solder the other two wires, coming from points X and Y (marked WJ18 and WJ20 on the PCB) to the end lugs of the new pot.

Now you have continuous control over your tone!

Just in case you've been wondering, the messy soldering joint on our amp's PCB (*) was done by its previous owner, who installed BitMo modification. That large capacitor in the bottom right corner is an add-on.

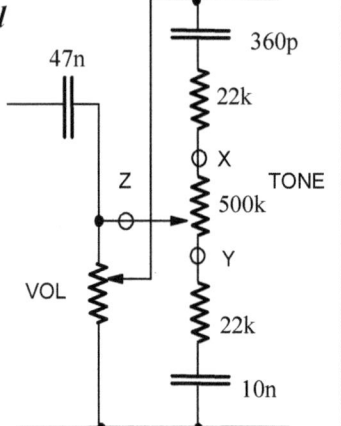

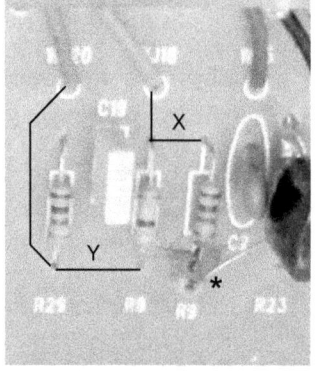

Removing the power supply sag or making it switchable (FIRM-SPONGY RESPONSE)

Fender installed two 270R resistors to simulate the voltage sag of tube rectifiers. If you want to eliminate it, and will never miss the Excelsior's original "spongy" response, simply bridge each resistor with a piece of insulated wire, solder the two on top of the resistors, and you are done. This will make the high voltage power supply stiffer, will give you more clean headroom at higher power levels and slightly higher maximum output power levels.

This mod is fully reversible; if you change your mind later or sell the amp to someone who wants the old response back, simply unsolder the two jumper wires.

If you want the versatility of having both firm and droopy response, install a DPDT (Double Pole Double Throw) switch, suitably rated in terms of voltage and current (250V_{AC} and 250mA rating is fine). In the upper position of the switch, the resistors are bypassed (FIRM response); in the lower position of the changeover contacts, the resistors are switched back into the circuit for a saggy HV supply.

The switch should be installed at the back, on the bottom chassis, next to the on-off switch.

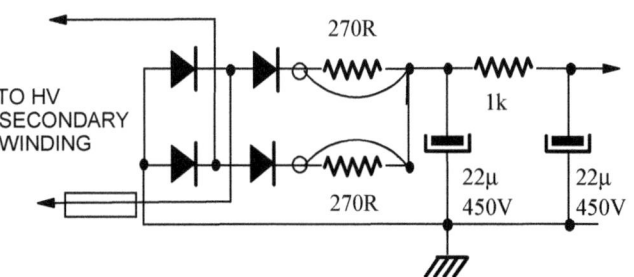

ABOVE: Partial HV power supply. Capacitors in parallel with rectifier diodes are not shown. To eliminate the sag completely, bypass each resistor with insulated wire.

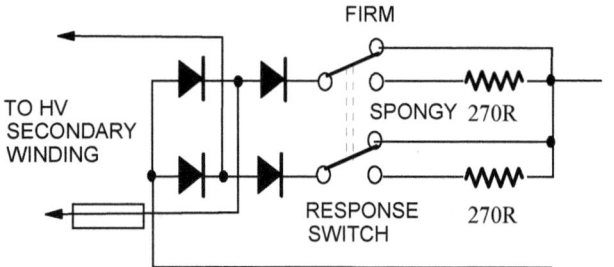

ABOVE: Switchable sag. The DPDT switch either bypasses the two 270R resistors (FIRM response, as shown) or switches them into the circuit ("SPONGY" or "DROOPY" response).

MODERN PUSH-PULL AMPS

Our mods: Step-down transformer, Firm-Spongy response switch and Triode-Pentode switch

The two 120V power transformer primary leads enter the PCB at (1). To install a step-down 240-120V transformer (2), they need to be removed and joined with the step-down transformer's secondary leads (3), and the step-down transformer's primary leads need to be soldered onto terminals (1)..

To install a small choke (4), with enough current capacity for screen grids and preamp stages, the 1k resistor (5) needs to be removed and the choke leads soldered there.

The Firm-Spongy switch (6) is wired as per instructions on the previous page.

The Triode-Pentode switch (7) requires the removal of the two 470R screen resistors from the PCB (8) and their installation between pins 4 (screen) and 1 (unused) or 4 and 6 (the pin also unused with 6V6 tubes), as illustrated at (9). Alternatively, you can install a small terminal strip between the two tube sockets.

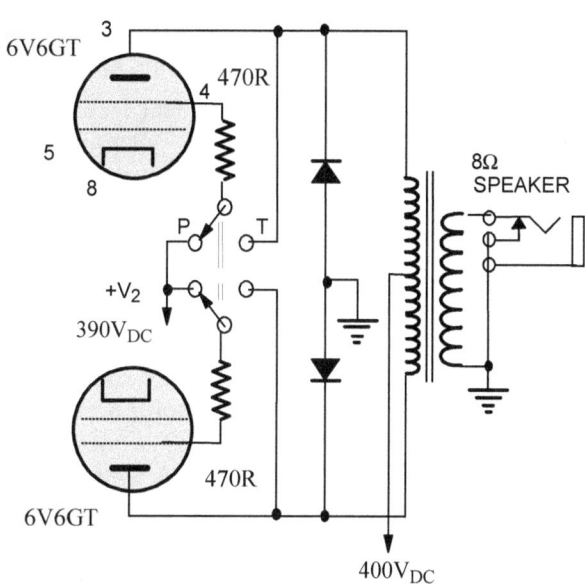

The addition of the filtering choke cleaned up the power supply (chokes provide better AC ripple filtering than resistors) and made the amp almost absolutely quiet.

The baby choke was salvaged from a Heathkit test instrument that used a dozen or so duo-triodes, so it should be good for at least 20-30 mA.

Both *Triode* and *Pentode* modes sounded very good indeed, but very different in output power level and tonal balance.

We could not detect much sonic difference between the *Firm* and *Spongy* response settings.

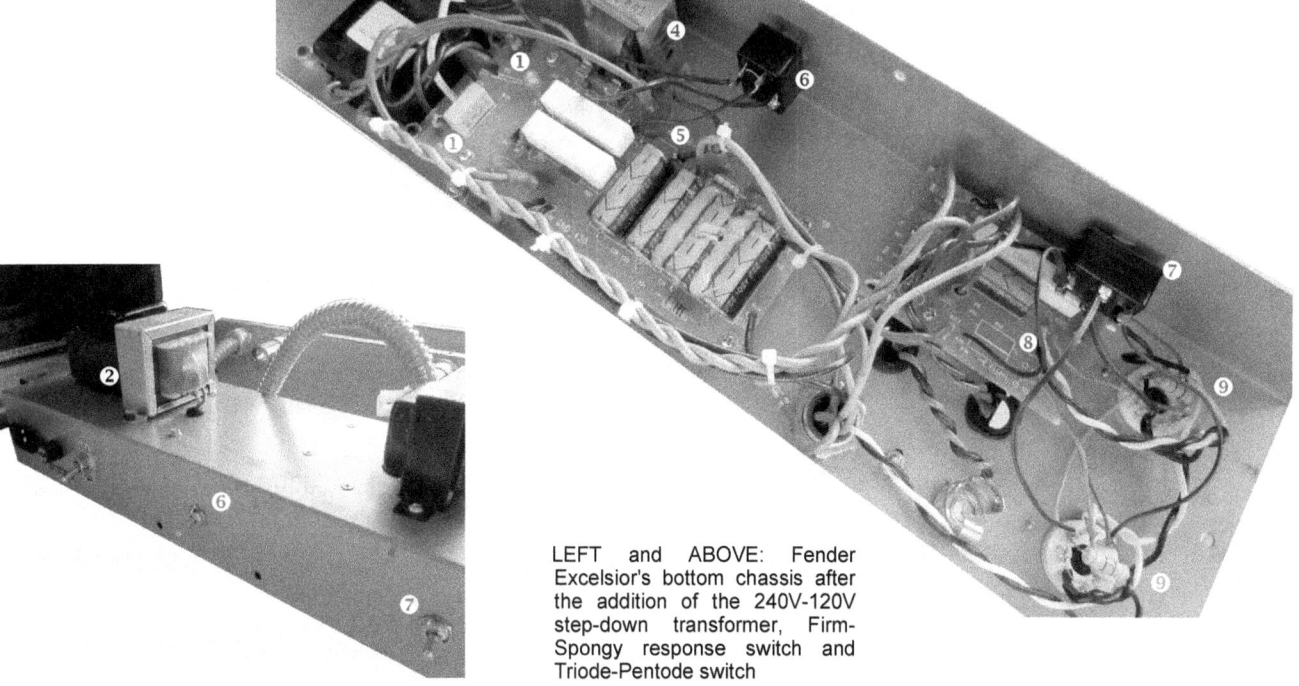

LEFT and ABOVE: Fender Excelsior's bottom chassis after the addition of the 240V-120V step-down transformer, Firm-Spongy response switch and Triode-Pentode switch

LORDEN TL-15R

Tube amp manufacturers and retailers must be getting desperate. Either that or the profit margins on the gear they sell is enormous, which I doubt.

In June 2016, I received a promotional e-mail from Muso City, a shop in Melbourne, Australia. This particular amp was reduced 40%, meaning its price was brought down from AU$399 to only AU$239.40. We didn't need or want another Chinese-made guitar amp as a case study, but being a sucker for specials, I could not resist getting my hands on one.

To put this pricing in perspective for our readers living in countries "that are not Australia," this price includes a GST (Goods & Services Tax) of 10% or AU$21.76, which goes to the Aussie federal government in Canberra, meaning the retailer will get AU$239.40- AU$21.76 or AU$217.64.

Also, the price includes free courier shipping within Australia, which for an amp of this size and weight is at least AU$37.64, meaning the shop will only get around AU$180!

The conclusion? Either the wholesale price of this amp is way below AU$180 (at the time of writing, US$133) or, if it's higher, the retailer lost money on this sale. Also, since I paid by PayPal, that cost the retailer 2.6% + $0.30 per domestic sale (on AU$239.40) or around AU$6.52.

Why am I telling you all this? Because it is annoying to see retail prices of parts and components so high when many finished amps are so cheap! It makes us DIY constructors feel as we are being ripped off. For instance, the Celestion G10R-30 speaker inside this amp has a retail list price of US$35 or AU$47.

The five Ruby tubes in this amp will set you back AU$86 or US$64 ($22 for a pair of EL84s and $42 for three 12AX7 duo triodes). For some inexplicable reason, the preamp tubes are more expensive than power tubes?

Anyway, just the tube set and the speaker in this amp would cost you AU$133. So you get the rest of the amp (power & output transformers, cabinet, PCB, and the rest) for only AU$107!

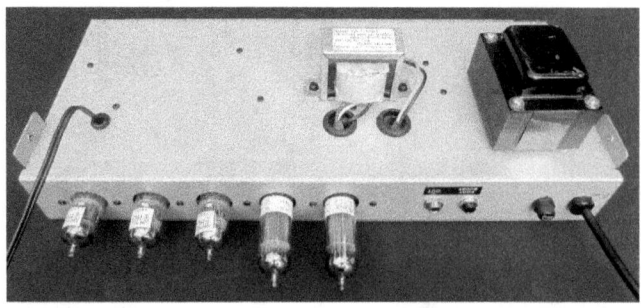

Lorden TL-15R

- 15W PP combo amplifier
- Construction: PCB
- Rectifier: solid state diodes
- IC-driven spring reverb
- 3-band EQ (Bass-Middle-Treble)
- Foot-switchable gain boost
- Controls: Gain, Volume, Reverb level
- 10" 8Ω Celestion G10R-30 speaker
- Tubes: 2 x EL84, 3 x 12AX7
- Dimensions: 460x430x250mm

This brings us to my other point. It is much cheaper to buy a commercial amp and improve it than build an amp of the same level of quality from scratch - primarily due to the extortionary cost of parts.

A similar rip-off is noticeable in the motor vehicle industry. If you wanted to buy genuine parts and assemble a $20,000 car yourself, it would cost you more than $200-300,000!

As with most currently-produced PCB-based amps, to perform any mods, the PCB must be removed, and to do that, all six potentiometers and the guitar jack must be detached from the top control panel.

MODERN PUSH-PULL AMPS

Luckily, all the connections are via various types of connectors, so there is no need to unsolder them to flip the PCB over. There is plenty of room on the top side of the chassis, should you wish to add a filtering choke or replace the questionable output transformer with a larger and better-sounding one.

The solid-state rectifier diodes (1) and filtering caps (2) are instantly identifiable. The primary (3) and secondary wires (4) from the output transformer are fed through their own holes in the chassis, which is proper practice.

All connections are marked on the PCB, either as colors of the wires (5) or the function, "Lamp" for instance (6), a nice and repair-friendly touch.

The spring reverb tank is driven by an integrated circuit (7), the two shielded cables to the reverb tank can be unplugged from the PCB if need be.

The output transformer (photo below) gives away the low-cost nature of this amp. It is not a custom designed or made audio transformer but a mains (power) transformer with a CT primary winding.

The voltage or turns ratio can be calculated as TR= 440V/13V = 33.85, so the impedance ratio is IR= TR^2 = 1,145, meaning the 8Ω speaker load will be reflected to the primary as anode-to-anode load of Z_P = 1,145*8 = 9.16kΩ impedance.

The DC resistances of the two primary halves are very different, 218 and 133 ohms, so there will always be an imbalance in the output stage and an increased distortion.

Due to the low-grade magnetic material used (non-grain-oriented steel), the primary inductance is a low (10 Henry), so the quality factor is also low, under 300!

LORDEN OUTPUT TRANSFORMER:
- EI57 laminations
- a=19 mm, stack thickness S=25mm
- Center leg cross area A=aS = 4.75cm^2
- Power rating: P=A^2= 22W
- Voltage (turns) ratio: TR= 440/13=33.85
- Impedance ratio IR=TR^2= 1,145
- Primary DCR (blk-white): R_{P1}=133Ω
- Primary DCR (red-white): R_{P2}=218Ω
- Primary inductance @1kHz L_P= 10H
- Leakage inductance @1kHz L_L=34mH
- Quality factor: QF= 10/0.034 = 294

EPIPHONE ELECTAR CENTURY

This amp was released as "Limited Edition 75th Anniversary Inspired by "1939" Century Amp." It certainly looks great (providing you like the art deco style). However, according to online reviews, there were numerous issues, ranging from its allegedly poor build quality and reliability to the loud hum (buzz) and tone/power-related issues.

Many guitar players claimed that other 7-9 watt rated tube amps were more powerful than this supposedly 18 watter! Indeed, 18 watts from two 6V6 in push-pull seems to be a far-fetched marketing claim, so we were intrigued.

The model was discontinued soon after, and many units flooded the market as "reconditioned," with a new serial number applied. While the list price of this amp was a relatively high US$665, these were offered at US$350-$399 mark.

As with most other case studies in this book, we were thinking, "if the electronics in this amp turn to be excrement, we can always gut it and rebuild it point-to-point," as a last resort, of course.

The power claim was debunked once we obtained the factory circuit diagram, which specified the maximum power as 13 Watts at 10% THD.

The chassis is relatively small for a push-pull amp, with the output tubes right next to the power transformer, mains IEC inlet, and the on/off switch. Since all connections, inputs, and outputs are on the same (the only) panel, the fascia is chock-a-block! One-quarter of it is occupied by three separate input jacks. More valuable space is taken by that huge pilot light and the "Electar" logo, not to mention the IEC power inlet and the on-off switch. Mods and additions are not easy for that and many other reasons (to be covered soon).

Instead of using three separate inputs, remove two input jacks and install a three-way "Bright-Normal-Dark" toggle switch!

The filtering choke (1) and bias pot (2) with its associated test points (3) also take some room on top. The interior (next page) is even more crowded. Most of the space on the larger PCB is wasted on the five large axial electrolytic capacitors. Notice the use of shielded cable despite very short runs of only a few centimeters, one between the footswitch and the smaller PCB (4), the other, longer run, between the two PCBs (5).

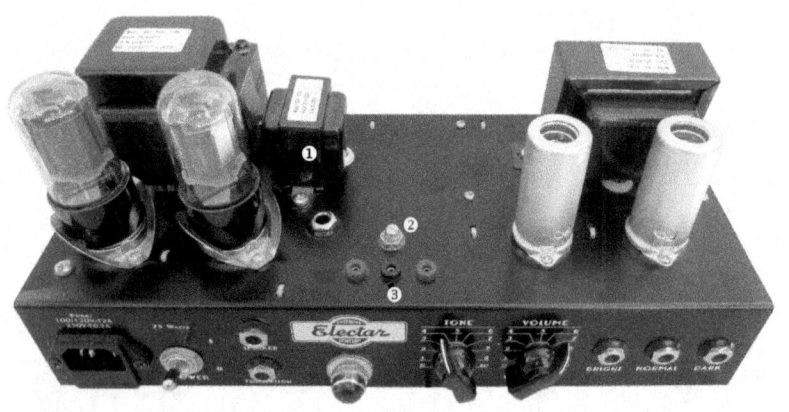

EPIPHONE CENTURY

- 2x12AX7 and 2x6V6
- Solid state rectification
- PCB construction
- Output power: 18 Watts claimed, 13 Watts in reality
- Volume with push/pull boost, Tone
- Inputs: Bright, Normal, Dark
- 12" speaker
- Dimensions: 36 x 22.5 x 40 cm
- Weight: 10.3 kg

ABOVE: The internal view of Epiphone Electar Century amplifier

Had the connector been positioned in location (6) instead of the faraway location (7), that second connection could have been a short piece of unshielded wire! To withstand initial high voltage before tubes warm up and start drawing anode currents, the two large 100μ/450V elcos (first in the filtering chain) are in series (making one 50μF/900V capacitor), but somebody grounded that middle point (8) with a jumper wire, effectively shorting the second elco. Actually, there are two wires, green and white, soldered at (9) and wrapped with a black insulating tape. What a shocker!

The circuit

The amp's circuit diagram is available online so it won't be reproduced here. Its topology is quite common, a push-pull 6V6 pentode stage driven by a long-tail phase inverter, preceded by two 12AX7 gain stages. There is a negative feedback loop from the 8Ω output to the 100Ω resistor in the cathode circuit of the phase splitter.

Tube heating is AC, with two 100Ω resistors to the ground and a 100Ω series resistor. Solid-state rectifiers are used (two 1N4007 diodes in CT arrangement), and the output stage uses fixed bias provided by a half-wave rectifier and the already mentioned 25k bias pot. The bias pot changes the bias voltage for both tubes (there is no balance adjustment).

Thus, matched pairs of output tubes should be used if you want maximum headroom and the cleanest power possible.

The tone control is a simple one-knob type, 470p-1M pot-4n7, but it's followed by a high pass filter formed by the 470p capacitor, bypassed by a 470k resistor, and a 100k grid resistor to ground.

Notice also a complete tone control stack, but without any adjustable pots, all fixed resistors. Very strange - a whole tone stack is included and then a simple one-knob tone control is tacked on?

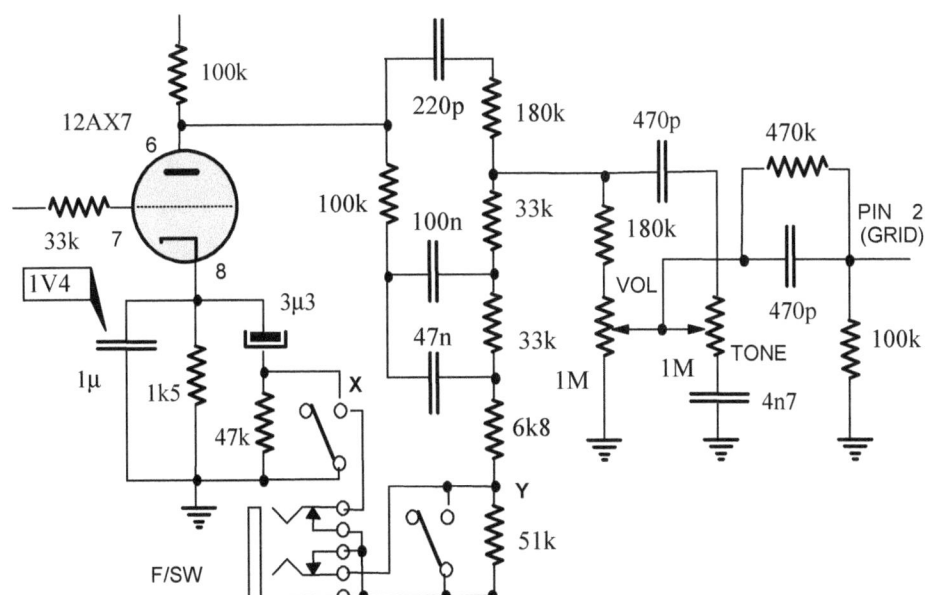

The input stage of Epiphone Electar Century, ©Epiphone Guitar Corp.

Positioned at the output of the first stage, before the volume and tone controls, the inclusion of this filter was obviously done for a reason, to provide a specific voice to the amp. However, the stack further attenuates the audio signal.

The two contacts of the switch on the VOL pot are in parallel with the two contacts of a footswitch.

One switches a 51k resistance at the stack's bottom in and out of the circuit (bypassing it when the contact is closed). The other contact, when open, leaves the 47k resistor in series with the 3μ3 cathode bypass capacitor. When closed, it short circuits that 47k resistor, effectively connecting the negative end of the elco to GND.

The output transformer

Rated at 30 Watts, the output transformer inside this Epiphone amp seems to be a quality unit. The two primary halves have an almost identical DC resistance of around 201Ω. While the primary inductance of around 18 Henry is fine but nothing to write home about, the leakage inductance is very low (11mH), making the quality factor a respectable 1,718!

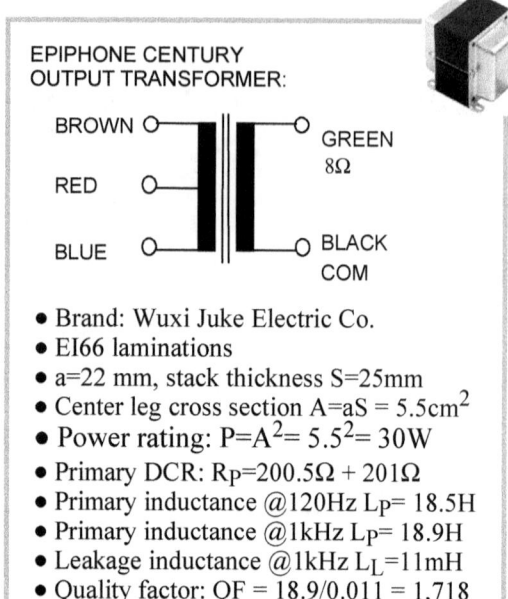

EPIPHONE CENTURY OUTPUT TRANSFORMER:

BROWN — GREEN 8Ω
RED
BLUE — BLACK COM

- Brand: Wuxi Juke Electric Co.
- EI66 laminations
- a=22 mm, stack thickness S=25mm
- Center leg cross section $A = aS = 5.5 cm^2$
- Power rating: $P = A^2 = 5.5^2 = 30W$
- Primary DCR: $R_P = 200.5Ω + 201Ω$
- Primary inductance @120Hz $L_P = 18.5H$
- Primary inductance @1kHz $L_P = 18.9H$
- Leakage inductance @1kHz $L_L = 11mH$
- Quality factor: $QF = 18.9/0.011 = 1,718$

Sonic impressions

The location of the controls at the bottom of the cabinet and at the back is the worst possible location from the user's point of view, and for that reason abandoned decades ago, so releasing a modern take on flawed cabinet design is questionable.

The amp sounded decent, but it was extremely noisy, with not just a low-frequency hum present but also an even more annoying high-frequency hiss. No recording with this amplifier, mate!

It never ceases to amaze me how and why a supposedly reputable manufacturer of musical instruments such as Gibson/Epiphone would release a product that has not been well thought through or has been poorly designed or built and jeopardize their reputation by doing so!

DIY PROJECTS: CONVERTING SOLID STATE GUITAR AMPS TO TUBES

- EVALUATING POTENTIAL CABINET DONORS
- DIY PROJECT: THE FIFTH AVENUE
- DIY PROJECT: THE FOURTH FORCE
- DIY PROJECT: AGENT ORANGE
- DIY PROJECT: THE SCORPION STING
- DIY PROJECT: THE FRONTMAN

Perhaps paradoxically, the most expensive and the most challenging part of a guitar amp for a DIY enthusiast to build is not the actual electronics; it is the speaker cabinet or enclosure, the metal chassis, and the control panel. Small solid-state guitar amps are cheap yet contain all those parts and look professionally made!

The idea is thus to remove the tonally lousy solid-state innards of these amps and build a tube amp inside, retaining as much functionality of the existing controls, ideally all of them, so the switches and knobs do what the fascia says they'd do.

EVALUATING POTENTIAL CABINET DONORS

Epiphone Studio Acoustic 15C

This cute little acoustic amp has an XLR input for a microphone and a 1/4' jack for a guitar and looks good in a cream Tolex finish.

Although small, the speaker is positioned higher up towards the top-mounted chassis, so its magnet presents an obstacle for tubes that would be hanging down on its sides. Plus, it is not centered in the sideways direction either.

This means there is not enough room for two transformers and at least two tubes on the speaker's sides. The transformers would have to be mounted on the sides of the cabinet, or a bottom chassis would need to be used, housing the power supply and perhaps the output stage as well. That would complicate things immensely, so for that reason, we didn't proceed with this conversion.

Epiphone Snakepit 15G

This small combo amp looks cool in black Tolex, although it would look even better in brighter colors such as red, orange, or blue. We got one from the USA for only $30, so we shouldn't complain too much. Our amp didn't have any serial number on it, so perhaps it was a factory second that was somehow released onto the market.

The top-mounted chassis seems large enough for easy conversion to tubes. Although a decent-looking and OK sounding 8-incher, the speaker would still be the main limitation of the finished amp. As always, you can easily upgrade it to a better one of your choosing.

RIGHT: Despite its weird name, the solid state Snakepit 15G combo looks cute, similarly styled to Epiphone Valve Junior.

The clean channel has volume control only, while the overdrive channel has both gain and volume controls. The 3-band equalizer is common for both channels, which are manually switchable only (no footswitch provision), but one can easily be added at the back.

In fact, except for the 24 V_{AC} power input socket, there are no jacks or anything else at the amplifier's back, which is good news; you can add whatever you like.

Apart from one PCB and what looks like a voltage regulator chip or output power transistor bolted to the chassis, nothing else is inside. The IEC power inlet socket would be installed at (1), instead of the DC power in jack, the power transformer at (2), the on-off switch stays as is at (3), as does the input jack (4), and all control pots.

Of course, their resistance values are wrong for a tube amp, so they cannot be reused; different pots need to be installed.

The external speaker jack, 1/2 power (triode/pentode) switch, negative feedback on-off switch, and similar controls (if implemented) that do not require constant access from the front during performance can go to the back panel (5).

Verdict: Perfect for tube conversion!

VOX Pathfinder 10

When converting solid-state amps to tubes, the topology of your tube amp will be determined by the existing control panel. So, look for donor amps without too many knobs and functions. Of course, you can always leave some of them inoperative, but that is the last resort. Ideally, all knobs and functions should remain and perform their intended functions. That way, when you decide to sell your masterpiece, you'll get the highest possible price for it.

In this case, we only have Gain, Master Volume, Bass & Treble tone controls, Overdrive/Clean push button switch, and the combined headphone/line out jack, which is excellent news since even the simplest and lowest power tube amp should have all of those features.

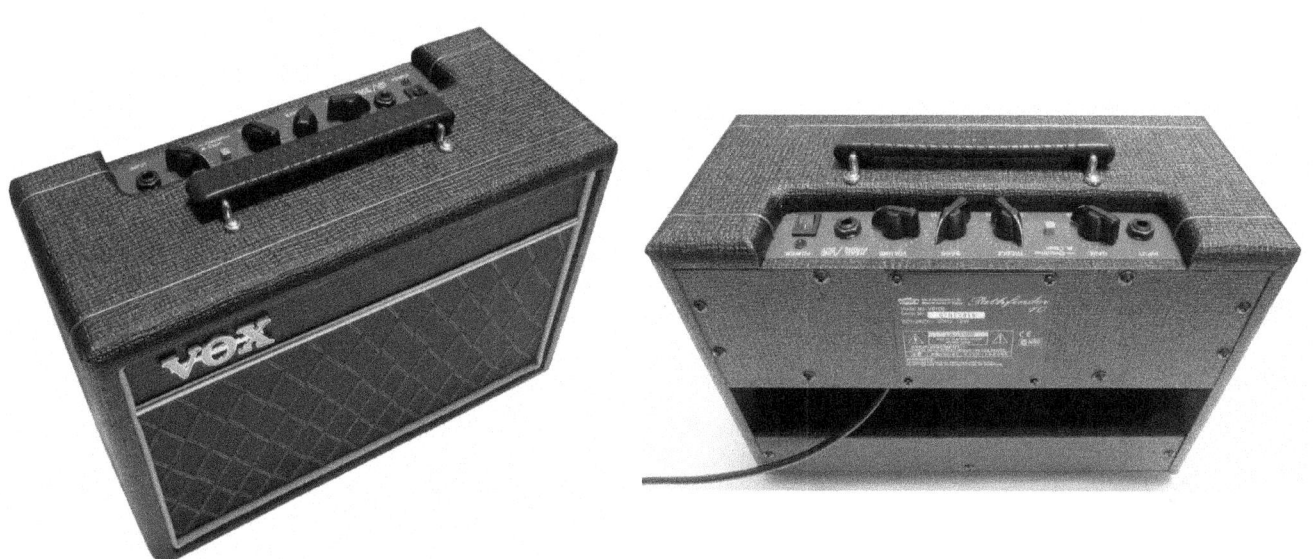

As seen from the photo of its internals, Pathfinder 10 uses integrated circuits in its preamp stages (1) and LM1875T, a 20 Watt integrated power amp chip IC (2) on a heatsink.

The cabinet is sturdy and well made, and since the inbuilt "heavy-duty" (??) speaker is so tiny (3), there is plenty of room around it for the rectifier preamp and power tubes.

There is no external speaker out jack, half-power switch, or attenuator, but you can mount those controls on the back cover.

The cabinet is wide enough for a 10" speaker, but it isn't high enough. However, an 8" speaker can fit height-wise, so if you don't mind removing the front baffle and the fabric and enlarging the cutout, that is the first upgrade you should make.

Verdict: Not much space, but tube conversion is possible!

Fender Frontman 15G

This small closed-back combo solid-state amp from Fender can be had for $20-30 second-hand, so we got one to see if anything can be done with it.

It hasn't got many controls, which is one of the main criteria when choosing a body donor for tubing. There are three volume controls, Normal Volume, Drive Volume, and Gain, and three tone controls, Bass, Middle, and Treble, so a proper tone stack should be implemented.

There is also a Drive Select pushbutton switch, Aux. Input and headphone output. An external speaker output jack would need to be added somewhere at the back and a ventilation grille or two, one at the bottom of the backplate (1), the other on top of the amp behind the handle (2).

The chassis was designed and mounted awkwardly, making it difficult to use as a platform for a tube amp.

The metal plate (3) sits tight against the top of the cabinet so nothing can be mounted on that side. It is held in place by the two bolts that attach the handle to the cabinet.

The PCB (4) is held in place (at an angle) by the potentiometers and the heatsink for the output power IC chip (5). Another metal profile would have to be added to hold the tube sockets and the terminal board (tag strip).

Notice two round forward-facing openings in the front baffle, presumably to act as bass reflex ports for improving the bass response from its tiny speaker.

Verdict: Not the easiest of conversion platforms, unless you want to flaunt Fender's logo around or are an incurable Fender fanatic. There are better options, such as VOX Pathfinder 10 or 15 models and Epiphone Snakepit 15G.

Orange Crush 12

There are physically larger cabinets in the Orange Crush series of solid-state amps, but as they get larger, their price rises steeply, and the number of knobs, digital controls, and gadgets (such as tuner) increases significantly. For a tube conversion, we want a body donor with as few knobs as possible.

The smallest in the Crush series, Model 12 already has three volume control knobs (Volume, Overdrive, and Gain), and three-tone controls (Treble, Middle, and Bass).

The L-shaped chassis is not the best configuration for tube placement. Tubes will either have to sit upwards (in the photo), just as the existing mains transformer (1) does, or we would need to add a bracket at its bottom (2), and tubes would then be parallel with the PCB. Once the chassis is returned to the cabinet, in the first case, tubes would sit horizontally; in the latter option, they would hang down towards the bottom of the cabinet.

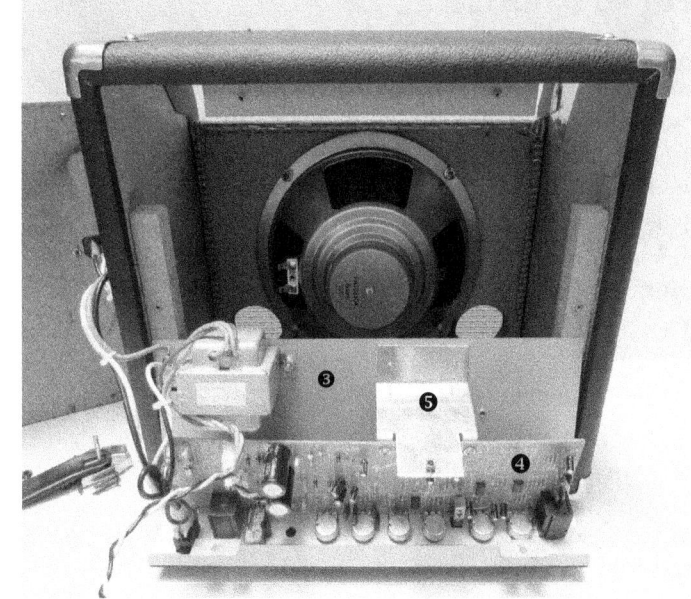

ABOVE: The knobs on Orange Crush are so close together that only miniature sized potentiometers can be used.

In the space now occupied by the PCB we'd need to fit a terminal board. A new power transformer would go in the same position as the current one, and the output transformer would sit at (3), opposite the input guitar jack (4).

Assuming a tube design with one preamp and one small output tube and no tube rectifier, there could be enough ventilation through the four frontal ports. If needed, a rectangular cutout could be made in the bottom panel, and a ventilation grille installed at (5).

LEFT: The L-shaped chassis is held in place by four bolts, making it easily removable.

FOUR FORCE EM-1

It was hard to assess this amp as a potential body donor from its web photos. It seemed small, but it's actually quite large. Secondly, it was hard to tell what its cabinet was made of. We assumed a timber (plywood, MDF, etc.) construction. Wrong again. The body is made of thick black plastic.

You may or may not like the Darth Wader-ish looks of this unusual amp. While ours had a nicely contrasting silver metal front speaker grille, the white-bodied unit also looks cool.

The simple control panel has only five knobs (Gain, Volume, Bass, Mid, Treble), input and external speaker output jacks, and an on-off switch.

Apart from the IEC power inlet (1) and two smallish round acoustic ports, nothing else is on the back panel. Interestingly, it isn't plastic but made of MDF board. This was most likely done to increase the rigidity of the enclosure and prevent unwanted vibrations and resonances.

There is no mains transformer. Instead, an AC-DC converter (2) takes any mains voltage 100-240V and supplies the required low DC voltage for the solid-state amp on a single PCB (3). The power amplifier is an IC chip (integrated circuit) mounted on a small heatsink (4).

The metal chassis is a simple L-shaped affair, not ideal for tube conversion. A bottom plate with mounted tube sockets needs to be added, from which tubes will hang downwards on both sides of the speaker.

The two back vents are positioned near the bottom of the enclosure, so they would serve as an intake of cold air entering the enclosure and rising upwards.

A small metal ventilation grille needs to be installed at the top panel (5) or towards the top of the back panel at (6).

Verdict: Tube conversion possible but fiddly!

Washburn Bass King VBS-30

This bass amp has a 10" speaker driven by a simple solid-state circuit. It's quite rare since it was only made for a year or so, which isn't surprising. A low-powered solid-state circuit and a feeble 10" speaker do not make a good bass amp! Many other competing models were equally useless as bass amps. However, their cabinetry was not as sturdy or nice looking as Washburn's.

We got this amp from the USA for its lovely tweed-covered cabinet and a simple control panel with only five knobs. They are Volume, Treble, Middle, Bass, and Presence, facing the amp's back and going from the input jack to the left. Thus, a push-pull circuit with a full tone stack and presence in the long-tail phase inverter stage would fit nicely into this beautiful body.

Unless you want to keep the solid-state preamplifier and just add a tube power stage, the turret board will now sit where the existing PCB is (1).

A filtering choke can fit inside the chassis in place of the existing power transformer (2), with a rectifier tube (if used) hanging down at (3) and all preamp and power tubes lined up on the other side (4).

The output transformer would be mounted on top of the chassis at (5) and the power transformer at (6).

As with most solid-state body donors, especially with closed cabinets, the heat extraction issue must be resolved. Install a ventilation grille on top of the amp (7) or on top of the back cover. As the hot air from the cabined rises, it will suck in outside air through the two front ports (8), acting as ventilation holes.

Verdict: Apart from the ventilation issue, perfect for tube conversion!

Pignose Hog 30

A battery-powered solid-state bass amp with a simple control panel (Gain, Volume, Bass, Middle, and Treble controls and "Funk Bass" switch), Headphones and Line Out jacks at the back, and plenty of room inside the reasonably large cabinet. The space vacated by the left battery would be used to mount a power transformer, and the output transformer would sit exactly where the right-hand battery is now.

All in all, this amp's chassis and cabinet make it an ideal candidate for tube conversion!

ABOVE: While the dark orange ("burned ochre" sounds better) cabinet looks appealing, we didn't like the protective metal grille over the speaker, somehow it looks dated.

DIY PROJECT: THE FIFTH AVENUE

The body donor: Washburn Bass King VBS-30

A rare solid-state bass amp with a nice tweed-covered cabinet, a 10" speaker, and a simple control panel with only five knobs, Volume, Treble, Middle, Bass, and Presence. Thus, a push-pull circuit with a full tone stack and presence in the long-tail phase inverter stage would fit nicely into this gorgeous cabinet.

The power tube: 6AR6 (6098)

6AR6, an octal beam power tube, was developed for service as a horizontal deflection ("sweep") amplifier for black & white TV sets and oscilloscopes. It was also used in radar systems, which means it passed stringent JAN (Joint Army Navy) controls.

From our experience in hi-fi amps, 6AR6 is a very reliable, robust, long-lasting, and, above all, great-sounding tube!

Bendix 6384 are electrically equivalent to Tungsol 6AR6 but are grossly overpriced. While 6AR6 can be found for US$10-20 (in 2017), 6384 rarely sold for under US$100, yet it delivered no sonic advantages.

6384 has an anode power rating of 30 Watts, so in a guitar amp, it could deliver higher output power than 6AR6 (19 Watts) or 6AR6WA (21 Watts). However, it has the same maximum anode current of 125mA, so even that higher anode power claim is questionable since the anode is of the same size as 6AR6WA.

BELOW: Sylvania 6L6GB on the left and Tungsol 6AR6WA on the right

6AR6

TUBE PROFILE: 6AR6

- Indirectly-heated beam tetrode, octal socket
- 6.3V, 1.2A heater
- $V_{AMAX}=565V$, $V_{G2MAX}=300V$
- $I_{AMAX}=115mA$, $P_{AMAX}=19W$, $P_{G2MAX}=3.2W$
- As a pentode: gm=6mA/V, $r_I = 21k\Omega$, $\mu=126$
- As a triode: $\mu= 6.1$, $r_I = 1.0k\Omega$

TUBE PROFILE: 6AR6WA (6098)

- $V_{AMAX}=400V$, $V_{G2MAX}=315V$
- $I_{AMAX}=125mA$, $P_{AMAX}=21W$, $P_{G2MAX}=3.5W$

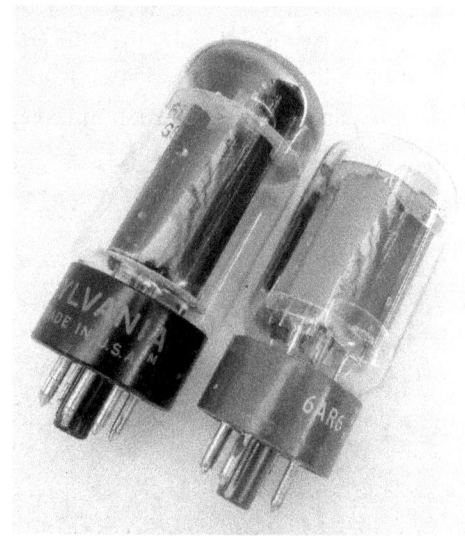

Topology

We have one input, a single channel, a three-knob tone control stack, single volume control, and presence adjustment, so the topology is pretty simple. Instead of three amplification stages and a split-load inverter, we will have two amplification stages and a long-tail phase inverter, providing gain. That way, we can implement presence control in its tail.

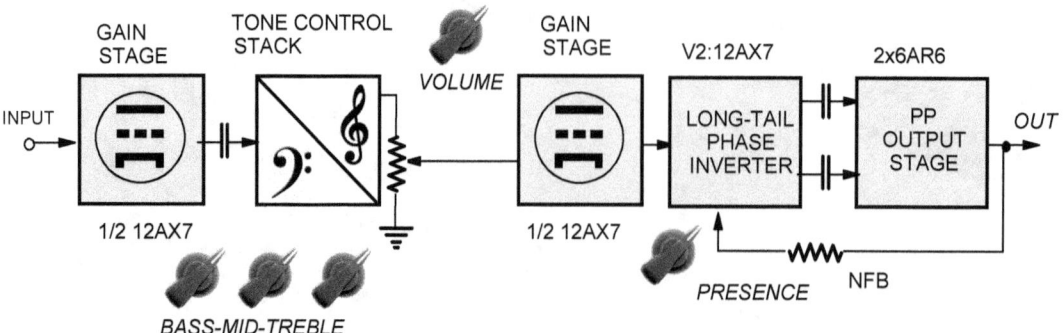

The composite curves for pentode operation

Since the preamp stages are standard, let's focus our attention on the output stage. If we choose 350V as the quiescent DC voltage and position the operating points Q1 and Q2 at 20mA idle current, very close to Class B operation, the composite graphs of two 6AR6 in pentode mode yield the maximum current swing of 400mA and the voltage swing of 600V. The composite load impedance is R_{COMP} = 600V/400mA=1k5. The output transformer's primary impedance (anode-to-anode) needs to be four times higher, or Z_{AA}= $4R_{COMP}$ = 6kΩ!

We can expect the maximum output power $P_{MAX}=\Delta V_A * \Delta I_A / 8$ = 600*0.4/8 = 240/8= 30 Watts.

With an idle current of 20mA, the bias is around -50V, so with 40mA through the common cathode resistor, its resistance needs to be 50/0.049 = 1,250Ω or a standard 1k2 resistor.

Option #2: If we bias the tubes a bit closer to Class A operation (idle current of 45mA), the bias is around -40V, so we need a common cathode resistor of 40/0.09 = 444Ω resistance, close to the standard value of 470Ω.

Option #3: Finally, as per the data sheet (-22.5V bias, 77mA current through each tube), the common R_K must be 22.5/ = 146Ω, so use a 150Ω resistor. This would bias tubes deep in class A, but would significantly reduce the maximum achievable output power.

Unless you order a power transformer that will provide AC voltages needed to get the precise DC voltages used in your composite curves, your finished amplifier will usually work in a different regime of that envisaged in your estimations.

The estimations that we have just done are just a starting point anyway, so don't lose your sleep over such discrepancy between theoretical "predictions" and practical results.

We ended up with 360V on 6AR6 anodes (circuit on the next page), and, using a 390Ω common cathode resistor, got 39V on the cathodes and 100mA of combined idle cathode current, close to option #2 above.

We now need to check 6AR6's anode power dissipation. With 39V on the cathodes of output tubes, the current through the 390Ω cathode resistor is exactly 100 mA. The voltage drop of 1V on 560Ω screen grid resistors means the screen current of each tube is 1/560= 1.8mA, so at idle, the anode current of each tube is 50-1.8 = 48mA.

The anode-to-cathode voltage across the 6AR6 tubes is 360-39=321V, so finally, the power dissipated into heat on each anode is thus P = 0.048*321 = 15.4 Watts. For a 19 or 21 Watt rated tube, this is very conservative, so these lovelies will have an easy and stress-free life.

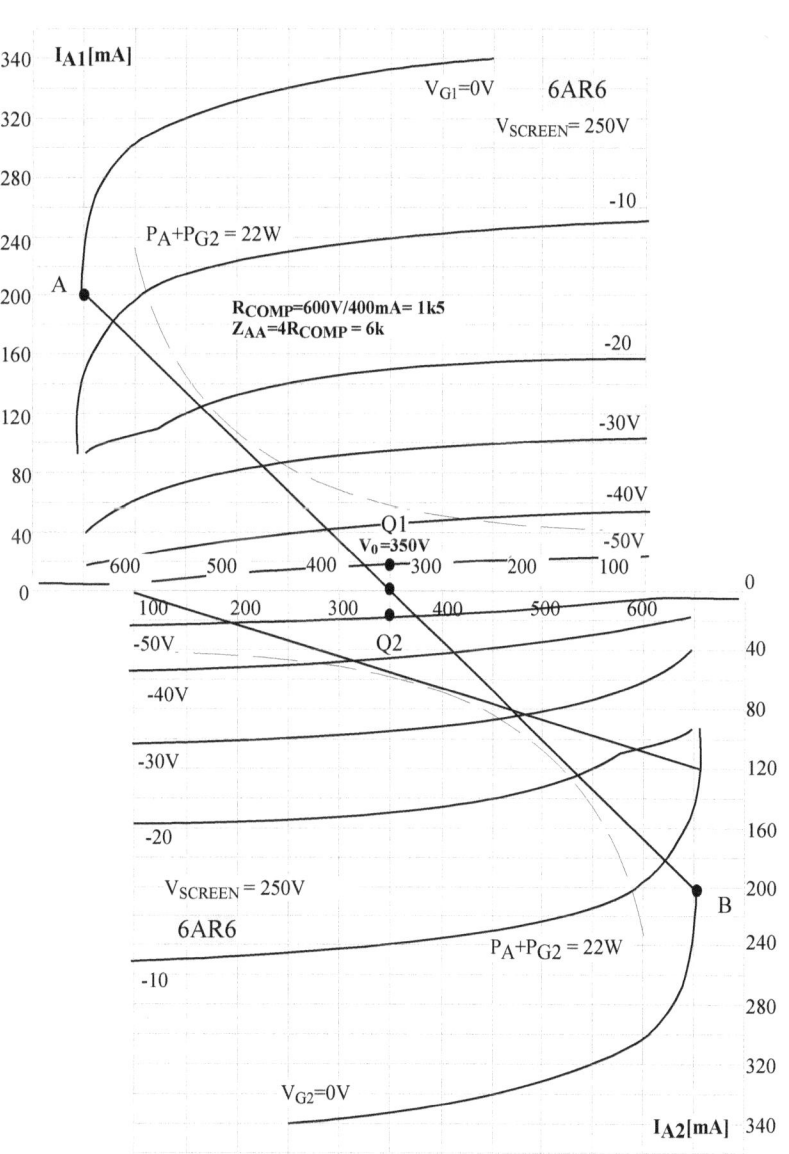

Circuit diagram

The most straightforward power supply possible: AC heating and two fast recovery silicon diodes in a CT arrangement. A chain of four DC voltages, as with most push-pull amps, supplying the CT of the output transformer (power tubes' anodes), power tubes screen grids (V2), anodes of the phase inverter (V3), and anodes of the first two gain stages (V4).

After the initial tests, two issues became apparent. Sonically, the Presence control didn't do much. The amp was very loud and "clean" - the preamp stages would only start distorting at the highest power levels (too late).

Thus, the Presence was replaced by the Master Volume control, and all was perfect; early preamp distortion was now possible at low overall power levels.

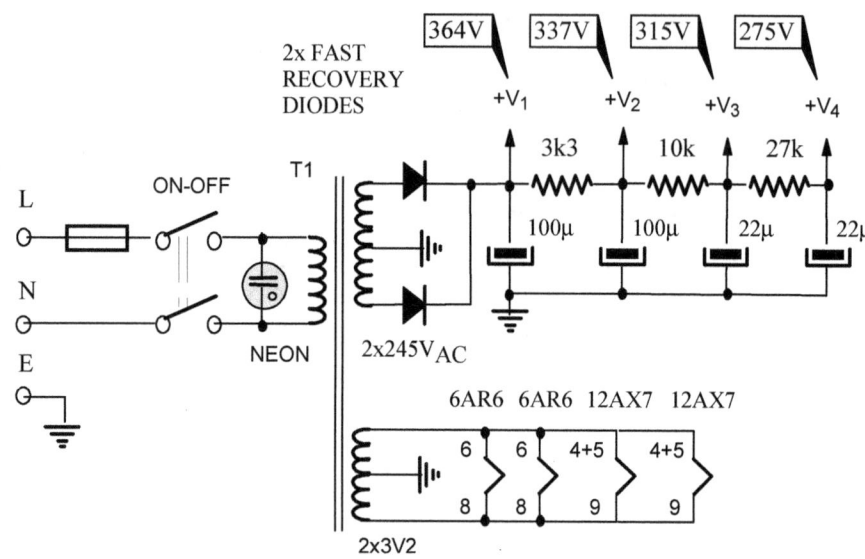

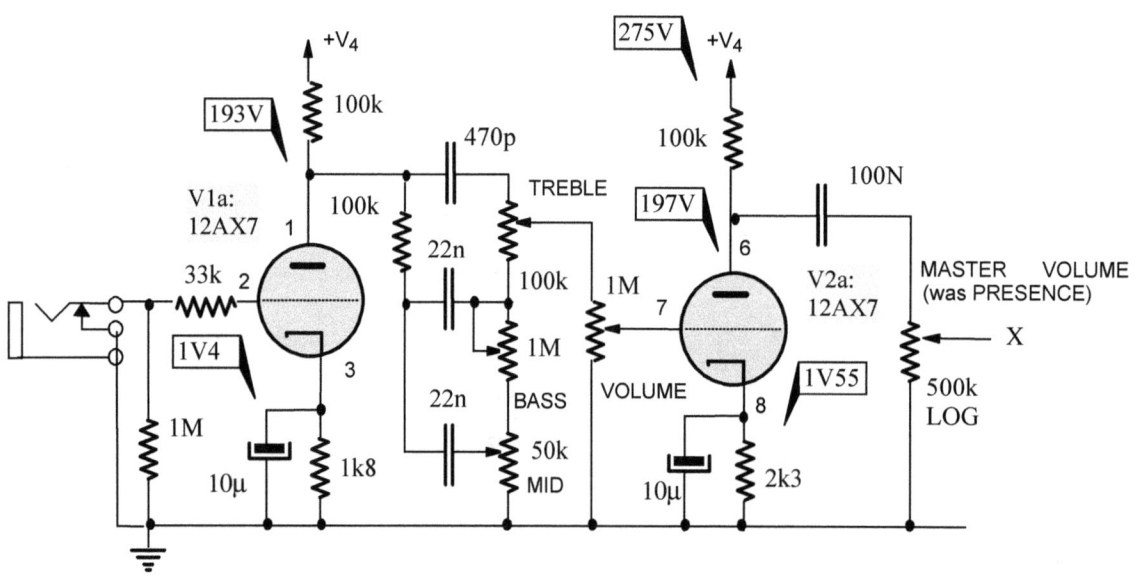

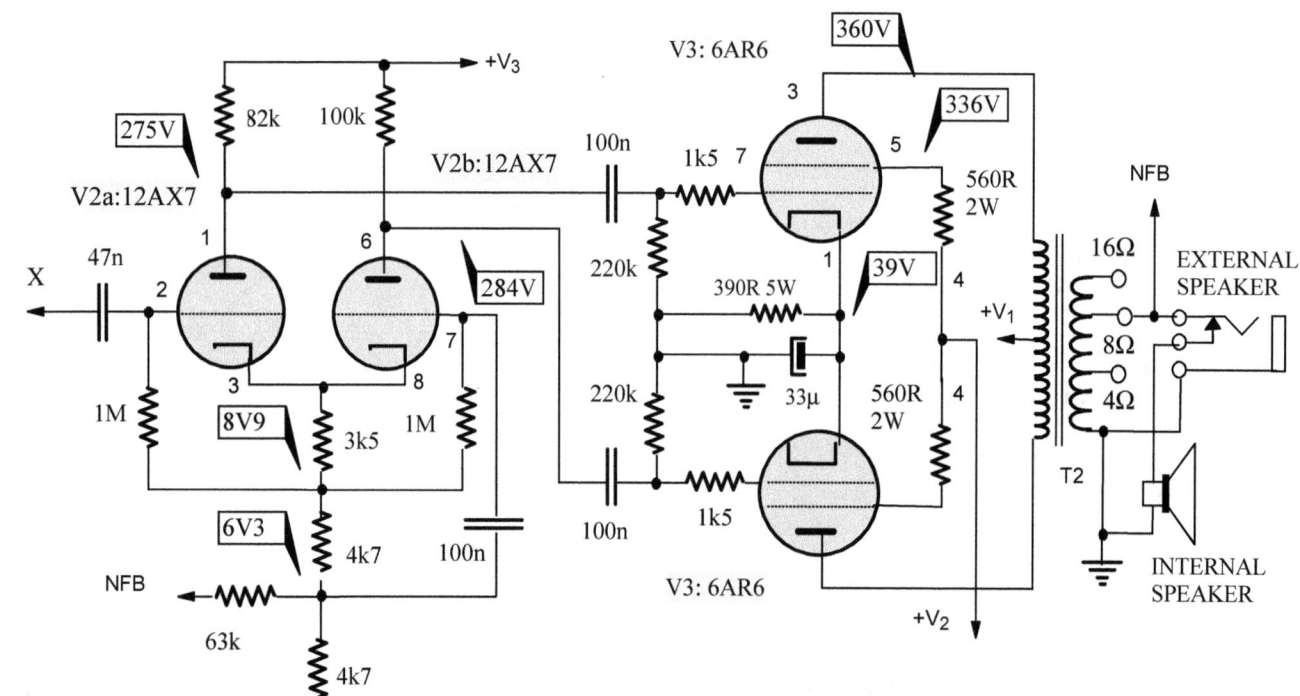

DIY PROJECTS: CONVERTING SOLID STATE GUITAR AMPS TO TUBES

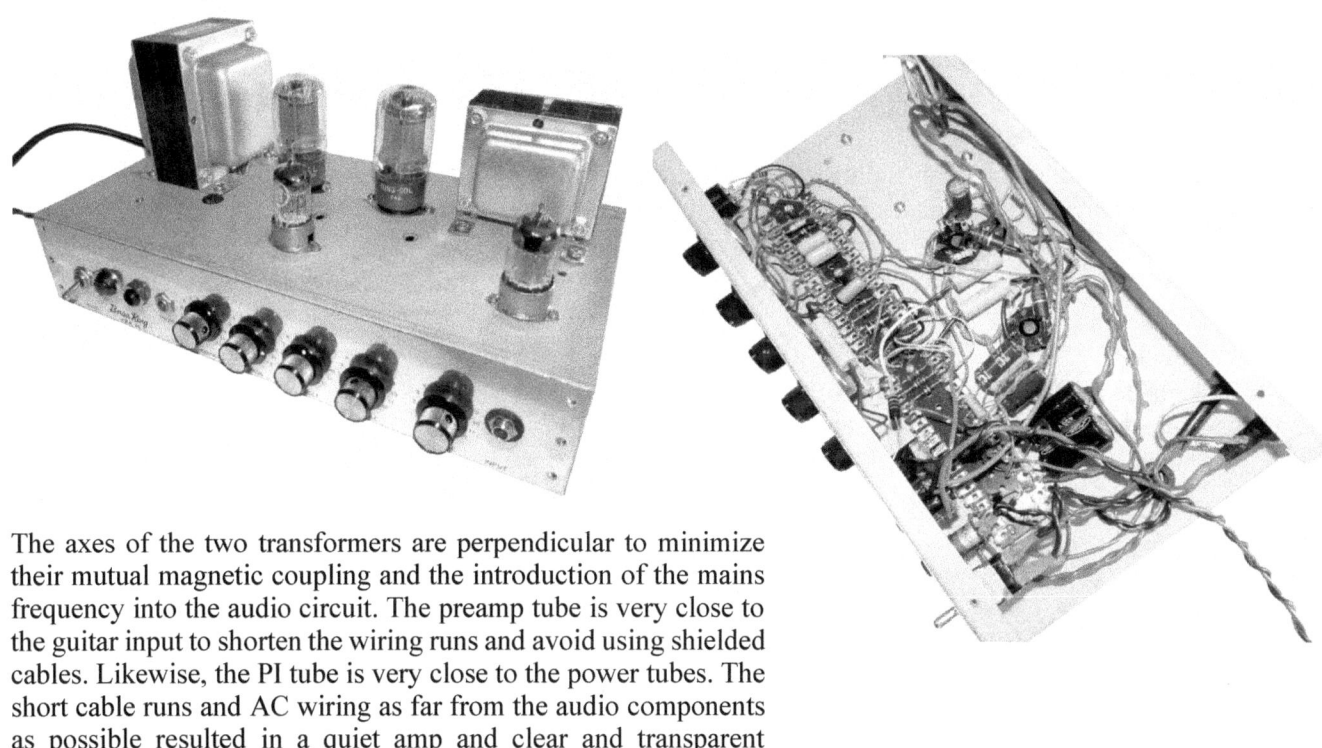

The axes of the two transformers are perpendicular to minimize their mutual magnetic coupling and the introduction of the mains frequency into the audio circuit. The preamp tube is very close to the guitar input to shorten the wiring runs and avoid using shielded cables. Likewise, the PI tube is very close to the power tubes. The short cable runs and AC wiring as far from the audio components as possible resulted in a quiet amp and clear and transparent sound.

Sonic impressions

6AR6 power tubes sound magnificent in single-ended hi-fi amplifiers, so I was anxious to hear the verdict on The Fifth Avenue's tonal aspects. Sure enough, the amp sounded full-bodied and engaging. Was that due to 6AR6 lovelies, our design skills, the large and sturdy Washburn bass amp cabinet, its speaker, or "all the above"? It's hard to tell. In any case, one of the best-sounding designs in this book!

DIY PROJECT: THE FOURTH FORCE

The body donor: Four Force EM-1

The simple control panel has only five knobs (Gain, Volume, Bass, Mid, Treble), the input jack, and the on-off switch. Despite its relatively large cabinet, the chassis is small, and the speaker's magnet is close to it, so we had to position the tubes on each side of the chassis as far from the speaker as possible.

The output transformer

M1115 is an audio transformer for 100V Line PA speakers, of unknown origin and manufacturer, sold by Australian DIY electronics chain Altronics (AU$14.50 or US$10.50 in 2017).

The secondary is 8Ω only, with primary taps marked 0-1.25-2.5-5-10-15W. These figures indicate the amount of power delivered to the speaker from a 100-volt public address line. Its frequency response is specified as 30Hz - 20kHz (±3dB) and the primary inductance is a healthy 7.7 Henry. Its magnetic stack (core) is rated at 15 Watts.

It reflects an 8Ω load as 8.7kΩ anode-to-anode load onto the primary side. The 5 Watt lug is the CT (Center Tap), while "0" and "1.25W" lugs are anode connections.

Is a two-tube push-pull amp possible?

ECL82 (6BM8) is a triode-pentode developed for audio use in radios, tape recorders/players and low cost stereo systems. The 300mA heater version, PCL82, was used in the audio stages of domestic TV sets. The successor, ECL86 (6BW8), was released in 1962.

The voltage amplification factors of the triode sections are high, µ=70 for ECL82 and µ=100 for ECL86. Their internal resistances are also high, in the order of 30kΩ and 60 kΩ respectively, but still lower than that of 12AX7, so these triodes will drive tone stacks with lower losses and more overall gain.

The pinouts of these two otherwise similar tubes are different (a serious design faux pas, in my opinion), so, unfortunately, the two tubes are not pin-compatible; otherwise, they could be interchanged in the same amp for two different voicing options.

With a 100k anode load, the triode section of ECL86 achieves a voltage gain of 60; with 220k anode load and 250V anode voltage, a gain of 75 is possible!

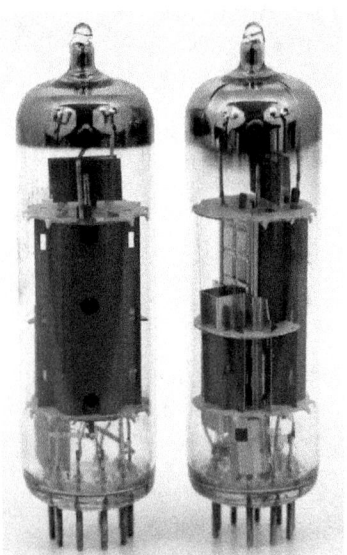

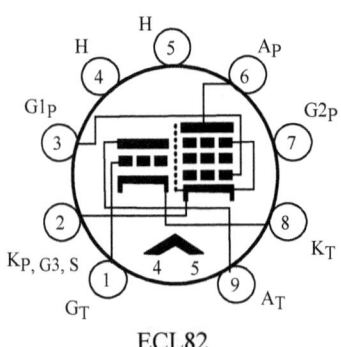

ABOVE: ECL82 pinout
LEFT: ECL86 from two angles

TUBE PROFILE: ECL82 (6BM8)

- Indirectly-heated audio triode-pentode
- Noval socket, heater: 6.3V/780 mA
- TRIODE: P_{AMAX} = 1W, I_{AMAX} = 15mA
- V_A=300V, V_G=-1.3V, I_0=1.1 mA
- gm = 2.5 mA/V, r_I=28kΩ, µ=70
- PENTODE: P_{AMAX}=7W, I_{KMAX}=55mA
- gm = 7.5 mA/V, r_I=16kΩ, µ=7.5x16 =120

TUBE PROFILE: ECL86 (6GW8)

- Indirectly-heated audio triode-pentode
- Noval socket, heater: 6.3V/700 mA
- TRIODE: P_{AMAX} = 0.5W, I_{AMAX}= 4mA
- gm = 1.6 mA/V, r_I=62kΩ, µ=100
- PENTODE: P_{AMAX}=9W, P_{SMAX}= 1.8W
- I_{KMAX} = 55mA, V_{HKMAX}= 300V
- gm = 10 mA/V, r_I=48kΩ, µ=480

To confuse things even further, the other members of this tube family include ECL83 (5W pentode's rating and 3.5W triode with µ=85), ECL84 (4W pentode and 1W triode with µ=65) and ECL85 (9W pentode and 0.5W triode with µ=50). ECL83 and ECL84 have the least powerful pentodes, while ECL85 has triodes with the lowest mju, so overall, ECL82 and ECL86 are the two highest merit tubes.

Suitability for a self-splitting arrangement

The only commercial amp we know of that uses a self-splitting output stage, Kustom 15H, uses 6BQ5 pentodes with gm=11.3 mA/V and µ=450. In a self-splitting arrangement we are looking for tubes with high mutual conductance and high amplification factor µ. Since there is a significant difference between ECL82 (gm=7.5mA/V and µ=120) and ECL86 (gm=7.5mA/V and µ=480), ECL86 is better suited to this application. Its parameters are in the same ballpark as those of 6BQ5, so it should work well in the same type of circuit.

The topology

Two ECL86 triode preamp stages, with Gain after the first stage and Volume after the 2nd stage, are followed by the self-splitting push-pull output stage. Since we hadn't tried ECL86 in this type of circuit before, if they failed to work satisfactorily in a self-splitting stage, we thought our fallback position was to rewire the two output pentodes in a parallel single-ended mode. Obviously, a SE output transformer would need to be substituted.

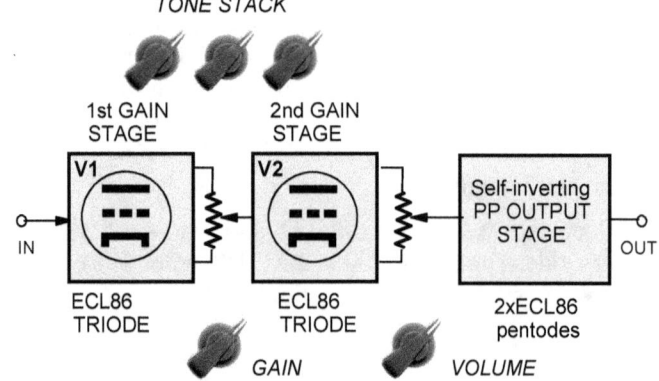

Circuit diagram

Due to the minimal space inside the chassis, the solid-state power supply is as simple as possible, with only two HV filtering stages and AC heating of both tubes.

In the cathode-coupled self-inverting output stage, the common cathode resistor must not be bypassed by an electrolytic capacitor. That would shunt the audio signal developed on V1a (the driving tube) cathode to the ground and deprive the driven tube's (V2b) input (its cathode in this case) of the signal so that the output stage would work in a single-ended mode.

Since the driven tube works in the common grid (CG) mode, its grid is grounded, and the signal input is its cathode, with the output taken from the anode, just as in the common cathode (CC) V1b.

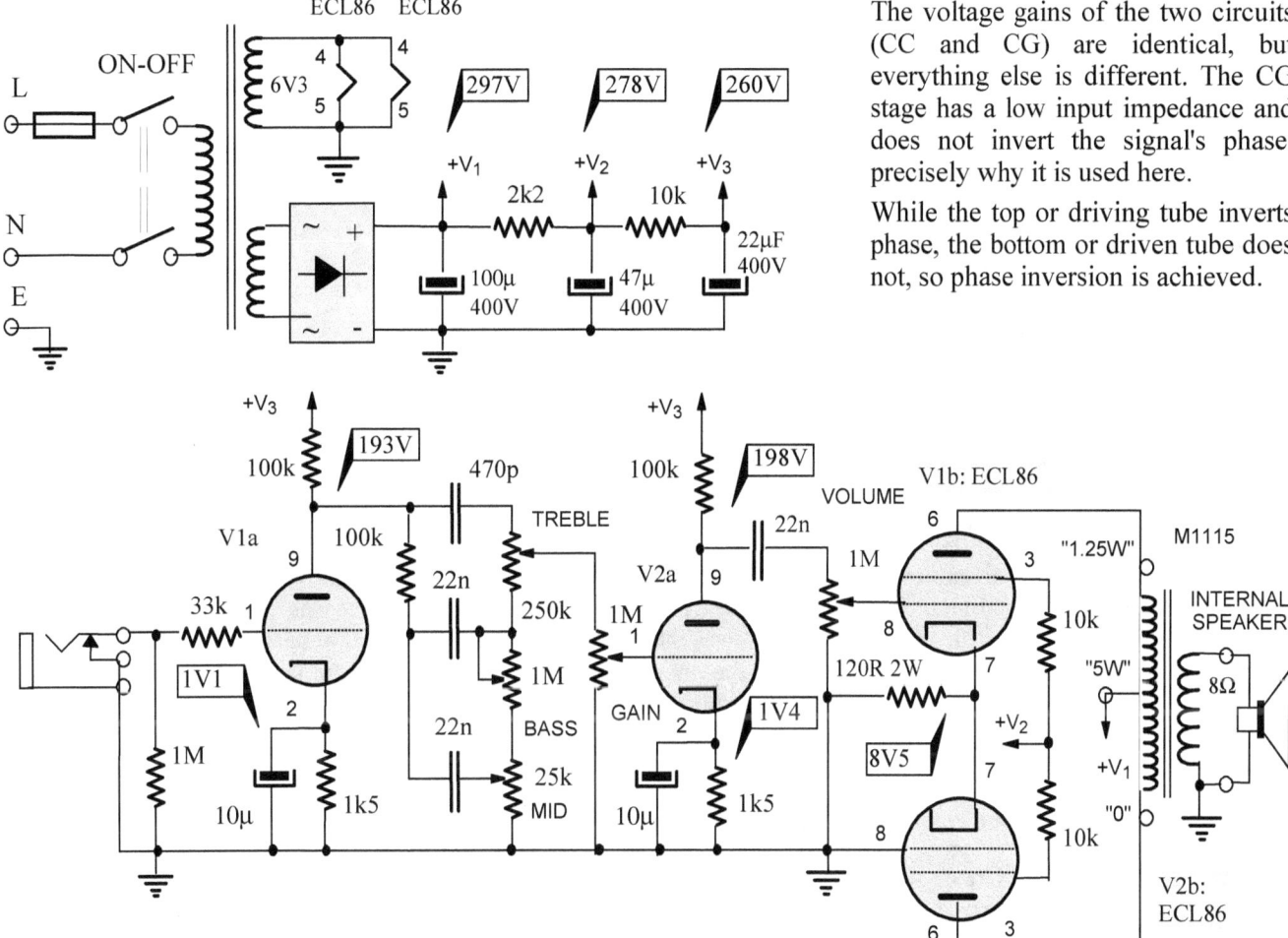

The voltage gains of the two circuits (CC and CG) are identical, but everything else is different. The CG stage has a low input impedance and does not invert the signal's phase, precisely why it is used here.

While the top or driving tube inverts phase, the bottom or driven tube does not, so phase inversion is achieved.

The voltage drop on the 10k HV filtering resistor was 15 Volts, so the first two stages drew 1.5mA together. The cathode current of both output tubes together was $2I_K=8.5/0.12=70.8$mA or 35.4mA per tube. With around 283V between their anodes and cathodes, the anodes of output tubes dissipate P=283V*33mA=9.3 Watts, which is slightly over the maximum allowed (9W), but these tubes are cheap and plentiful, so no significant loss even if they last only a few months.

The wiring and construction details

The L-shaped metal chassis (next page) is held in place by three bolts (1). We installed the power transformer (3) onto the existing bracket that held the switch-mode power supply module in the original amp (2). The transformer mounting bracket fit the two holes perfectly, but it protruded just a few millimeters too far to the back, so the back cover could not be put back on. So, it had to be moved and mounted onto the sidewall.

The U-shaped aluminium bracket (4) onto which tubes were mounted was actually a drawer handle from Ikea, an ideal budget solution.

The three main (AC) wires, phase, neutral and earth (5) come bundled to the on-off switch (6) straight from the IEC inlet mounted on the back cover. The output transformer (7) is right next to the amp's guitar input (8). Alternatively, we could have mounted it in location (4) on the U-bracket, in between the two tubes.

The sonics and playabilty assessment

Since we had not included an external speaker jack in this retrofitting project, we could not evaluate the sound of this amp with a larger speaker cabinet, but what we heard with the Four Force stock speaker was quite impressive. It was hard to ascertain if the deep, full and punchy sound was the result of the plastic cabinet, of the fact that it is of a closed back type with the back firing bass reflex ports, the quality of the stock speaker, or the uniqueness of our design. Most likely, it was the fortunate synergy of all these factors that voiced this amp.

DIY PROJECT: AGENT ORANGE

The body donor: Orange Crush 12

Even the smallest in the Crush series of solid-state amps (40x40x25cm) had more knobs than we would like. Luckily most were standard functions, Volume, Gain, Bass, Middle, and Treble, so the only question was implementing the Overdrive control.

Instead of switching a fixed resistor between the bottom of the tone stack and ground for a gain boost or overdrive and bypassing it for the gain attenuation of the tone control stack, we had an idea of adding a control pot to vary that resistance continuously. When the slider is in the + or CW position, the whole 100k is inserted, and the gain is boosted, partially disabling the tone control stack. As the slider is moved upwards (CCW), more and more of that resistance is bypassed, the normal operation of the tone stack resumes, and the gain is reduced.

Nothing could be added to the tiny control panel, so the external speaker jack was placed on the back panel.

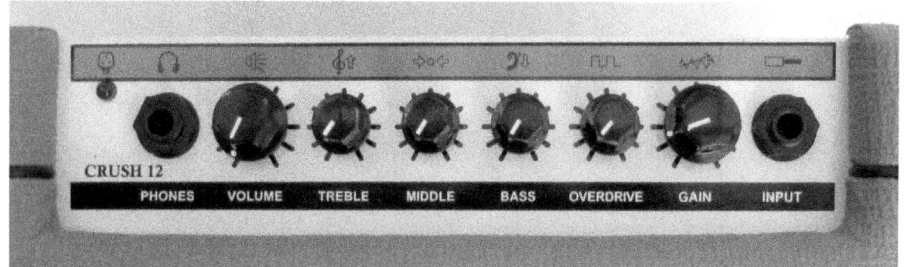

ABOVE: The control panel of Orange Crush 12 solid state amplifier
RIGHT: ECLL800 is the most unusual triode+dual power pentode tube

The output tube: ECLL800

Instead of one output tube driving the other in a self-inverting push-pull output stage, some innovative soul in early 1960s had an idea of placing both tubes in one glass envelope. For good measure he added a driver triode as well, and all that within the confines a small Noval base body, the same size as EL84 pentode!

Two pins are "lost" to the heater, so three triode pins plus four pins for each beam tube (anode, screen, control grid and cathode) bring us to 11 pins, so the only way out was to share four pins, to bring 11 "active" pins down to 7.

The cathodes and screen grids are shared between the power tubes, saving of two pins. Beam forming plates are also shared and internally strapped to the common cathode, so that does not save us any pins. The other two pins are saved by connecting the triode's cathode to the cathodes of power tubes, and also by strapping its control grid (where the signal enters the whole tube-cum-output stage) to the grid of one of the power tubes.

The short-form catalog tells you (almost) everything you need to know.

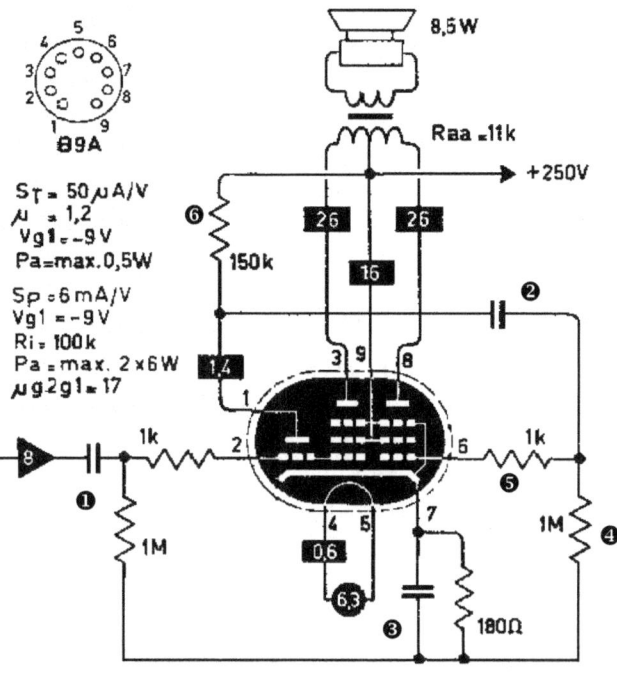

The little triode's gain is only 1.2, its power rating is measly 1/2 Watt, and its mutual conductance ST (from German Steilheit meaning "slope") is only 0.05mA/V. However, its only job is to provide phase inversion for the other beam tube (pins 8-6-7, while the first power tube's grid is driven directly by the signal from the triode's grid (pin 2).

The beam tetrodes have a very modest power dissipation, only 6 Watts each (half that of E84), but their bias of -9V and triode mju of 17 are in the same ballpark as those of EL84. Since their 6 mA/V transconductance is also approximately half that of EL84, one could speculate that by paralleling two of these, we would get one 12 Watt tube with a slope of 12mA/V, an EL84 indeed!

Since a pair of EL84s can easily produce 15 Watts, one of these weird tubes should be good for at least 7-8 Watts, and sure enough, the diagram specifies 8.5 Watts with an 11 kΩ load.

Notice that the values of two coupling capacitors (1) and (2) aren't specified; that is left to the designer to choose, depending on the amps' service, as a guitar or as a hi-fi amp -ditto with the cathode bypass cap (3). 22-47nF are the usual coupling cap values in a guitar amp; for a hi-fi amp 100-220n are needed.

The 1M grid leak resistor (4) and 1k grid stopper (5) aren't critical either. The same applies to the 1k and 1M resistors in the grid circuit at pin #5.

However, the other component values, the 150k anode resistor (6) of the driver triode and the 180Ω cathode resistor, have been selected for optimal operation of the whole stage and should not be changed!

Circuit diagram

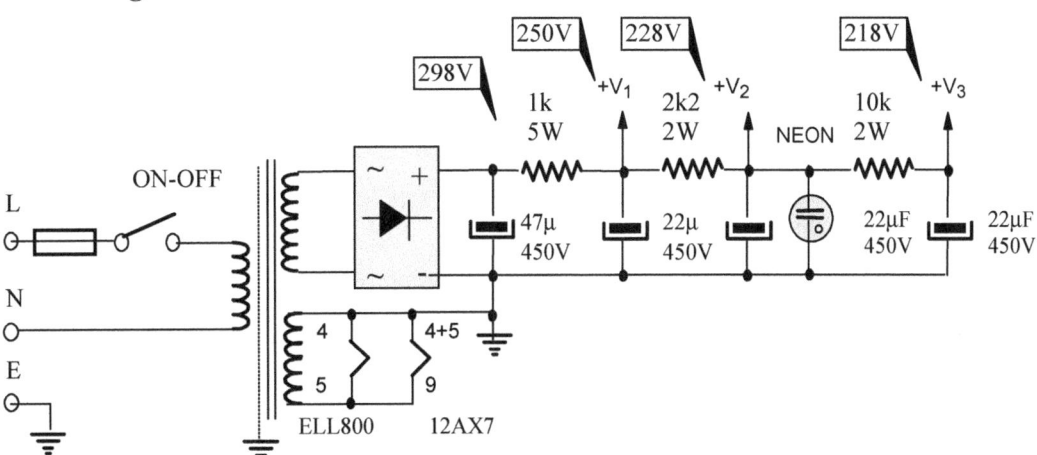

There is no need to draw a block diagram; the topology of this little beast is elementary, a preamp stage followed by a full 3-knob tone stack and Gain control, then the driver stage and its Volume control, and the ELL800 PP stage.

The bias voltage for both preamp stages was identical, -1.3 Volts, but the anode currents were different since the cathode resistors have different values. The 1st stage anode current was 1.3/2k = 0.63mA, while the 2nd stage anode current was even lower, 1.3/3k = 0.43 mA or around 1 mA for both stages together.

The 8.7V cathode voltage of ECLL pentodes with 191W cathode resistor (180R marked) meant that the total cathode current was 8.7/191= 46mA.

The power supply voltage chain (298-250-228-218Volts) shows that the total HV current draw through the 1k resistor was (298-250)/1=48/1=48mA. The current through the 2k2 resistor was (250-228)/2.2 = 22/2.2=10mA, which means that the screens drew 9mA (since we know that the preamp stages only drew 1 mA together). Indeed, the current through the 10k resistor was (228-218)/10 = 10/10 = 1mA!

Now we can determine the anode currents, which are I0=(48-10)/2 = 38/2 = 19mA each.

Finally, we have to check the idle anode power dissipation on each output pentode. The anode voltage was 231 V, so the voltage between the anode and cathode was V_{AK}= 231-8.7= 222.3V, and the power was $P=I_0*V_{AK}$ = 0.019*222.3 = 4.2 Watts, which was comfortably below the 6 Watt maximum.

The overdrive control worked very well, with clean voicing in the CCW position and a bit of crunch and fuzz when turned fully CW.

The *Line Ou*t voltage divider reduces the speaker voltage to 22/(22+82) *100% = 21.15% of its amplitude. With the maximum outptu power of around 5 Watts on 8Ω load or 6.3 Volts, the line out voltage would be 0.21*6.3 = 1.32 Volts.

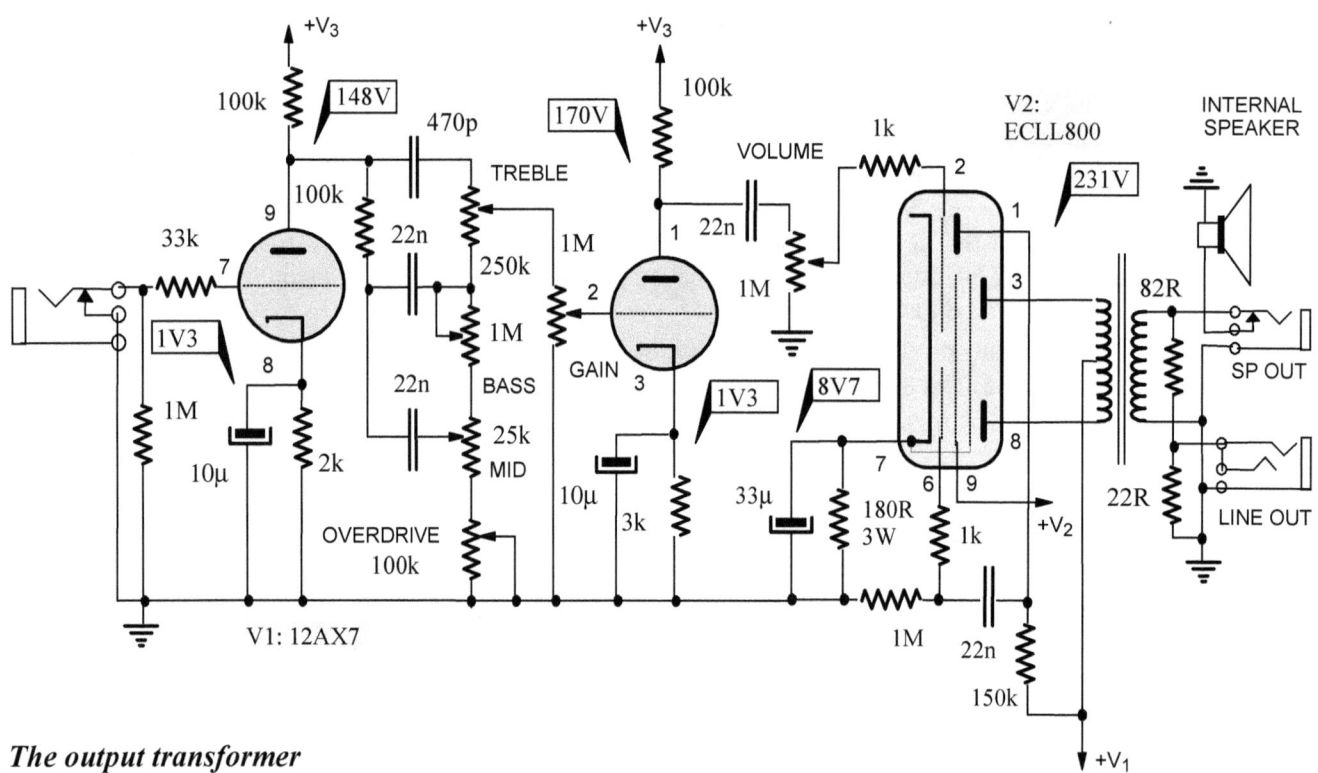

The output transformer

We could not find a push-pull output transformer for this project, something with a relatively high primary impedance of 10-15kΩ, so we had to design and wind one.

Since we didn't sectionalize it, the leakage inductance was relatively high at 91mH. Still, since the primary inductance was also very high (153 Henry!), the quality factor was much higher than similar China-made transformers, a whopping 1,681!

This transformer can reproduce bass frequencies like few others, something you only appreciate when you plug in a large external speaker box. The tiny internal Orange speaker cannot do it much justice.

AGENT ORANGE OUTPUT TRANSFORMER:
- EI60 laminations
- a=20 mm, stack thickness S=30mm
- Center leg cross section $A=aS = 6cm^2$
- Power rating: $P=A^2= 36$ Watts
- Primary impedance with 8Ω load: 15k
- Primary DCR: R_P= 1,005 Ω + 897 Ω
- Primary inductance @1kHz L_P= 153H
- Leakage inductance @1kHz L_L=91mH
- Quality factor: QF = 1,681

DIY PROJECTS: CONVERTING SOLID STATE GUITAR AMPS TO TUBES

The wiring and construction details

The terminal board had just the right width to fit between the potentiometers and power inlet (4). It was mounted on 1" hexagonal distance holders, one at each end (1). It also rests on the plastic tip of the input jack (2) and the headphone/line out jack (3).

The power inlet is very close to the input tube, but we could do nothing about it. There was a low residual hum, so a steel divider was installed (not shown in the photos).

The power transformer (5) and the power supply board are bolted to the bottom of the cabinet so that no add-on screws can be seen. The heater wires (7) from the power transformer were soldered directly onto the terminal board. The output transformer (8) was quite large, but it just fit on the metal chassis with not a millimeter to spare.

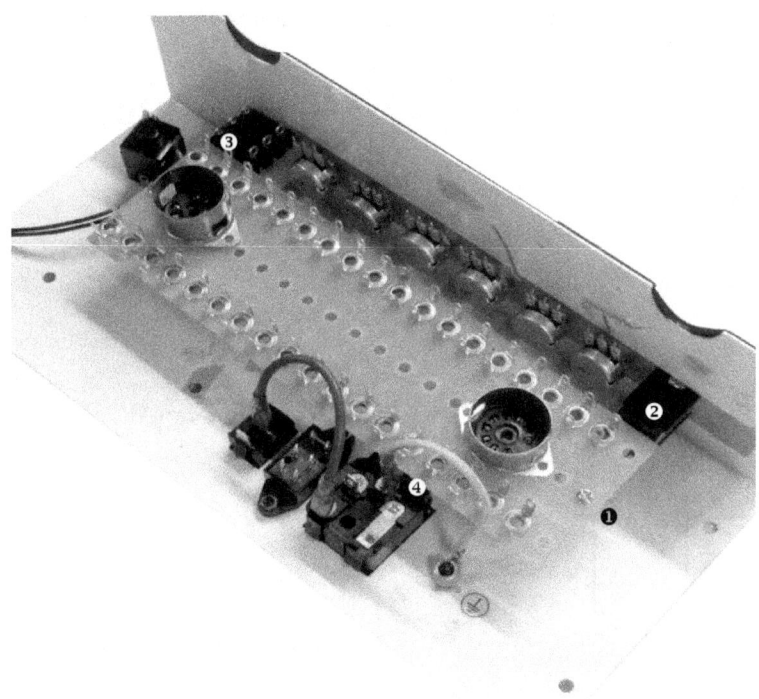

The external speaker jack (9) was very close to the headphone-line out jack and the output transformer, which was very fortunate because it minimized the length of interconnecting wires between them and the internal speaker.

The wiring is easy to follow and troubleshoot later on. In this book, most small amps we used as body donors for retubing projects had push-on knobs (to fit split-shaft pots), which we could not reuse with our solid shaft potentiometers. Small diameter knobs with a tightening screw were needed.

The ventilation grille (1) and external speaker jack (2) were added, the mains voltage selector switch was disconnected but left in place for esthetic reasons (3). You could wire it up with a bypass capacitor to provide normal/bright voicing or as a triode/pentode switch.

The triode pentode/switching could not be done here since the two ECLL800 pentodes share the same screen grid.

ABOVE: All switches, connectors and pots mounted onto the chassis, together with the terminal strip and tube sockets, ready for soldering of resistors and capacitors and final wiring.

LEFT: The small cabinet is almost completely filled up with componentry.

BELOW: The amp's rear after re-tubing

DIY PROJECT: THE SCORPION STING

Body and chassis donor: Epiphone Snakepit 15G

Compared to the relatively small size of its cabinet (only an 8" speaker), the chassis of this solid-state amp is quite spacious, suitable for a small push-pull amp. There are two channels, Channel 1 (Overdrive) and Channel 2 (Clean), selectable by the Channel Select Button.

The clean channel only has Volume control, while the Overdrive channel has Gain, Volume, and Bass-Mid-Treble tone controls. There are also "Aux-in" and Headphones jacks.

The chosen topology

Channel 1 (Overdrive) will pass through an additional gain stage and the two preamp gain stages of the Clean channel. With the Channel Select switch in the Clean position, the input stage will be bypassed, and the signal from the input jack brought directly to the 2nd stage.

ABOVE: Epiphone Snakepit solid state donor amp

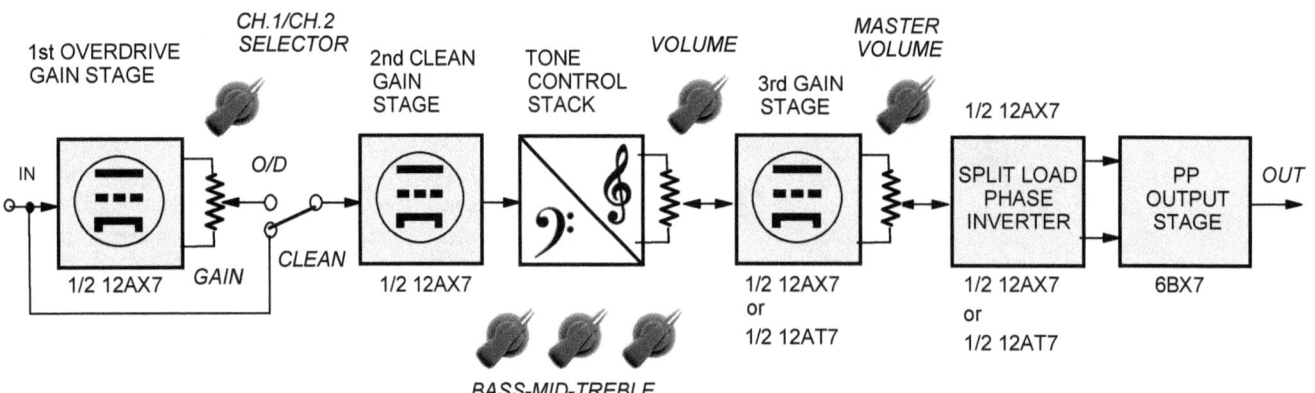

6BX7 duo-triode

The 10 Watt rating of each 6BX7 triode is promising for a low-power push-pull amplifier. The anode curves seem linear and equidistant, forecasting low to medium distortion. The good-looking tube is roughly the size of 6V6 and uses the standard octal socket. However, it isn't pin-compatible with any other octal tube, so there will be no substitutes if chosen for this design.

If we position the operating points at $V_A=250V$ and $I_A=25mA$, the required bias is only -21V. The current swing from A to B is 2*55mA=110mA and the voltage swing is 2*(250-75)= 350V, so the composite tube's load is $Z_{COMP}=350/0.11=3.18k\Omega$. The real anode-to-anode primary impedance is always four times higher than the composite figure, or $Z_P=4*Z_{COMP}=12k7$!

The maximum output power in class AB_1 is $P_{OUT} = \Delta I_A \Delta V_A/8 = 0.11*350/8 = 4.8$ Watts.

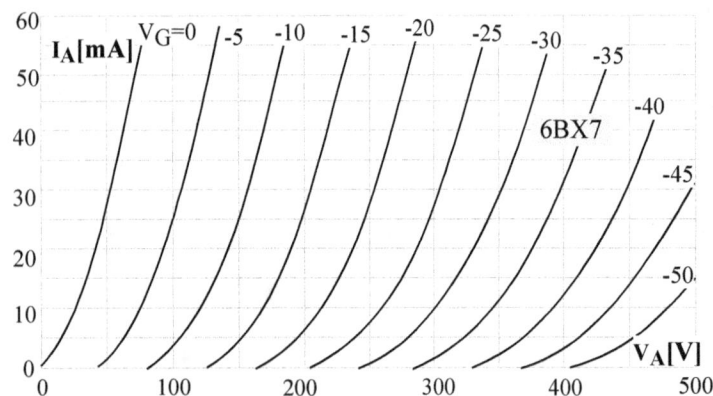

TUBE PROFILE: 6BX7GT

- Indirectly-heated twin triode
- Socket: octal, heater: 6.3V, 1.5A
- Anode dissipation: 10W each, 12W both
- $V_{HKMAX}= 200V$
- Typical operation:
- $V_A= 250V$, $V_G=-16.4V$, $I_0=42$ mA
- gm=7.6 mA/V, $\mu=10$, $r_I = 1,300\ \Omega$

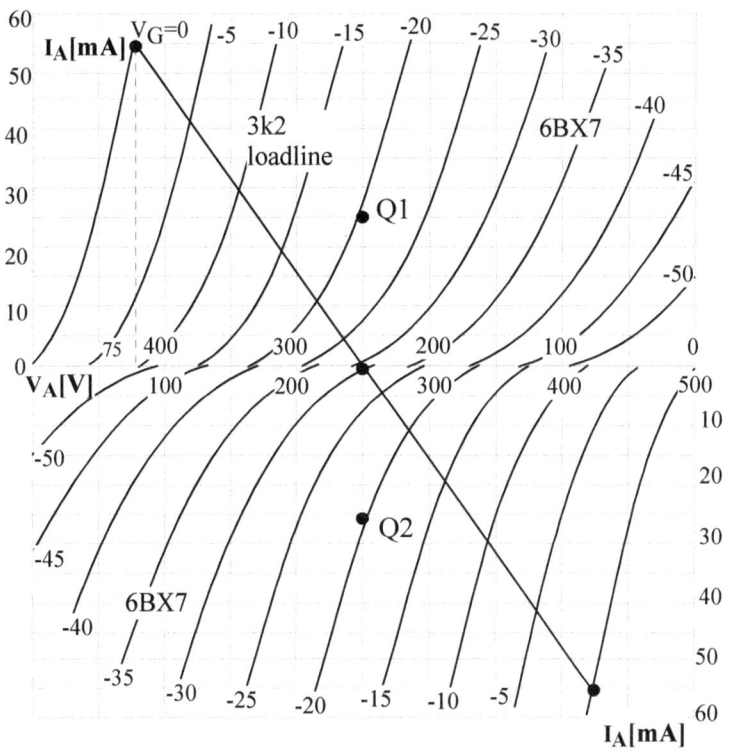

ABOVE: One of many starting points - 3k2 plate-to-plate load impedance and Q-points chosen so the amp operates in class A most of the time

ABOVE: 6BX7 is a good-looking, very linear and inexpensive octal duo-triode.

The good news is that we only need one tube, so it is a much cheaper proposition than trying to find and buy matched pairs or buying five or six tubes and trying to match them. NOS USA-made 6BX7 of vintage production, most commonly branded RCA, GE, and Sylvania, are available and inexpensive.

The circuit diagram

The clean channel has two amplification stages. Since its first stage has an un-bypassed cathode resistor, a local NFB that reduces distortion but also lowers its voltage gain, we increased the anode resistance from the usual 100k to 150k. That will partially compensate by increasing the gain somewhat. The overdrive stage does not need much gain, so the 110k anode resistor is fine.

In this design for V2 you can use 12AX7, or, if you need less gain but prefer a more robust and creamy sound, use 12AT7. The DC voltages in key nodes with 12AT7 tubes are marked in rounded frames.

Due to the low anode supply voltage of the power stage (only around $220V_{DC}$), the bias of the output tubes is also low, 12.3V, instead of 16-22 V_{DC}. The total cathode current is 12.3/390 = 31.5mA or 16mA per 6BX7 tube. With around 200V between anodes and cathodes, the idle power dissipation on each tube is 200*0.016 = 3.2 Watts. This is very low, meaning the 6BX7 will last a long, long time, but our maximum output power will also be low.

Our main culprit was the power supply. We didn't have a suitable power transformer at the time of construction, so we adopted some guerilla tactics, two mains transformers with low voltage secondaries connected back-to-back, as described elsewhere in this book.

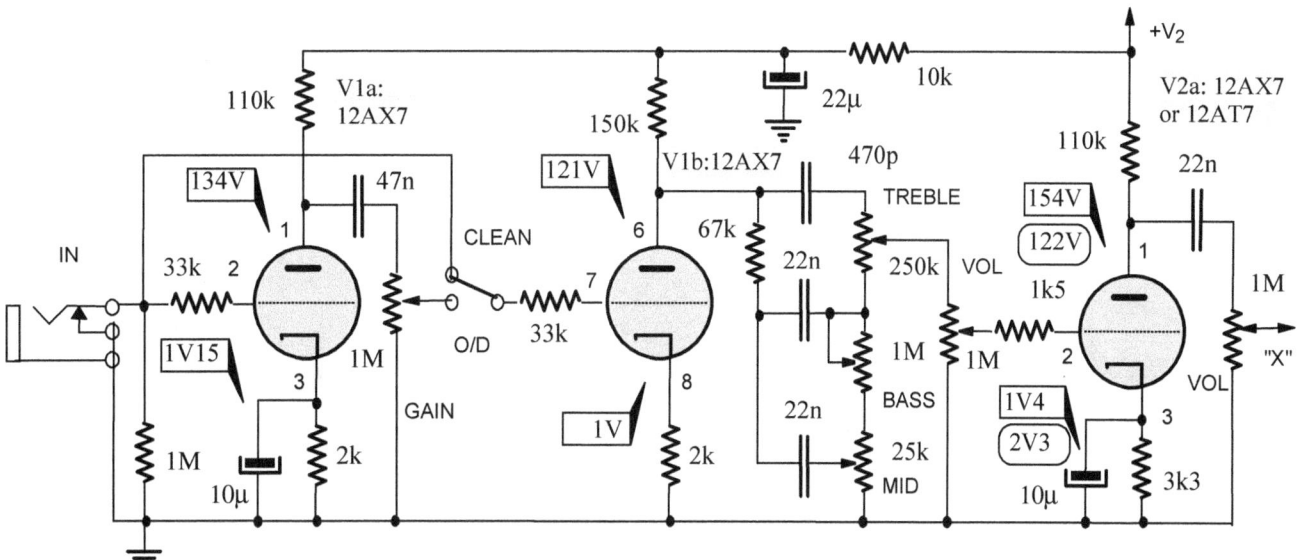

The larger transformer (TR1) had various secondary voltages, 5V, 6.3V, 7.5V, 9V, 12V, and 15V. Transformer TR2 (the smaller one) only had 7.5 and 15V CT secondaries (15V and 30V in effect), so we had no choice but to connect a 15V tap on TR1 to 7.5V CT secondaries of TR2. Should you wish to build this amp, choose a power transformer that will give you 320 to $350V_{DC}$ in point Y (the first elco).

We could have taken the anode supply for the power stage from point "Y", as is usually the case with push-pull amplifiers. The argument is that the relatively high AC ripple superimposed on the DC voltage at that point will cancel out due to the push-pull nature of the output stage.

That is true, but only at idle and only if the output triodes are perfectly balanced, meaning completely identical, which is seldom the case.

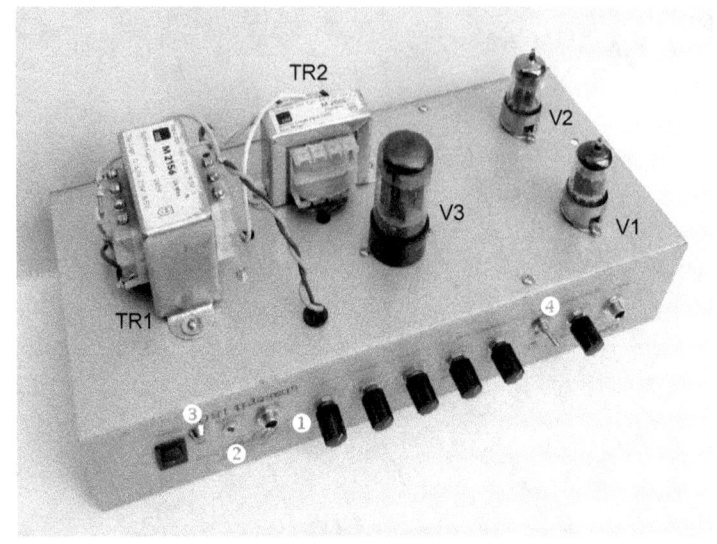

More importantly, as soon as the signal comes through, the output stage becomes dynamically unbalanced and the ripple mixes with the signal and creates all kinds of unwanted harmonic and intermodulation distortion artifacts.

By taking V1 after the first filtering stage (the other end of the 1k series resistor), a much cleaner DC supply, the unpleasant distortion is eliminated. The difference is relatively subtle, but noticeable by a trained ear!

The two mains transformers TR1 and TR2 don't have to be oriented 90° to each other; we've done it simply so the terminal strip and the V3's octal tube socket could comfortably fit without the need to push the transformers all the way to the edge of the chassis.

The AUX In socket was removed, its hole enlarged for the External Speaker socket (1); the Headphones socket remained (2), and instead of the PCB-mounted LED, a neon Power On indicator was substituted (3).

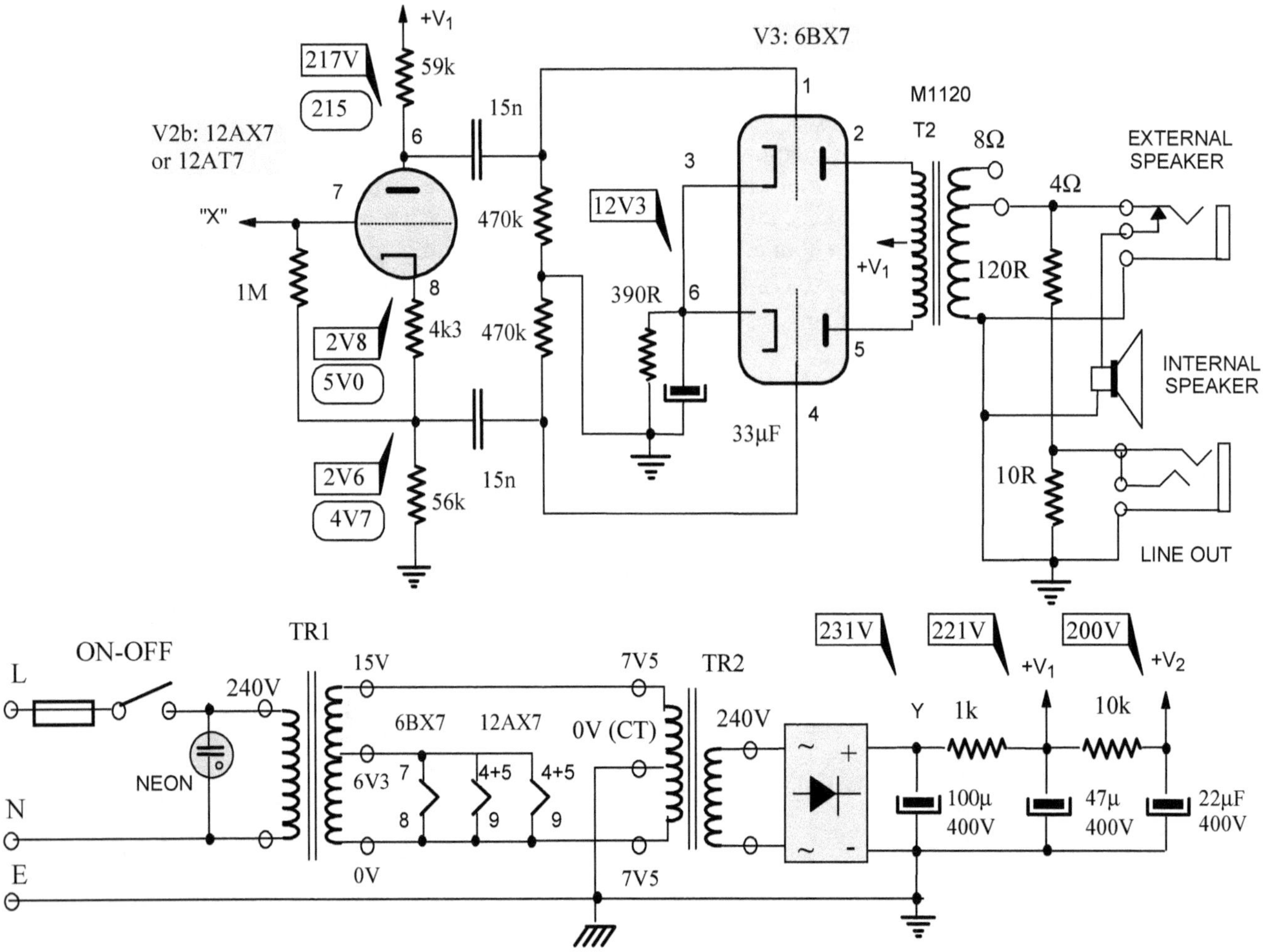

ABOVE: The plasticky Epiphone logo was screaming "no self-respecting *adult* guitar player would touch me" and was replaced with a chrome-plated metal scorpion emblem.

ABOVE: The back "cover" was nothing but an MDF frame and wasn't covering anything. A steel mesh was added to prevent contact with the amp's insides and to make the back look classier.

The channel selector push-button switch was also PCB mounted, so it was replaced by a toggle switch (4). At the back, an IEC power inlet with an integral fuse was added (5).

Since the terminal board had only 12 pairs of lugs, not all components could be accommodated. Since tube sockets were so close to the terminal board, a few resistors were wired in a true point-to-point fashion (6). The two voltage divider resistors for the headphones output (7) and two capacitors in the tone stack (8) were joined "in the air," but the mechanical integrity hasn't been compromised; there is no way they could move or short circuit anything!

The cathode resistor of the power stage and its bypass elco (9) are on the smaller power supply terminal board, which is also "chock-a-block", with not a single lug to spare.

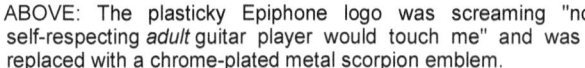

DIY PROJECT: THE FRONTMAN

6EM7 dissimilar duo-triode

6EM7 was designed for service in TV sets. The first triode (μ_1=68) was an oscillator working at the mains frequency of 60Hz, and the second triode (μ_2=5.4) was employed as a frame power amplifier, driving the vertical deflection plates of cathode ray tubes.

TUBE PROFILE: 6EM7
- High μ triode + low μ power triode
- Octal socket, heater: 6.3V, 925 mA
- TRIODE #1: P_{AMAX}=1.5W
- V_{AMAX}=330V, V_{HKMAX}=200V_{DC}
- μ=68, r_I=40k, gm= 1.6 mA/V
- TRIODE #2: P_{AMAX}=10W
- V_{AMAX}=330V
- μ=5.4, r_I=750Ω, gm=7.2 mA/V

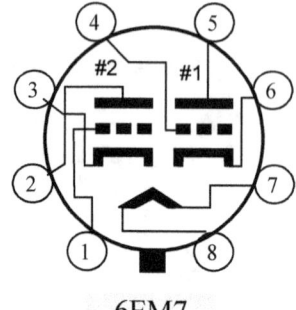

6EM7

Loadline analysis

Assuming a 350V_{DC} anode supply voltage and choosing a 180k anode resistance, the quiescent point needs to be at exactly 1mA of anode current and around -2.4V grid bias.

The cathode resistance must be $R_K = V_K/I_K$ = 2.4/1 = 2k4!

The voltage gain is $A_1 = -\Delta V_A/\Delta V_G$ = -(305-50)/4.8 = -255/4.8 = -53

The power stage

Since small output transformers of around 5kΩ primary impedance are the most common, let's position a 5k2 load line on the 6EM7 curves.

The -40V bias seems optimal, and since the idle current is 37mA, the cathode resistance needs to be $R_K = V_K/I_K$ = 40/37 = 1k1, so use a 1k2 standard resistor. It will dissipate 1.65 Watts of heat, so a 5 Watt rated resistor would be perfect.

With V_{MIN}=50V and V_{MAX}=385V we have ΔV=335V and ΔI=70-7=63mA, so the output power will be around $P_{OUT} = \Delta V \Delta I/8$ = 335*0.063/8 = 2.6W, a perfect level for practice and recording.

The anode dissipation is $P_{IN}=I_0 V_0$ =0.037*230 =8.5W, perfect for a 10W triode.

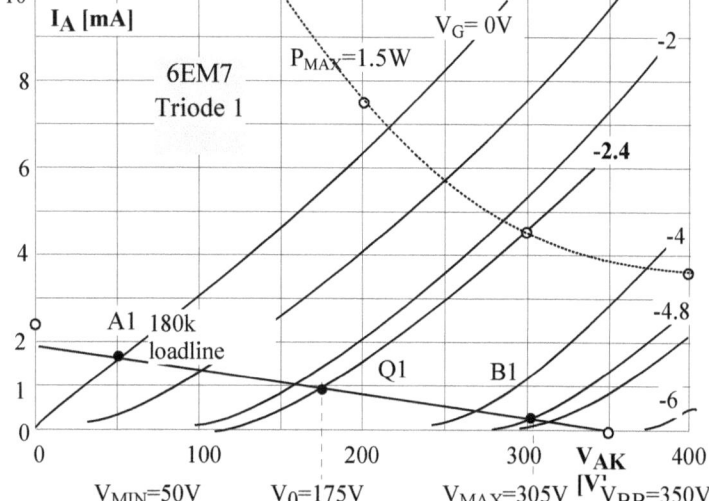

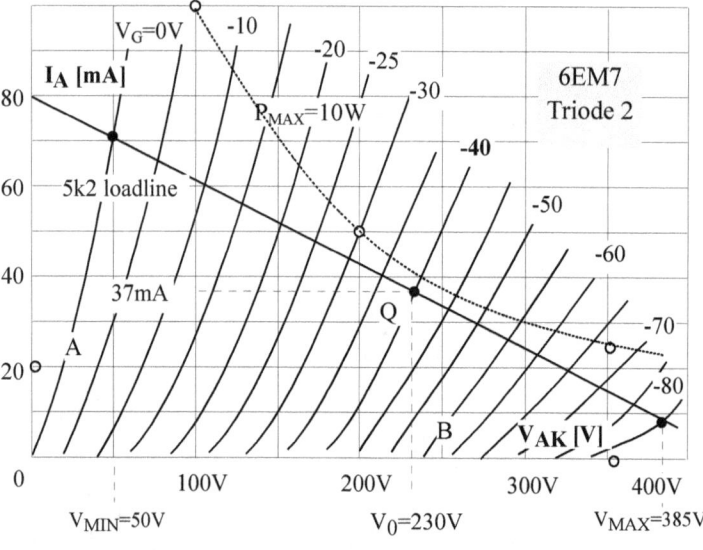

The body donor: Fender Frontman 15G

This small SS combo amp isn't the easiest platform for conversion; its chassis is of an awkward shape, and the internal space is minimal. However, we will only have two small tubes and a tiny output transformer, which should be manageable. The Bass, Middle, and Treble tone controls require the inclusion of a classic Fender tone stack.

The topology

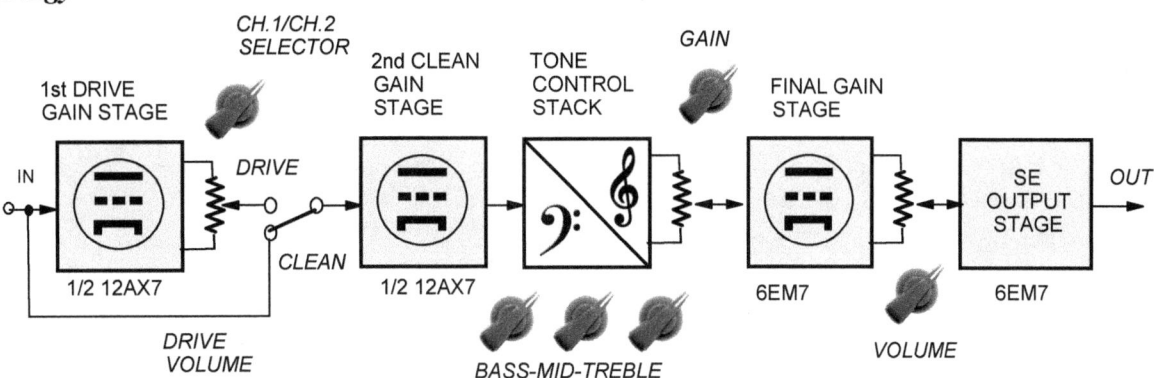

Since the final gain stage, using 6EM7's high mju triode, will have a voltage gain of up to 50, the 2nd clean gain stage with a similar gain (using one half of 12AX7) will be sufficient. The *Drive* gain stage will be switched in and out of the circuit and does not need a very high gain. Plus, its lower gain will avoid sudden jumps in loudness levels when the *Clean/Drive* switch is operated.

Circuit diagram

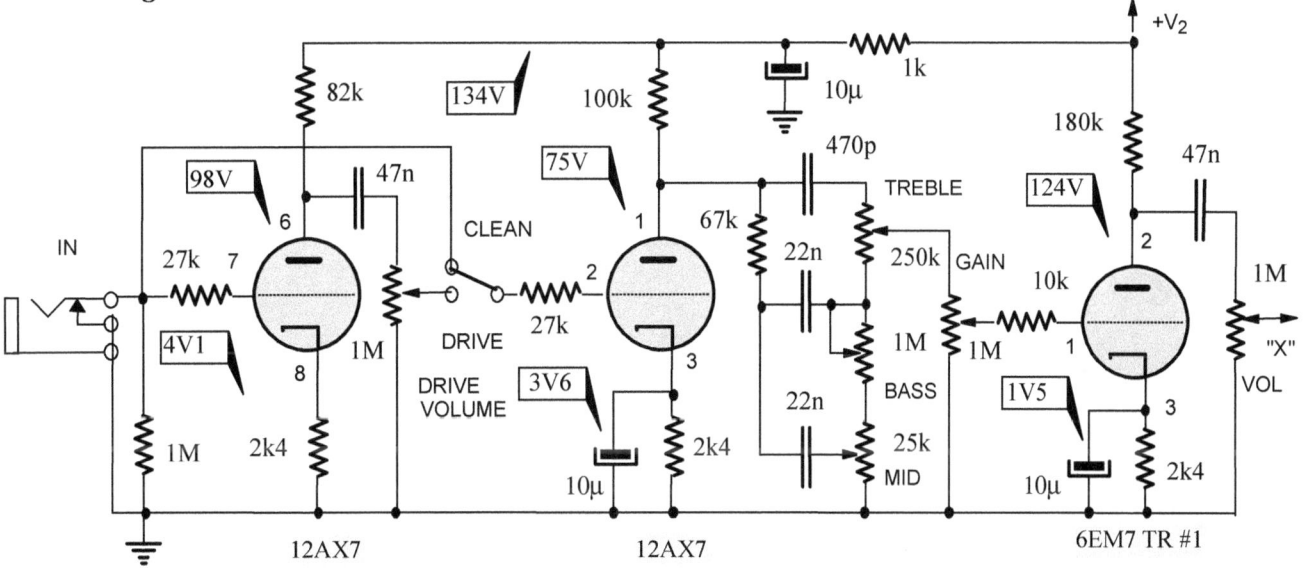

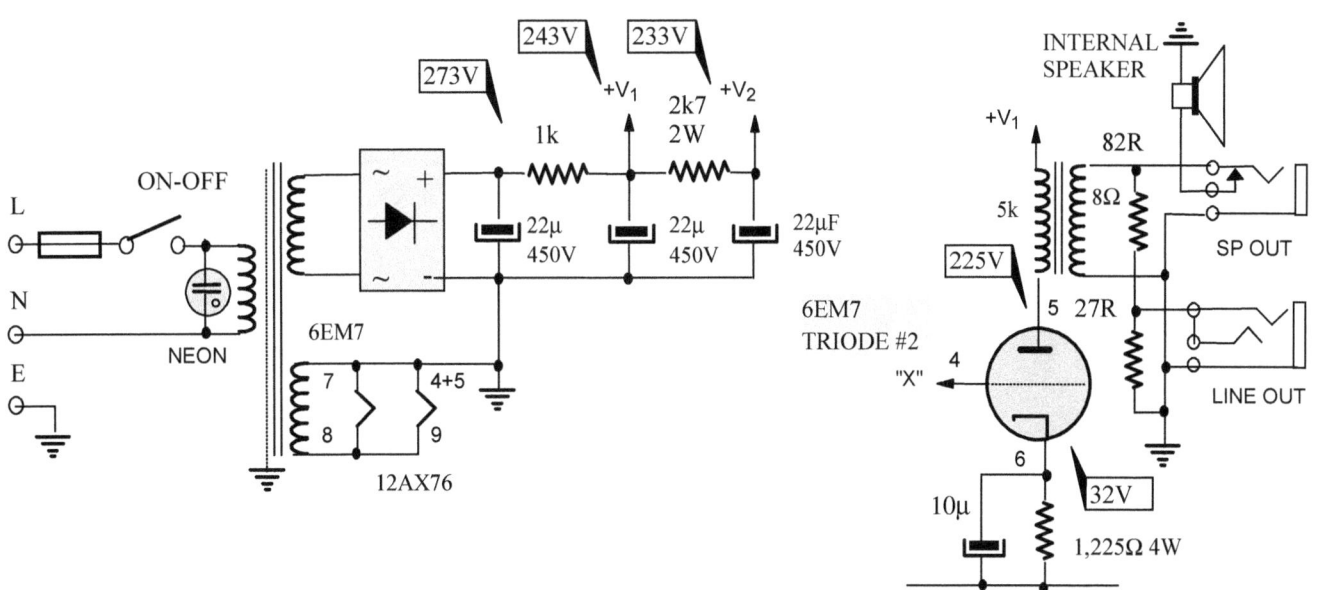

The 1,225Ω cathode resistor of the output stage (next page) was made up of two 2,450Ω 4W low inductance resistors with gold-plated leads. All three stages use them as cathode resistors, resulting in relatively high bias values for the first two stages of 4.1V and 3.6V, respectively, meaning their anode currents are 1.7 and 1.5 mA. High current (hot biased) preamp (and power) stages sound better!

With 39V on the cathodes of output tubes, the current through the 390Ω cathode resistor is exactly 100mA. The voltage drop of 1V on 560Ω screen grid resistors means the screen current of each tube is 1/560= 1.8mA, so at idle the anode current of each tube is 50-1.8 = 48mA.

The anode-to-cathode voltage across the 6Em7 power triode is 225-32=193V and the anode current is $I_A = V_K/R_K$ = 32/1,225+26mA. The heat that the anode must dissipate is P=0.026*193 = 5 Watts. This is very conservative for a 10 Watt rated triode; these lovelies will have an easy and stress-free life.

The wiring

The power transformer and the power supply terminal board are bolted onto the bottom of the cabinet, so no unsightly add-on screws or bolts can be seen from the outside (1). The power transformer leads were not long enough, so we had to use a terminal strip (2). The mains and heater AC wiring are bundled together (3). The preamp tube is mounted as close as possible to the guitar input jack (5).

The output transformer (6) is right next to the 6EM7 tube, and the added External Speaker jack (7) is right in front of it, next to the Line Out jack (8), thus minimizing the output stage and speaker wiring.

The back cover was cut with a jigsaw, and a small steel grille was added to allow some air circulation (9).

The playability and listening impressions

Initially, the cathode resistors of the input stage (only active in Drive mode) were bypassed by a 10μF capacitor. However, there was too much overall gain with three cascaded stages in Drive mode; at high Drive Volume levels, the amp started to howl (oscillate).

To reduce gain, the bypass elco was removed, and the anode resistor of the first stage was reduced from 100k to 82k. That brought the gain right into the desired range. The little amp sounded more like a baby Marshall than a Fender, but we kept the Fender emblem to add a bit of harmless confusion to the fore.

DIY PROJECTS: ULTRA-SMALL AMPS

- COMMERCIAL BENCHMARKS: FLEA-POWERED AMPS
- DIY PROJECT: DIRTY THIRTY TUBED
- DIY PROJECT: EL COMANDANTE
- DIY PROJECT: TUBEFINDER

Someone once said, "All the world needs now is a great sounding 1 Watt amp". With a typical guitar speaker of 93-97 dB/Watt sensitivity, a one-watter would produce a deafening sound indeed. Due to its small dimensions and low weight, a low-powered amp would also be great for recording and transport.

Finally, and perhaps most importantly, while high power amps achieve that desirable distortion of the output stage only at very loud and thus impractical levels, the low powered amp's power tubes would distort at bedroom or studio levels, a huge benefit to many a guitarist.

In this section, we look at a few commercial flea-powered amps as benchmarks and then proceed to design and build three great-sounding baby amps, all capable of delivering between one and two Watts of the raunchiest tube power ever.

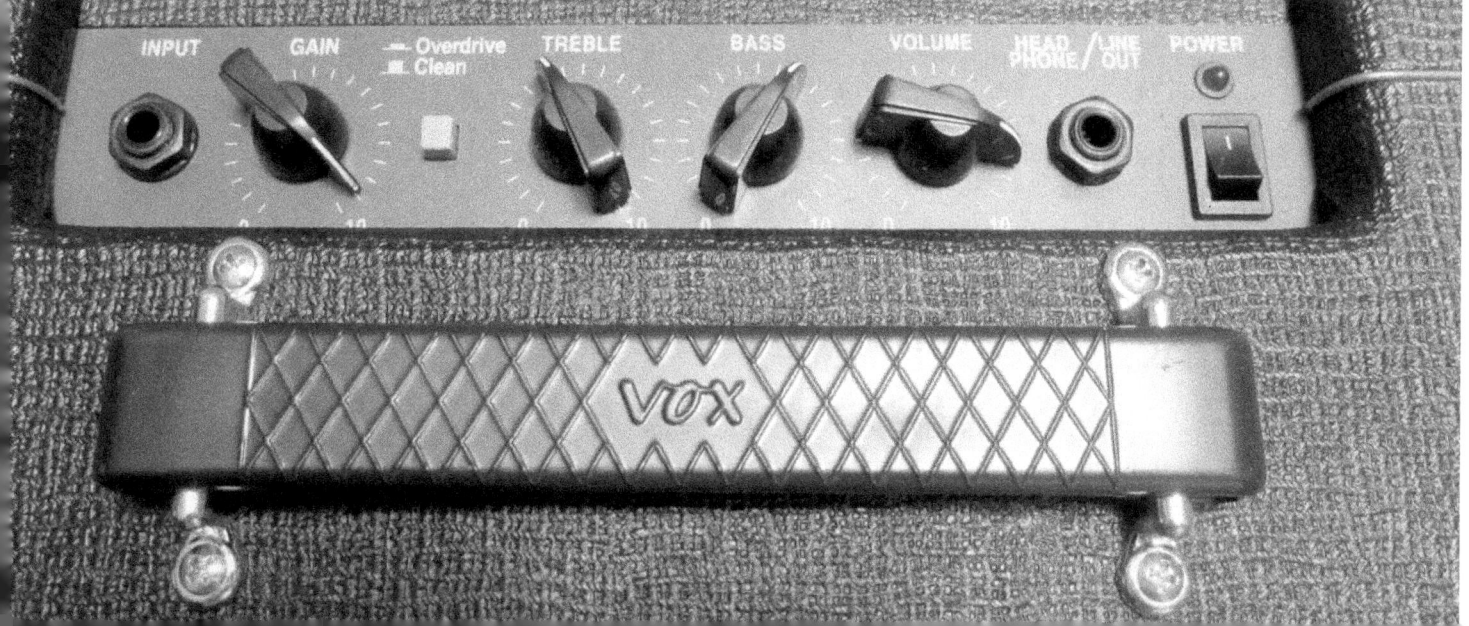

Somebody relatively famous in hi-fi circles once said that all the world needed was a great 1-Watt amplifier. He talked about highly efficient horn speakers, which would produce 100+ dB of sound pressure with a 1-watt input. Guitar speakers do not quite reach 100dB SPL but come pretty darn close; 93-96dB/W sensitivity levels are pretty common). There has recently been a trend of using microamps (power levels between half and one Watt) for practice and recording.

COMMERCIAL BENCHMARKS: FLEA-POWERED AMPS

Marshall JTM-1C

This classic-looking but brazenly expensive 1Watt combo with a 10" Celestion speaker used two ECC83 (12AX7) preamp tubes and one ECC82 (12AU7) duo-triode in the output stage. It featured loudness (volume) and tone controls, and, as if one watt was too loud, the low power switch brought the output power down to 0.1 Watts!

Most of the componentry (including the output transformer) was on a tiny PCB. The retail price was very steep £549 in the UK (€815 in Europe and street price of US$799 in America), probably the main reason the amp soon went out of production. There was a head version as well, JTM-1H, which retailed in Europe at €700!

Biyang (Wangs) VT-1H

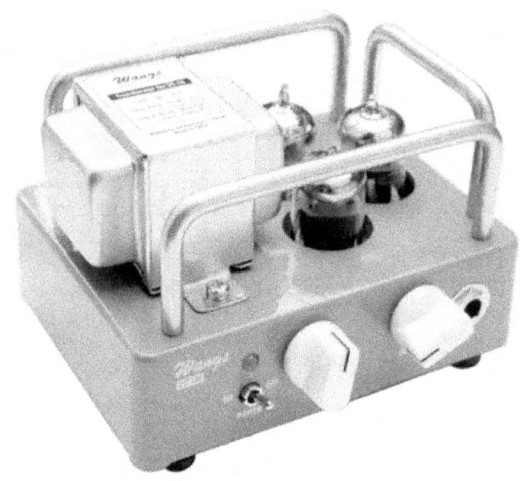

This Chinese-made 1 Watt head comes under two names, Biyang and Wangs. It measures tiny 13x8.5x10 cm and weighs 1.5kg. The same tube complement as JTM-1.

Since the specs say that preamp tubes are 1.5×12AX7 and that power section tubes are 0.5×12AX7 and 1×12AU7, that would indicate three stages of voltage amplification, a phase inverter as a 4th stage and a push-pull stage with 12AU7.

The fuse is soldered to the tiny PCB, and the amp is so cramped that mods and additions are impossible. The retail price of around US$180 seems attractive, but if you opt for the DIY route, that money can buy your much better parts for a design with more features and more power.

Z.Vex Nano

If you thought the Biyang amp was small, Nano head by Z.Vex Amps is even tinier. It uses two sub-miniature 6021W dual-triodes and produces a maximum of 0.1 Watt of "clean" power and 0.5W of distorted power. An external power supply of 12V and at least 1.5A capability is needed. An internal DC-DC converter provides the 230VDC anode voltage.

The array of controls is quite extensive for such a microamp. Apart from Volume control, there is a "Bright" switch, a "Thickness" switch with Normal, Thin, and Fat positions, and a "Mellow" switch, whose positions are Mellow 2, Normal and Mellow 1, with "Normal" being the brightest voicing.

There is even a fan that keeps the overcrowded internals cool. The price? Incredibly expensive US$519! I guess those four chrome handles must cost a lot ...

Mooer Little Monster

Mooer Little Monster is a 5 W guitar head with an above-average array of controls for such a tiny amp: "Normal/Bright" channel selection, "Thin/Mellow" tone control, gain and volume controls, plus 8W and 16W outputs. Tube complement includes 1xECC83, 1xECC81, and 1x 6V6GT. In 2016, this British-made all-tube head sold in the UK for £179.99 and in the US for $249.-

Blackstar HT1

British-made HT1 and HT1R (with digital reverb) are 1 Watt hybrid amps with nine dual op-amp integrated circuits. The reverb version also has a special reverb IC. Tube complement includes 1xECC83 and 1xECC82. In 2016 HT1 combo sold on Amazon USA for US$249.- while HT1R combo had a street price of US$329.-

The first dual op-amp is used in the first two input stages, which are followed by two cascaded ECC83 stages, and another op-amp stage driving the tone stack with Blackstar's patented ISF control (Infinite Shape Feature). More on ISF in the "Tone Controls" chapter in Volume 1 of this book.

The driver stage is a MOSFET push-pull arrangement with a BC184 bipolar transistor in the common source circuit as a CCS (Constant Current Sink). Thus, the source and drain currents are kept constant, maximizing the stage's gain and minimizing distortion.

This is a common approach in hi-fi amplifiers; without a detailed on-the-bench test of this amp, it is unclear why the designer went through all this trouble and why this stage was necessary at all.

Cleverly, the LED in the base circuit is used as an indication that the power is on.

The 22nF capacitors couple drains of the driver MOSFETS with the control grids of the output ECC82 duo triode. The output stage is cathode-biased.

BELOW: HT1 output stage © Blackstar Amplification

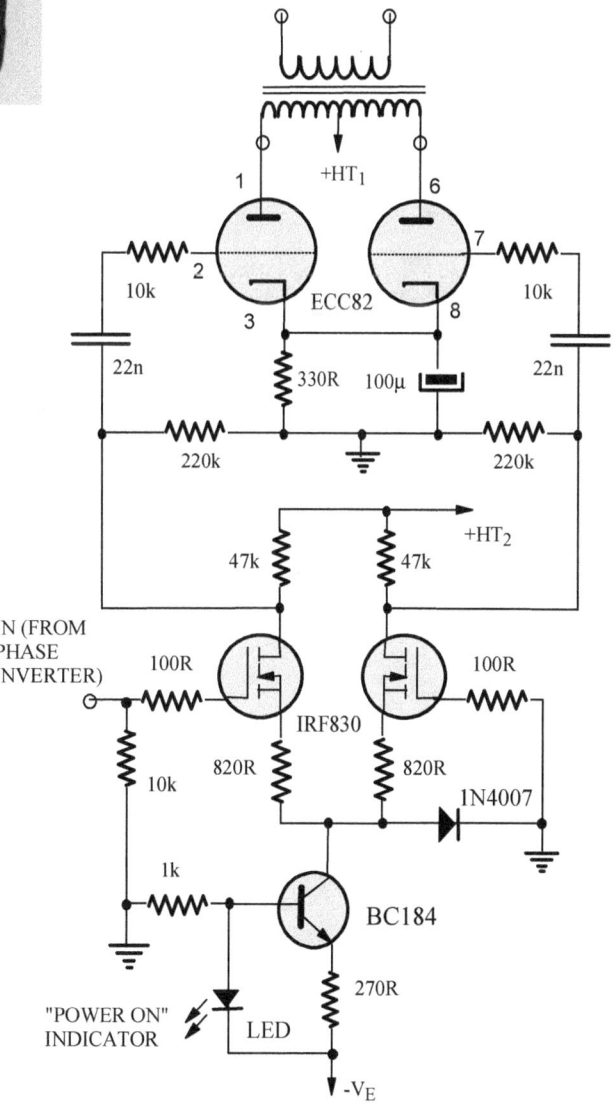

Mahaffay HiLo Watt Plexi

This tiny (8"x8"x7" and only 10 lb. of weight) hybrid is made in the USA and lists for US$449.- The design goal was to emulate the distorted sound of a 50 Watt Marshall Plexi amp cranked to the maximum but run at a reduced mains (line) voltage of only 80VAC (through a variable transformer). Such early and severe distortion but at much lower loudness levels has been the design goal of many amp designs, but how many (if any) achieve it?

The input stages use the low gain 12AU7 duo-triode, followed by 12AX7. 33 Watts into an 8W load from the solid-state output stage is claimed. The soldering is sloppy; most joints aren't done on terminal strips but simply "in the air." Insulation sleeves aren't used either; long component leads are laid bare.

Blackheart Killer Ant BH1H

Two 12AX7 duo-triodes are used in this PCB-based little amp, three triodes in the preamp stages and one in the SE output stage, producing 0.25 Watts. The first gain stage is followed by a volume control pot. The 2nd common cathode stage is DC-coupled to a cathode follower, which drives the fixed tone stack and the output stage. Interestingly, there are 4, 8, and 16Ω speaker outputs, but only Volume control, no tone controls of any kind.

With 12AT7, the maximum output power should double to 0.5 Watts, and with 12AU7 to almost 1 Watt, so plenty of tone and power tweak options here.

Fender Greta

Two 12AX7 preamp stages and a single-ended parallel 12AT7 output stage deliver up to two Watts of triode power. Incapable of handling even that power level, the 4" toy speaker lets the amp down. It breaks up very early and does not sound appealing even before it starts struggling.

The insides look better, with decent transformers and a power supply filtering choke, which isn't really needed in low power class A amps, whose anode power draws are constant.

The VU meter adds a nice touch but still does not compensate for the amp's daggy looks. Fender should have spent those few bucks on a larger (5-6") speaker and a sturdier cabinet with a carrying handle.

Its street price off around US$199.- was too high, especially considering that for only a hundred bucks more, one could have bought Fender Excelsior, Greta's bigger brother, and a respectable amp in every sense.

VOX Lil' Night Train

The first 12AX7 duo-triode is used in the two input gain stages, followed by the second 12AX7 as a phase inverter and a 12AU7 in a push-pull output stage. 2 Watts of class AB power with a 16Ω load or 1.5 Watts with an 8Ω speaker is claimed.

Apart from the Gain, Volume, Treble, and Bass controls, the "Bright / Thick" switch changes the link between the first two stages. In the "Bright" position, the output of the 1st stage is routed through the "Top Boost" tone control circuit, while "Thick" switches the signal directly to the second gain stage.

The retail price was US$400, but in 2016 the "certified refurbished" heads sold for US$199.- on Amazon.

DIY PROJECTS: ULTRA-SMALL AMPS

COMMERCIAL BENCHMARK: Randall Diavlo RD1H

This 1-Watter head looked cute and we haven't had any Randall amps in this book, so let's have a closer look at it. In 2016 its street price was around US$250.-

A modular design is adopted, presumably to save on design and production costs. The circuit is spread over four PCBs.

The preamp/driver circuit board is shared with two other models, RD5H head and RD5-112 combo amp. Both of these use a single-ended 6V6 producing around 5 Watts of output power.

There are two Noval tube sockets for 12AX7 duo-triodes (1) and two integrated circuits (2), so the preamp circuit is of a hybrid kind.

Two lots of four discrete diodes provide rectification (3), one for the high voltage, the other for heater voltage. A 3-terminal IC (mounted on a heatsink) regulates the DC voltage for ICs (4).

Some components are marked on the board but were not installed; perhaps they are for the 5 Watt model only (5).

The thin PCB (6) carries only three red LEDs that provide the devilish glow inside the chassis.

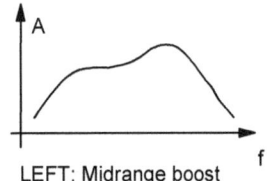
LEFT: Midrange boost

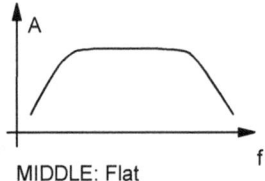
MIDDLE: Flat

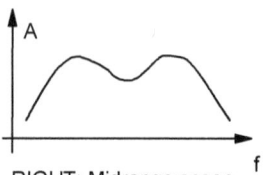
RIGHT: Midrange scoop

ABOVE: The 3-way Tone Stack Shift switch is a preset with three midrange voicing choices.

The 12AU7 output tube is on its own board (7), which gives you the flexibility to install a different tube on a different socket, should you wish to go through such trouble. For instance, an octal socket would enable you to use 6SN7 duo-triode or 6BX7. In stock form, you can replace 12AU7 with 12BH7 for slightly more power.

All output connections (Send, Return, Speaker Out, and Speaker Emulator) are on a separate PCB (8).

The 3-way Tone Stack Shift switch is preset with three midrange voicing choices. In its left position, it boosts the midrange, with flat bass and rolled-off treble. The middle position is the stock configuration with a flat frequency response curve, and the right position is a midrange scoop or attenuation, with boosted bass and treble.

The mains transformer had only a 120VAC primary, so a step-down transformer was needed for 220 or 240VAC mains countries; luckily, there was plenty of room next to the power transformer.

Unfortunately, we could not find the amp's circuit diagram online, and to trace its complicated circuit via hard-to-reach components would take many hours, so our analysis will stop here.

We measured DC voltages at tube pins using our Noval jig. The cathodes of 12AU7 output tubes are at 10.5V (the photo below); one anode is at 280.0V, and the other at 282.5V, indicating unequal resistances between the two halves of the output transformer's primary winding. The anode supply voltage of the power stage is 289V.

The anode voltages of stages 1 to 4 are $191V_{DC}$, $181V_{DC}$, $199V_{DC}$ and $163V_{DC}$, while their cathodes are at $1.45V_{DC}$, $1.55V_{DC}$, $2.17V_{DC}$ and $0.9V_{DC}$ respectively. The heater voltage is $6.12V_{DC}$.

However, despite DC heating and the use of tube shields, the hum at the amplifier's output (with both Gain and Volume at maximum and no guitar plugged in) was an incredibly high 1.5VRMS! Without a guitar plugged in, the input is shorted to the ground. Thus, that hum voltage is generated by the amplifier itself.

Furthermore, when the Tone control is turned CW, a loud high-frequency hiss appears, even more irritating than the low-frequency hum just mentioned. All in all, a complete disappointment of an amplifier!

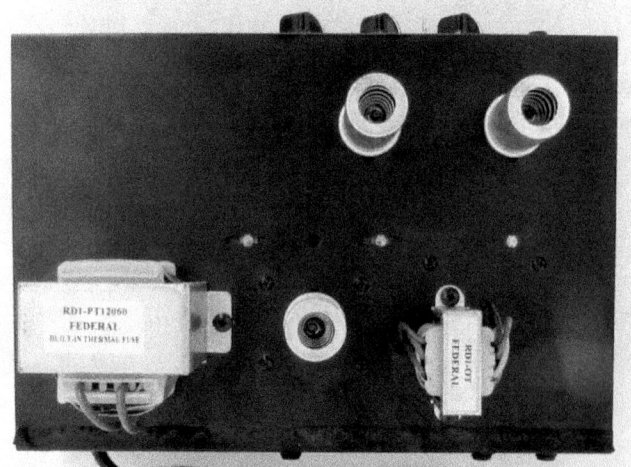

RANDALL DIAVLO RD1H

- 2x 12AX7, 1x12AU7 tube
- Solid state rectification
- PCB construction
- Output power: 1 Watt
- Controls: Tone, Gain, Volume
- 3-way switchable tone stack shift
- FX loop, speaker-emulated XLR output with dummy load
- Dimensions: 310x220x176 mm
- Weight: 5.5 kg

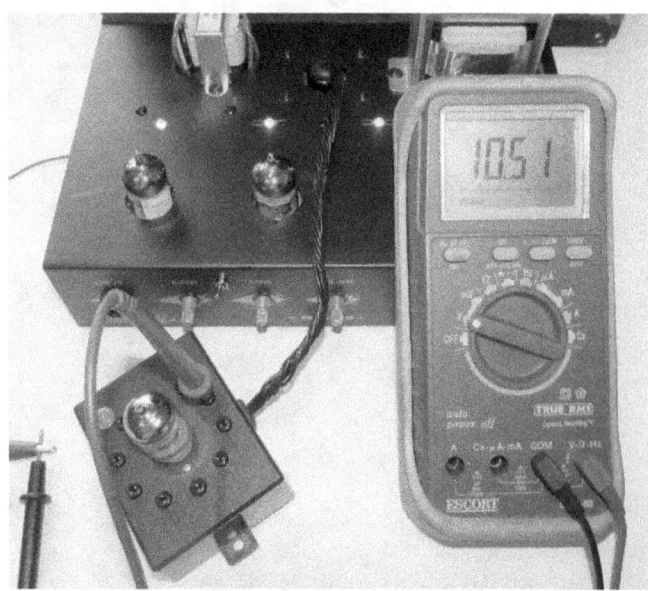

ABOVE: Measuring DC voltage on 12AU7 output tubes' cathodes using our DIY Noval socket adapter.

DIY PROJECT: DIRTY THIRTY TUBED

Now that we have seen what your hard-earned money gets you from commercial amp makers, let's design and build an equally- or even better-sounding amp for much less money. Power reduction is not a necessary function, nor is the reverb, which would overcomplicate the design. It is next to impossible to fit more than two tubes in these small spaces.

Cabinet donor: Danelectro Dirty Thirty

We bought this tiny solid-state amp from a local shop for next to nothing. A pair of them just sat there on the floor, dusty and forgotten for seven years! A few months later, we bought the other one as well.

The enclosure is perfect for a flea-powered tube amp. The speaker is tiny, of course, but it will do for a 1 Watt bedroom practice amp. Just in case, we will include an external speaker jack. Since 1 Watt output into an 8-ohm speaker means the signal voltage is 2.83V_{RMS} ($P=V^2/R = 2.83^2/8 = 8/8 = 1$ Watt), the same jack can be used as a line or headphone output.

Candidates for the output tube

If paralleled or in a push-pull arrangement, many preamp triodes can handle 1-2 Watts of output power. Some of the choices are 12AU7, 6SN7, 5687, 6CG7, and 12BH7.

Audiophiles love 6SN7's creamy and seductive sonics. 6CG7 is its Noval (9-pin) version.

The anodes in 6SN7 are rated at 5 Watts each, or 7.5 W if both are paralleled, so we will draw the maximum dissipation curve for that power limit (next page).

The anode current in our chosen Q-point is 20mA, and since the bias is -12V, the cathode needs to be at +12V. These two numbers and Ohm's Law give us the value of the cathode resistor $R_K = V_K/I_0 = 12/0.02 = 600\Omega$, so a 590R standard value would be perfect!

The power dissipation on that resistor will be $P=V_K*I_A= 12V*0.03A = 0.36$ Watts, so a 1 Watt-rated resistor would be fine.

ABOVE: Danelectro Dirty Thirty looks cute & classy in two tone brown and cream.

BELOW: Top panel stripped, PCB and power transformer removed, the chassis ready for tubing.

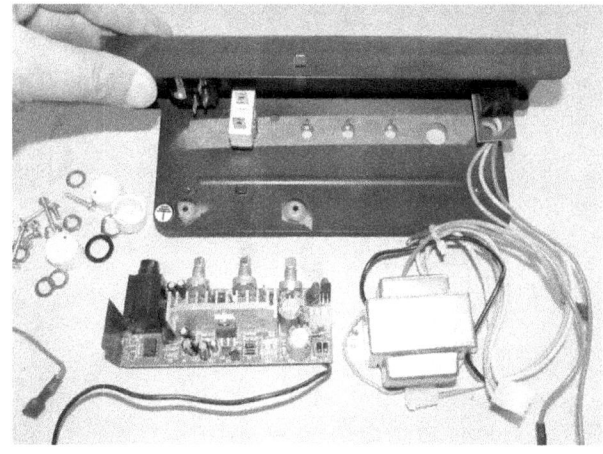

LEFT: 6SN7 & 12SN7 are good-looking, robust and great-sounding octal duo-triodes.

TUBE PROFILE: 6SN7GT - 12SN7GT

- Indirectly-heated octal duo-triode
- Heater: 6.3V, 0.6 A - 12.6V, 0.3A
- Maximum anode voltage 450 V_{DC}
- I_{AMAX} = 20 mA, V_{HKMAX} = 100 V_{DC}
- P_{AMAX}: 5 W each triode, 7.5 W both
- TYPICAL OPERATION:
- V_A=250V, V_G=-8.0 V, I_0=9 mA
- gm=2.6 mA/V, µ=20, r_I = 7.7 kΩ

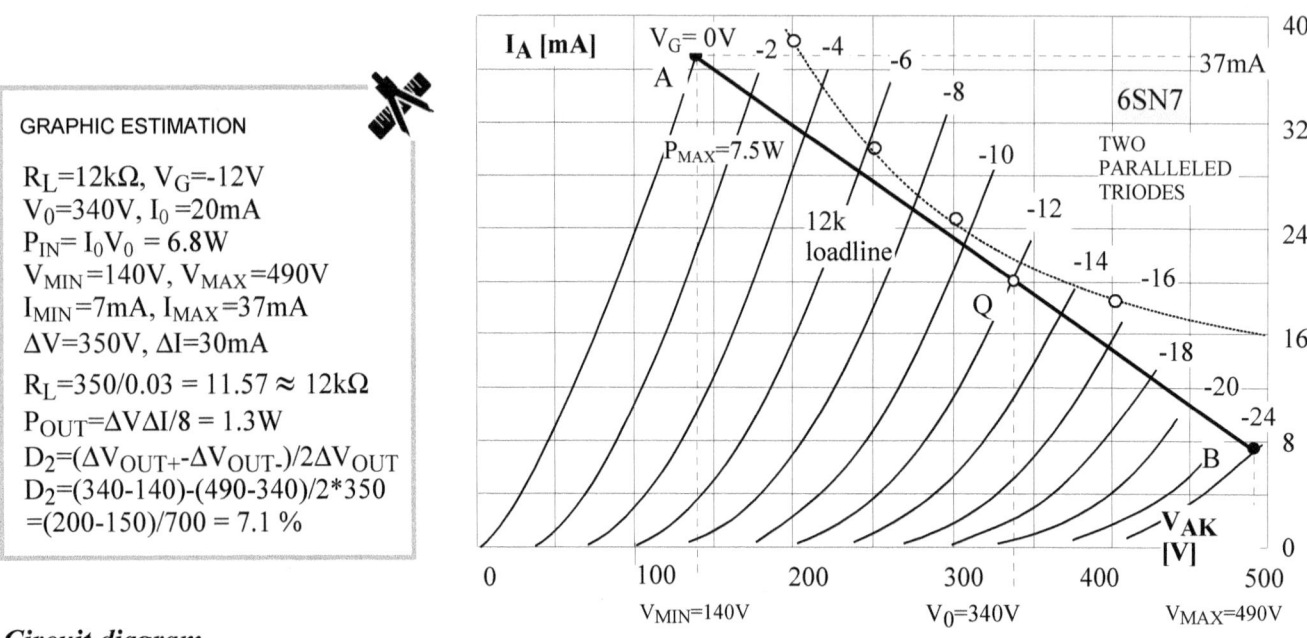

GRAPHIC ESTIMATION

$R_L = 12\,k\Omega$, $V_G = -12V$
$V_0 = 340V$, $I_0 = 20mA$
$P_{IN} = I_0 V_0 = 6.8W$
$V_{MIN} = 140V$, $V_{MAX} = 490V$
$I_{MIN} = 7mA$, $I_{MAX} = 37mA$
$\Delta V = 350V$, $\Delta I = 30mA$
$R_L = 350/0.03 = 11.57 \approx 12\,k\Omega$
$P_{OUT} = \Delta V \Delta I / 8 = 1.3W$
$D_2 = (\Delta V_{OUT+} - \Delta V_{OUT-})/2\Delta V_{OUT}$
$D_2 = (340-140)-(490-340)/2*350$
$= (200-150)/700 = 7.1\%$

Circuit diagram

Danelectro Dirty Thirty has three controls, Dirty Sweet, Level, and Tone, so our preamp will be simple, two 12AX7 gain stages driving two 6SN7 triodes (one physical tube) in parallel. The Dirty Sweet will become a Gain control, and the Level will be the Master Volume.

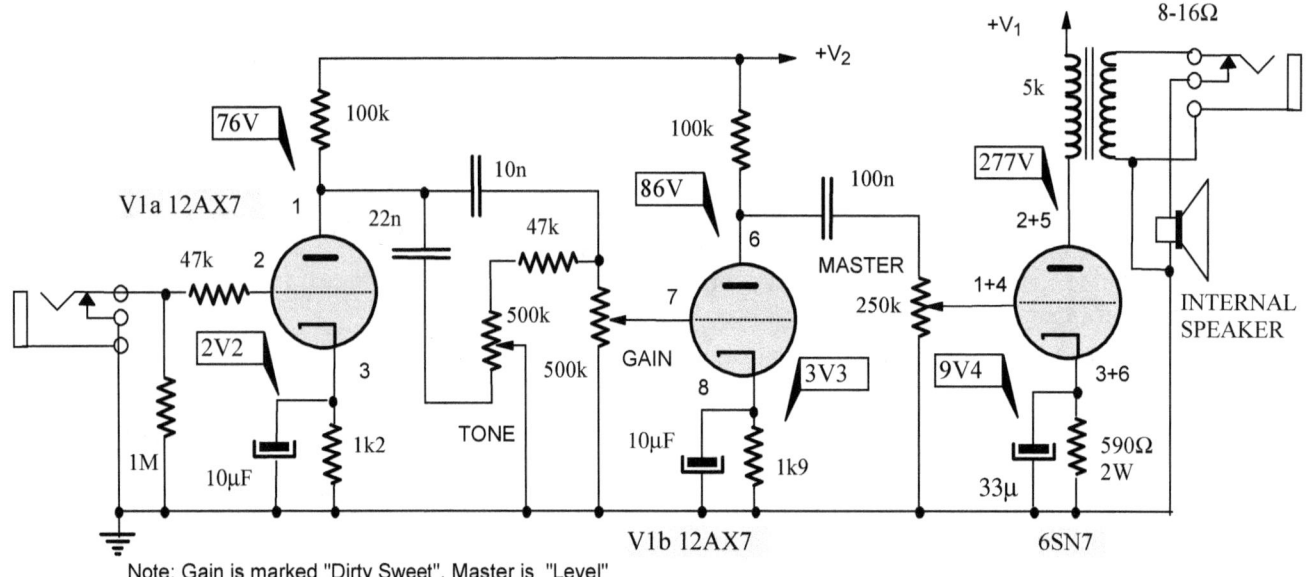

Note: Gain is marked "Dirty Sweet", Master is "Level"

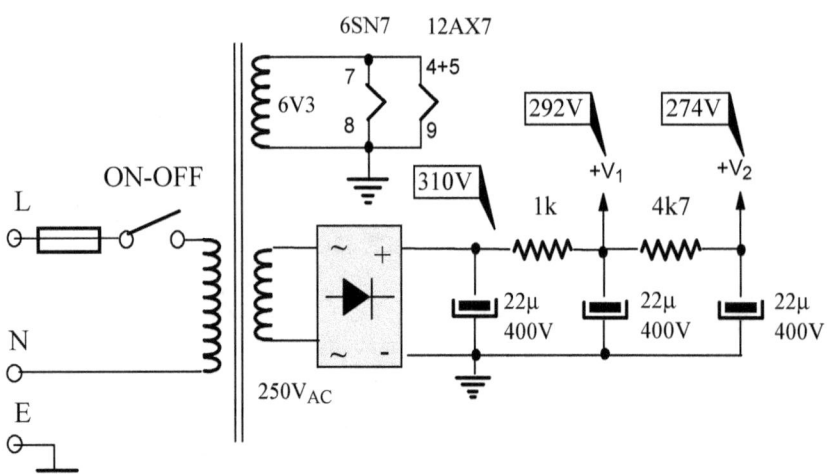

Sonic impressions

Just as in hi-fi, the 6SN7 lovelies did not disappoint as power tubes in this guitar amp application. The amp sounded much bigger and "confident" than its size would foreshadow, even through its tiny internal speaker.

With a large 12" external speaker cabinet, the bass and midrange expanded, the sound became fatter, full-bodied, and truly amazing! Sonically, the unnecessarily complex Randall Diavlo RD1H did not even come close to this little giant.

DIY PROJECT: EL COMANDANTE

Here is another option if you don't want to mess around converting small solid-state combos to tubes. Bought for a few bucks at surplus electronics sale, this small steel instrumentation case was designed for a test instrument to be built inside. The front, back, top, and bottom panels are removable, providing easy access.

There wasn't enough room to fit a large power transformer, so the amp was powered from a $12V_{AC}$ 1A power pack.

To keep things simple, let's stick to the omnipresent 12AX7 duo-triode as a preamp tube and choose its low μ cousin, 12AU7 duo-triode as the output tube.

This metal Che Guevara badge fits nicely between the two tubes and gave this mini head its name.

Topology

As we will soon ascertain, the 12AU7 triode needs to be biased at around -11V; its sensitivity is in the same ballpark as that of power pentodes and beam tubes of 6V6 and EL84 variety, so we will also need two voltage amplification stages to drive it.

Since a push-pull arrangement of output triodes would require adding a third tube, we will strap the two 12AU7 triodes in parallel single-ended triode arrangement (PSET). Each of the two triodes in 12AU7 has the anode power dissipation of 2.75 Watts, with two in parallel rated at 5 Watts, so they easily provide more than 1 Watt output power in a parallel single-ended circuit.

The good news is that another beauty, namely 12BH7, is pin-compatible with 12AU7. 12BH7's heater draws twice as much current than 12AU7 (600 mA and 300mA respectively at 6.3V) so we must ensure that the power transformer can accommodate the additional load on its heater winding. Each anode in 12BH7 is rated at 3.5 Watts, so with 7W on tap, if paralleled, we would get a maximum of around 2 Watts of audio power in a PSE output stage.

Now we need to estimate the power draw. Each preamp stage will draw less than 2mA of anode current, and the output stage will draw around 16mA in idle mode (we will go through detailed analysis soon), so 20mA total. At say 260V of anode voltage, that is 260*0.02= 5.2 Watts of anode power. In addition, at 12V, the two heaters will draw 0.15mA each or 300mA total (4 Watts), for a total of 10 Watts load on the transformer.

We will power the heaters directly from the power pack's output. Instead of the nominal 12.6V, we may have 12.0V, but that is fine; slightly underheated preamp tubes are quieter (less noise), and the output tubes won't be affected. But, how do we get the $250\text{-}300V_{DC}$ needed for the anode supply?

We had a stash of tiny surplus mains transformers, 240V input and 12.6V output at 500mA. If we connect one backward to a 12V input (the input will be at its secondary winding), on its primary we will get roughly $240V_{AC}$ which we will then rectify and filter to get $260\text{-}300V_{DC}$!

However, its feeble power rating is of concern. Our power draw is around 5.2 Watts, and these midget transformers are rated at 12.6V*0.5A = 6.3 Watts, so it seems they should be able to cope.

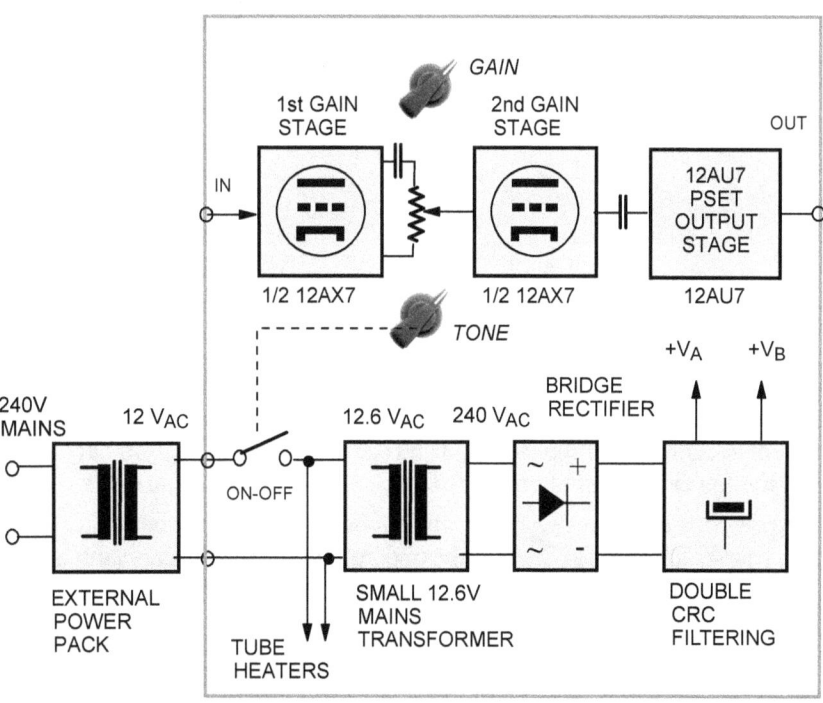

ABOVE: *El Comandante*'s block diagram

The power pack is rated at 12 Watts, so 4 Watts for the heaters and 5.2 Watts of anode power will load them at 9.2 Watts, so all is fine there. We are in the clear, but only for the 12AU7 output tube. Should you wish to use 12BH7, you must use a larger power pack and a more powerful transformer for the HV supply.

If you live in a country with 120V mains voltage, such as the USA, your power pack will be 120V in and 12V out. However, the small internal transformer must be rated at 240V-12V since 120V-12V transformer will only give you $120V_{AC}$, which is too low. In that case, you could use a voltage doubler instead of a bridge rectifier.

Biasing the output stage

ince the input stages are more-or-less standard, the choice of the operating conditions of the output stage is of most interest to us here. Again, there is more than one way to skin the cat, as that weird English proverb says, meaning you have lots of freedom.

The first step is to take anode characteristics, including the curve of maximum dissipation (very important), and choose the quiescent point Q. Since the curves go down to -30V bias, don't choose a curve below half that, or -15. So, Q will lie somewhere between -10V and -14V bias. In this first attempt, let's choose a low bias of -10V on the grid and place the Q-point at an anode current of 8mA. This determines that our anode-to-cathode voltage must be +260V!

Now we need to draw the load line. We have a wide range of choices, but remember that our load is not a resistor whose value you can choose at will, it is the output transformer's primary impedance, and these usually only go up to 10-12kΩ! The more horizontal the load line, the higher the anode impedance. Thus the 39k load line CD results in a larger voltage swing than the 14k load line AB, but its anode current swing is much smaller, meaning lower output power.

However, notice that in both cases, the positive swings AQ and CQ are much larger than the negative swings QB and QD, so there will be a high degree of distortion in this output stage. The graphic estimation table here is for the AB load line, and the 2nd harmonic distortion is a whopping 12.2%! As a practice exercise, estimate the distortion for the CD load line.

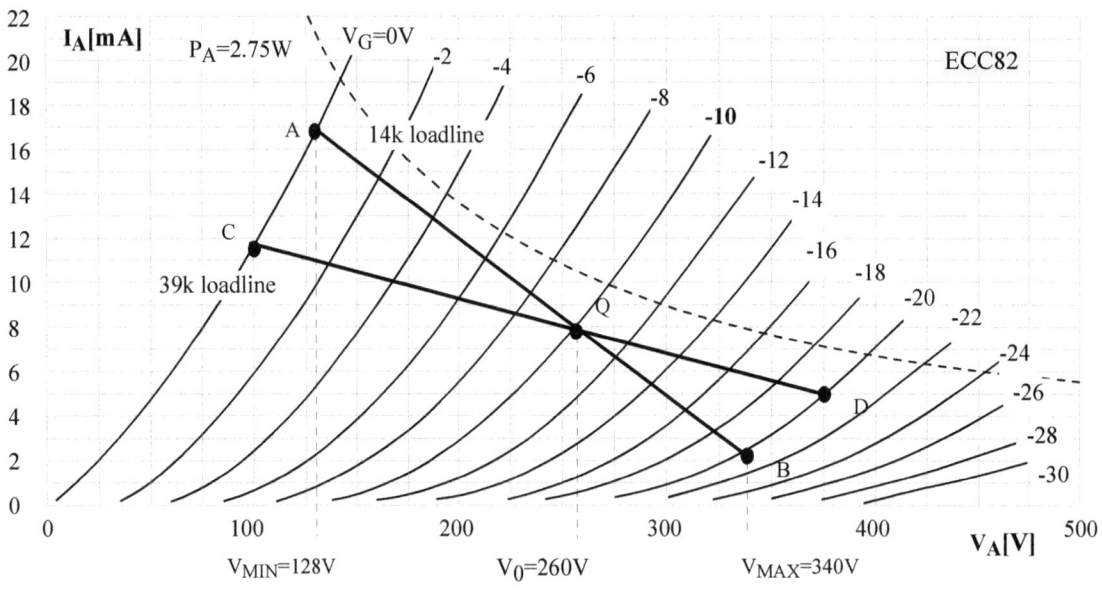

The answer? Around 8.2%, about 33% less than in the AB case. The rule-of-thumb becomes obvious now: the higher the load impedance, the flatter the load line and the lower the distortion, but also the lower the output power!

Our previous estimates were for one triode only, so if we connect both triodes inside one 12AU7 tube in parallel, everything stays the same, except the anode current values are doubled, and the load impedance is halved.

With two triodes in parallel, the AB load is 7kΩ, and the quiescent current is 16mA. Since the cathode needs to be at +10V to bias the grid at -10V, the common cathode resistor for both triodes must be R_K=10/0.016 = 625Ω!

GRAPHIC ESTIMATION

R_L=14kΩ, V_G=-10V
V_0=260V, I_0 =8mA
V_{MIN}=128V, V_{MAX}=340V, ΔV=212V
I_{MIN}=2mA, I_{MAX}=17mA, ΔI=15mA
P_{IN}= $I_0 V_0$ = 2.08W (76% dissipation)
P_{OUT}=ΔVΔI/8 = 212*0.017/8=0.45W
D_2=(ΔV_{OUT+}-ΔV_{OUT-})/2ΔV_{OUT}
D_2=(260-128)-(340-260)/2*212
=(132-80)/424 = 12.2%

DIY PROJECTS: ULTRA-SMALL AMPS

The problem with this exploratory design is that we don't know the voltage our power supply will produce, so it is prudent to allow for some leeway. What if our supply voltage is a bit higher, say 275V instead of 260V? How would that change the operating point? Well, let's consult the graph again.

Q1 now lies slightly above the curve of maximum anode dissipation. Also, the whole positive swing of the operating point (from Q1 to C) is way above the maximum power curve. The tube would work OK for a while, but its life would be shortened.

So, it would be better to position the Q-point lower, at the intersection of the 14kΩ load line with the -12V curve. This gives us the anode current of just under 7mA, meaning the idle dissipation on the tube is 275V * 7mA = 1.92 Watts, which is 1.92/2.75 = 0.7 or 70% of the maximum dissipation. Now the tube will cruise its way through its long and stress-free life.

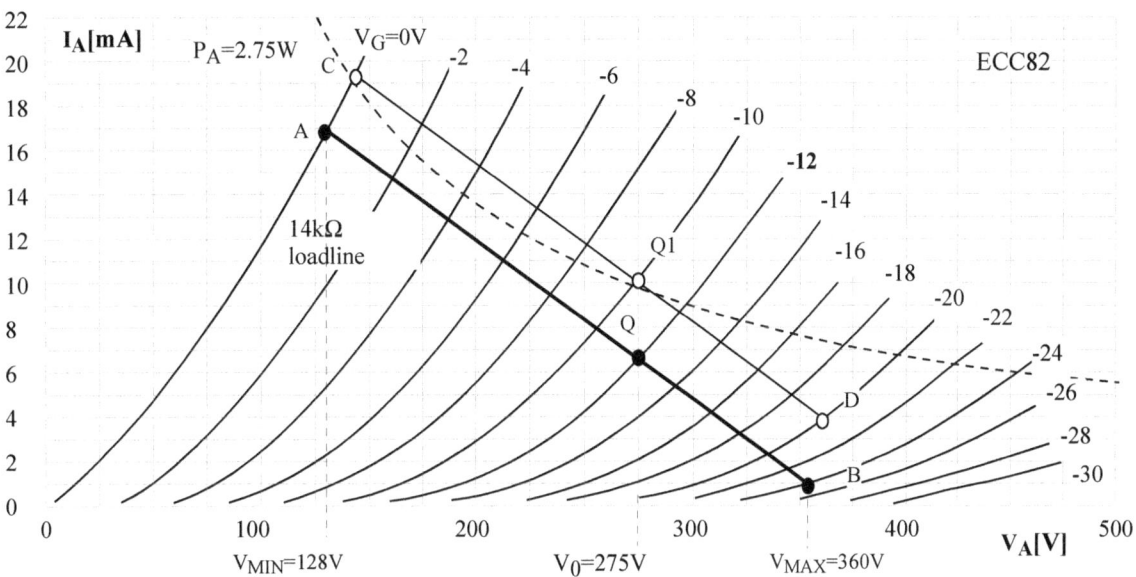

With two triodes in parallel, the quiescent current is 14mA, and since the cathode needs to be at +12V, the common cathode resistor now must be $R_K=12/0.014 = 857Ω$, a significant increase from 625Ω in the previous case.

Again, the rule-of-thumb is obvious. Despite a small reduction in idle current, as anode-to-cathode voltage increases, the bias must also be increased (made more negative). The cathode resistor's value must also be proportionally raised!

Notice that the negative swing now enters the area of low anode currents (around point B), where the anode curves are "squashed together," meaning high distortion. Indeed, our estimate indicates a 2nd harmonic distortion at a whopping 13.4%!

GRAPHIC ESTIMATION

$R_L=14kΩ$, $V_G=-12V$
$V_0=275V$, $I_0=7mA$
$V_{MIN}=128V$, $V_{MAX}=360V$, $ΔV=232V$
$I_{MIN}=1mA$, $I_{MAX}=17mA$, $ΔI=16mA$
$P_{IN}= I_0V_0 = 1.92W$ (70% dissipation)
$P_{OUT}=ΔVΔI/8 = 232*0.016/8=0.46W$
$D_2=(ΔV_{OUT+}-ΔV_{OUT-})/2ΔV_{OUT}$
$D_2=(275-128)-(360-275)/2*232$
$=(147-85)/464 = 13.4\%$

The output transformer

The tiny single-ended output transformer marked "April 1968" and "Rola" was salvaged almost two decades ago from a vintage Australian-made console with a tuner, turntable, and a small mono tube amplifier. Another example of the "don't throw anything away" rule, which continues "because you never know when you may need it in the future!"

The primary DC resistance was measured at 428Ω, which means it was wound using a very thin wire, 0.18mm diameter or even smaller. The primary inductance at 120Hz was a respectable 6.5H, dropping to around 5H at 1kHz. With shorted secondar,y the leakage inductance was incredibly high for such a small transformer, 110mH, meaning the windings were not sectionalized at all, the whole primary was wound first, and the secondary was wound on top of it. The quality factor was only QF=6.5/0.11=59.

With an 8Ω dummy load connected to its secondary, the primary impedance as measured by a digital LCR meter was 10.6 kΩ.

The "as built" circuit

The output stage of the finished amp deviated slightly from the theoretical estimate. We had around $250V_{DC}$ at the first filtering cap, but then we filtered it down to under 200V on the second capacitor. So, with $193V_{DC}$ on the anode, the output stage passed only approx. 10mA of current in idle mode (6/0.62= 9.7mA) or 5mA per tube. You could reduce the value of the 5k6 resistor in the power supply filter to raise the anode voltage and increase the cathode resistor's value from 620Ω (which gave us -6V bias) to 820Ω or even 910Ω.

Being lazy, we decided to leave the initially selected components in situ. More importantly, the amp sounded great and had plenty of grunt even at less than 1/2 Watt of output juice! As incredible as it may sound, even 1 Watt output may be too high. Feeding 1 Watt of electric power into a 90+ dB/Watt speaker would result in an ear-piercing sound pressure level of 90 dB!

The external power pack removes a significant source of EM radiation (and hum), namely the mains transformer, so no shielded cables were necessary. Furthermore, the steel instrumentation case used for the housing completely shields the circuit from the outside interference, contributing to a very quiet amplifier.

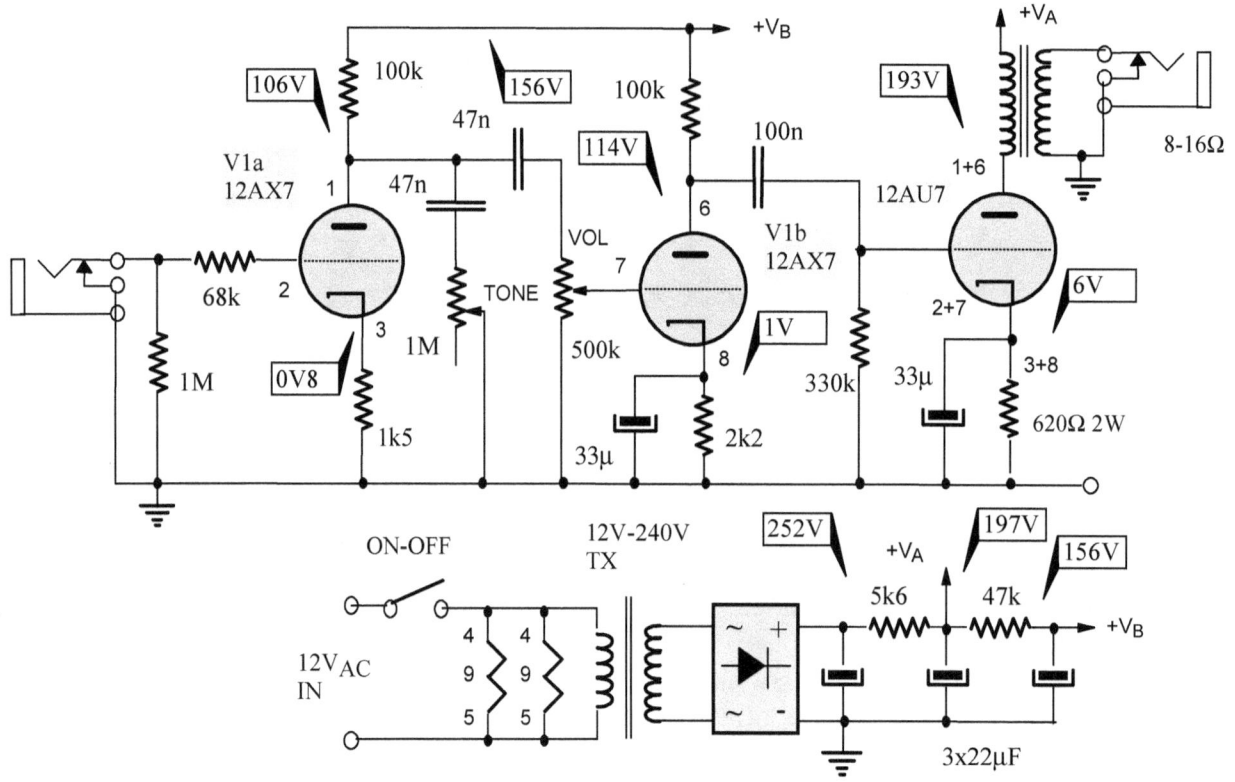

Construction details

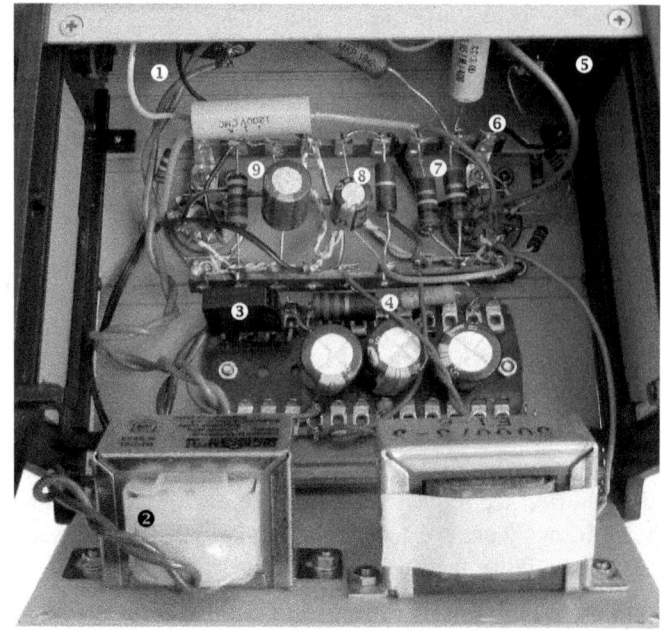

"El Comandante" from rear (ABOVE) and inside (RIGHT).

While the power supply was built on a terminal board, the audio section components were strung across two nine-lug terminal strips, bolted onto the two tube sockets (next page).Seven components, four resistors, and three coupling and tone capacitors were outside the terminal strips, with one end soldered directly to a socket pin and the other to either a ground lug, input jack, or one of the two potentiometers.

The on-off switch is at the back of the tone pot (1). Its output feeds the heaters and the HV transformer (2).The two wires in the foreground feed a solid-state bridge rectifier (3), whose positive terminal is on the right, where one end of the first series resistor (4) and the positive lug of the first filtering elco are connected.

The 68k resistor from the input jack (5) goes straight to pin 2 of the first triode, and the 1M resistor shunts the grid to the ground lug (6), one of the four already mentioned mounting lugs for tube sockets and terminal strips. All four were used as common (ground or earth) terminals, which is not a standard practice but resulted in no hum at all; the amp was very quiet.

The two 100k anode resistors (6) are flanked by the cathode resistor and bypass capacitor of the second stage (8) and the output stage (9).

The project case was too small to include a long fuse holder placed at the rear next to the $12V_{AC}$ input. Alternatively, install a fuse inside the AC power pack if it hasn't got one already.

Currently made products such as these "protective" cages often indicate a complete lack of critical (or any) thinking. The tubes are taller than the cages, so there is no physical protection at all! A solution would be to buy a third cage, take it apart and add one more ring and a set of spacers to the other two to make them taller.

DIY PROJECT: TUBEFINDER

6CG7 or 12BH7 duo-triodes as power tubes

Since Blackstar started using 12BH7 in their amps, the price of these duo-triodes has increased significantly. A few other guitar amp makers jumped on the bandwagon, for instance, Kendrick in their So-Lo 7 combo. "7" refers to the claimed 7 Watts of push-pull power, although I don't think it is possible to get more than 5 Watts out of one 12BH7! Indeed, both the Hughes & Kettner Tubemeister 5 and Blackstar HT5 are rated at 5 Watts maximum power, and even that is on a good day.

TUBE PROFILE: 12BH7

- Indirectly-heated Noval duo-triode
- Heater: 6.3V/600 mA or 12.6V/300mA
- V_{HKMAX}: $100V_{DC}$, P_{AMAX}: 3.5W each, 5W both
- Max. anode voltage & current: 300V/20 mA
- TYPICAL OPERATION:
- V_A=250V, V_G=-10.5V, I_0=11.5mA
- gm = 3.1 mA/V, µ= 16.5, r_I=5.3kΩ

TUBE PROFILE: 6CG7

- Indirectly-heated Noval duo-triode
- Heater: 6.3V/600 mA
- V_{HKMAX}: $100V_{DC}$, P_{AMAX}: 3.5 W each, 5W both
- Max. anode voltage & current: 300V/20 mA
- TYPICAL OPERATION:
- V_A=250V, V_G=-8V, I_0=9 mA
- gm = 2.6 mA/V, µ= 20 r_I=7.7kΩ

At -12.5V bias (grid to cathode voltage) and 12mA of anode current, the idle dissipation is just over 3.4 Watts, which is the absolute maximum since each triode is rated at 3.5 Watts.

0.8 Watts of single-ended output power at just over 8% second harmonic distortion is achievable, and two triodes in parallel can push 1.2 Watts due to required derating.

Notice that the load impedance is almost 18kΩ per triode. Such a transformer cannot be bought and would have to be designed and wound.

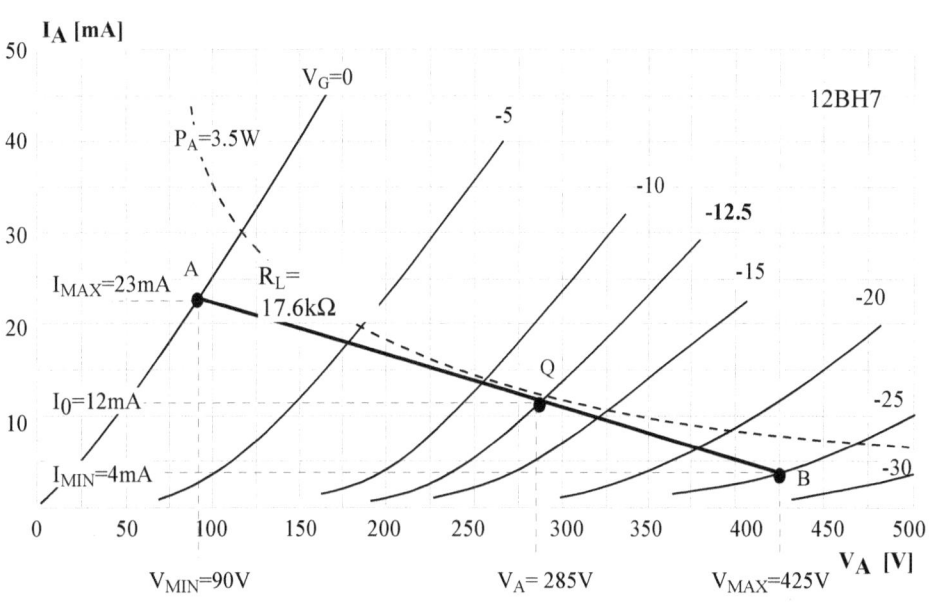

FAR LEFT: 6CG7 printed Penta Labs USA and Australian-made 6CG7 by Radiotron. Notice a much larger anode of the Australian tube. The Penta tube does not seem to have the power dissipation of 6CG7 at all, looks more like 12AU7!

LEFT: 12BH7 tube with larger and darker anode is by Radiotron Australia, the one with smaller anode is by Matsushita Japan.

CLASS A TRIODE (R_L=17k6):
V_{MIN}=90V, V_{MAX}=425V, ΔV=335V
ΔI=32-3=19mA
$P_{IN}=I_0V_0$=0.012*285 =3.42W
$P_{OUT}=\Delta V \Delta I/8$=335*0.019/8= 0.8W
$D_2=(\Delta V_{OUT+}-\Delta V_{OUT-})/2\Delta V_{OUT}$=
=(195-140)/2*335 = 8.2 %
$A=\Delta V_A/\Delta V_G$ = -335/25 = -13.4

Even with two triodes in parallel, the primary impedance would need to be 9kΩ, and even that is too high to be commercially available. There are some 10k output transformers for hi-fi amps, used for large HT transmitting tubes such as 211, 813, 845, and GM-70, but one of these large units costs more than a whole commercial guitar amp, so forget about those. Anyway, their sound would be too clean for a guitar amp.

There are a few other duo-triodes in this power range, so you don't have to fork out a fortune for vintage USA-made 12BH7. 6CG7 is electrically identical to 6SN7-GTB, but it is a more modern tube with a Noval socket.

12BH7 has the advantage that the heater can be configured for both 6.3 and 12.6V, while 6CG7 only works on 6.3V. The amplification factor of 6CG7 is slightly higher, but the internal resistance of 12BH7 is lower.

A better option: 5687 duo-triode as an output tube

5687 is a high perveance, high emission duo-triode. Each triode anode is rated at 4.2 Watts or 7.2 Watts together. Even the grid of 5687 can take 7 mA, which is higher than the anode of 12AX7! While μ is an par with 12AU7 (around 16), gm is high and r_I is 2.5 times lower than 12AU7, around 3 kΩ. 5687 was used in hi-end (meaning expensive) Audio Note hi-fi amps such as Ongaku, at some stage the most expensive stereo amp in the world.

By arbitrarily positioning the idle point Q at -12V bias (grid to cathode voltage) and 16mA of anode current, the idle dissipation is just under 4.1 Watts, which is the absolute maximum since each triode is rated at 4.2 Watts. Almost 1.2 Watts of sweet triode power is available, 50% more than with 12BH7!

CLASS A_1 TRIODE (R_L=11k1):
V_{MIN}=70V, V_{MAX}=390 V, ΔV=320V, ΔI =32-3=29mA,
$P_{IN}=I_0V_0$=0.016*255 =4.08W
$P_{OUT}=\Delta V \Delta I/8$=320*0.029/8= 1.16W
$D_2=(\Delta V_{OUT+}-\Delta V_{OUT-})/2\Delta V_{OUT}$=(185-135)/2*255 = 9.8 %
Voltage gain: $A=\Delta V_A/\Delta V_G$= -320/24 = -13.3

TUBE PROFILE: 5687
- Indirectly-heated medium μ dual triode
- Noval socket
- Heater: 6.3V, 0.9 A or 12.6V, 0.45A
- V_{AMAX}=300V_{DC}, V_{HKMAX}=90V_{DC}
- P_{AMAX}= 4.2 W, I_{GMAX}= 5 mA
- TYPICAL OPERATION:
- V_A=250V, V_G=-12.5V, I_0=12 mA
- gm=5.4 mA/V, μ=16, r_I = 3 kΩ

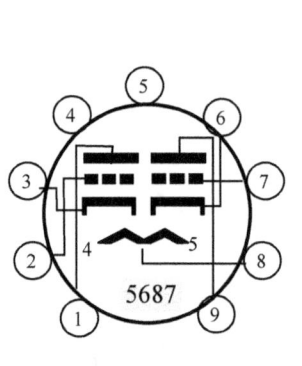

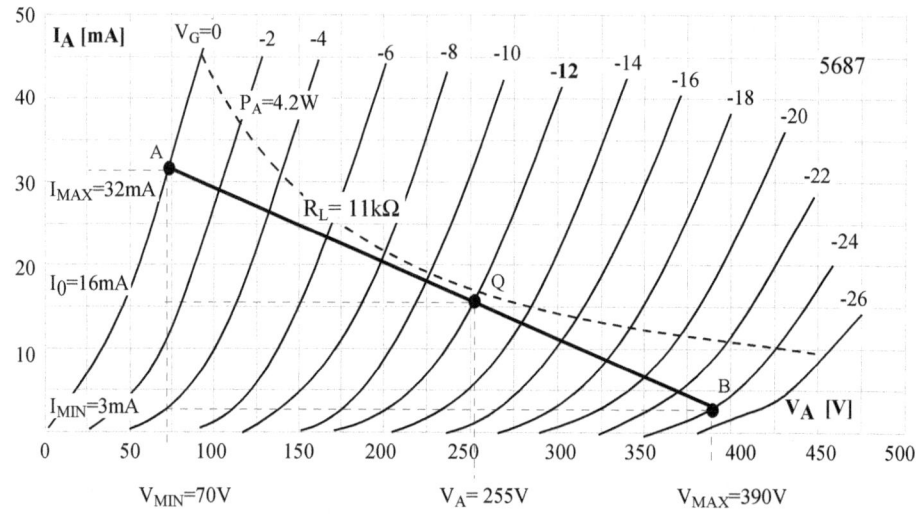

As already mentioned, the only difficulty is in obtaining a 10-12k SE output transformer. The most common primary impedances are in the 2-5kΩ range. However, since we have two triodes in one glass bulb, there is no point in using only one. By paralleling two triodes, we get an internal resistance of 1k5 and a very high transconductance of 10.8 mA/V. That translates into very dynamic sound. Now we can use 5-6k output transformers designed for 6V6 or EL84 pentodes.

Two triodes in parallel have to be derated from 2x4.2=8.4 Watts down to 7.2 Watts, so instead of 2x1.2=2.4 Watts of output power we can expect 1.8 to 2.0 Watts, which is still plenty for practice and recording purposes. Coupled with a very efficient speaker of 97-98 dB/W, this will be one incredibly loud little amp indeed!

Each paralleled triode will pass 16mA idle current, so with 32mA flowing through the common cathode resistor its resistance must be $R_K = V_K/I_K = 12/0.032 = 375Ω$. Use 390Ω (common value). The power dissipated into heat on that resistor will be $P = V*I = 12V*0.032A = 0.38$ Watts, so even a 1W rated resistor would be fine.

The circuit diagram

The transformer we earmarked for this project had no CT on its HV secondary winding, so we had to use a solid-state rectifier bridge.

The insertion loss of the two-knob tone control is not as high as that of the 3-knob tone stack, and the output tube's bias is relatively low at -12V, meaning we don't need a considerable amount of voltage gain from the preamp stages. We will still bypass both cathode resistors and won't use any negative feedback to keep the tone raw and dirty.

The *Overdrive/Clean* switch could be used in a few different ways. Assuming a two-pole switch, in the *Overdrive* position it could switch the two cathode bypass caps in and thus boost the gain of both stages.

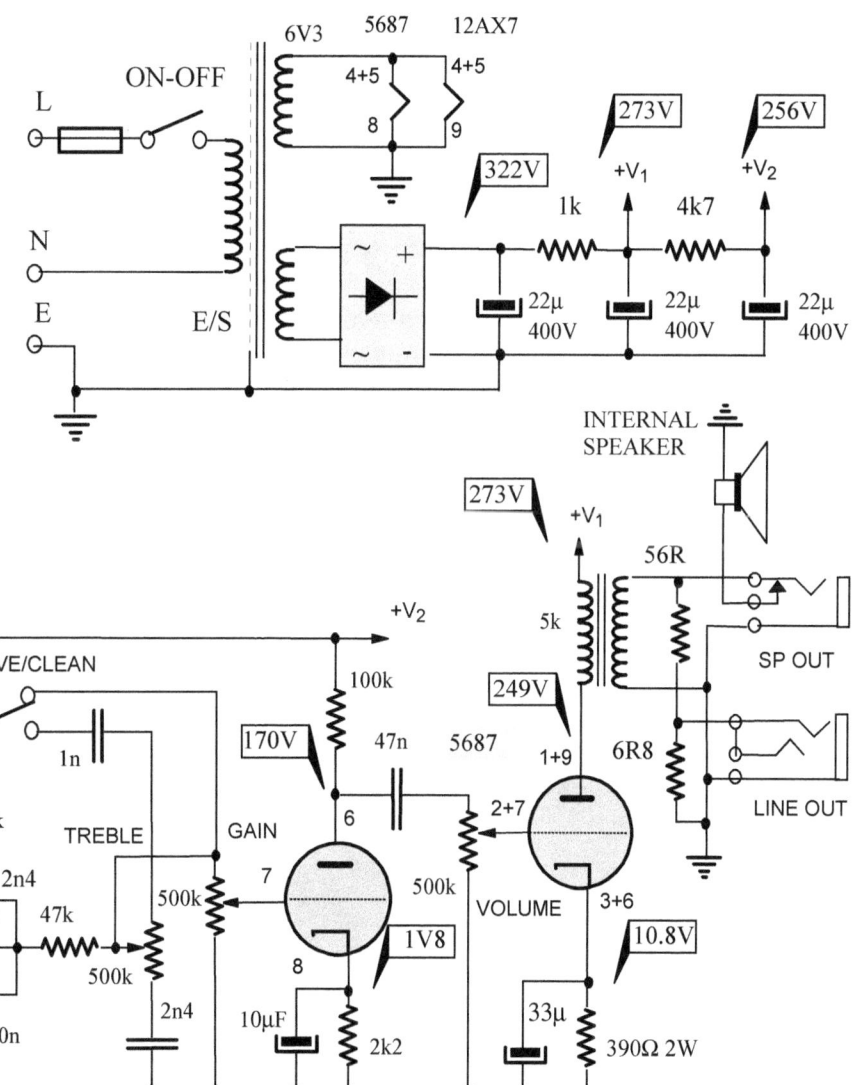

Or, it could switch off the negative feedback for more gain and more distortion. The default position (Clean) would have NFB operating.

Alternatively, it could increase the anode resistance of both stages by having two parallel 220k resistors in each anode (110k total), and then in the Overdrive position it would disconnect one 220k resistor in each pair and thus boost the gain of both stages. Remember, the higher the anode resistance, the higher the voltage gain of the triode stage.

We have chosen a simple single pole double throw (SPDT) switch to boost gain by bypassing the tone stack, a method used by Fender in their Vaporizer amp, for instance.

The body donor: VOX Pathfinder 10 solid state amp

VOX Pathfinder 10 control panel has separate Treble and Bass controls and Clean/Overdrive pushbutton switch, so we had to design our preamp circuit to incorporate these functions.

The Headphone/Line out jack was retained. There was no external speaker output, so we added one at the back panel.

Only a smallish mains transformer could fit under the chassis at (1), any larger units must be mounted on the outside, either on the top side of the chassis, behind (1) or hanging down at (2), next to the speaker. The input tube would be mounted at (3), the power tube would also hang down next to it at (4).

Construction details

Surprisingly, the small chassis accommodated the power and output transformers with plenty of physical separation. There was even some room for a choke.

The input tube (12AX7) is right opposite the guitar input jack and close to the output transformer, but that caused no issues. The 5687 output tube is closer to the power supply side, as is always the case.

However, inside the chassis, things got messier, bursting full of components, not all of which could fit on the terminal strip, so many had to be wired in a point-to-point fashion between various potentiometer, phone jack, and terminal board lugs.

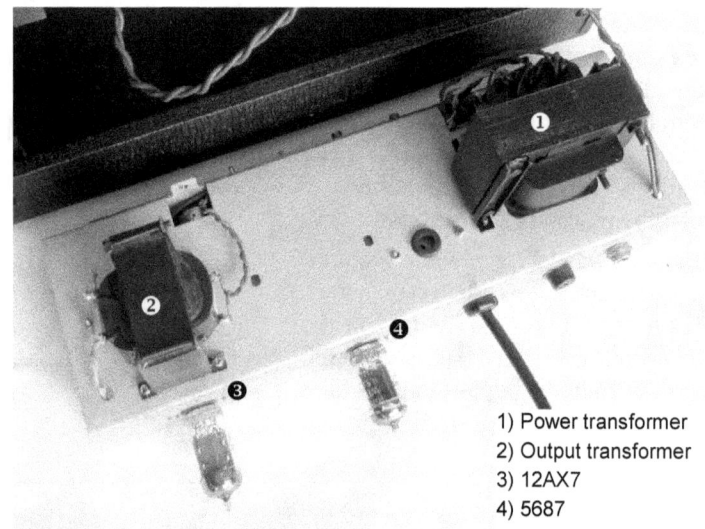

1) Power transformer
2) Output transformer
3) 12AX7
4) 5687

1) Speaker out
2) Fuse holder
3) Pilot light
4) Rectifier bridge
5) Filtering elcos
6) Input

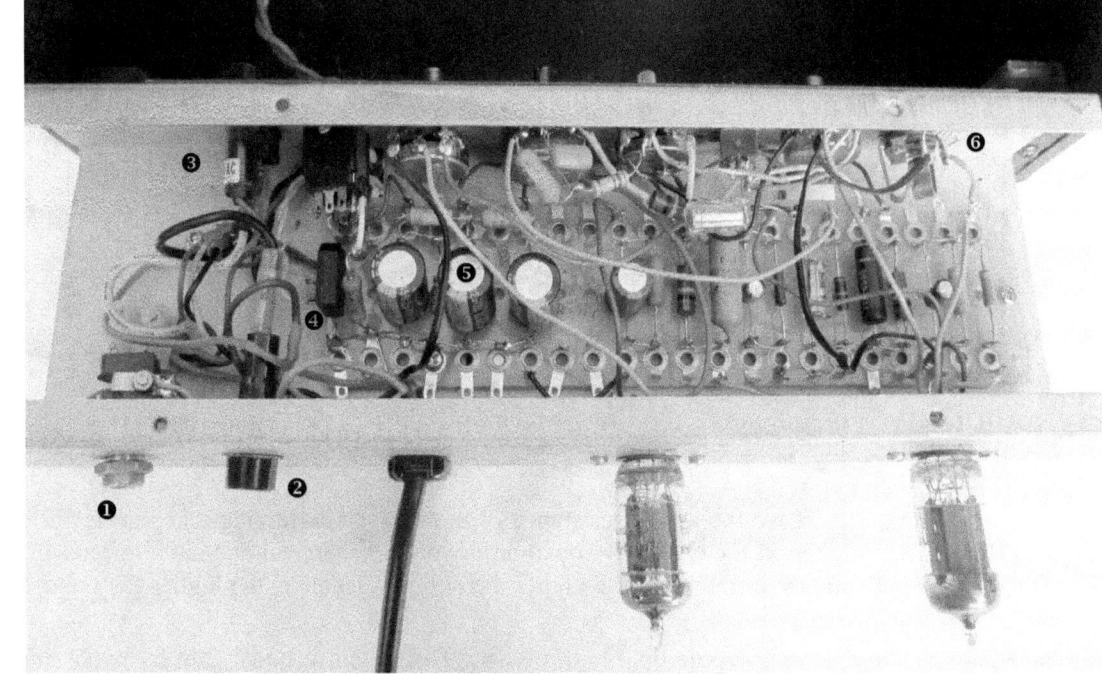

REBUILDING COMMERCIAL TUBE AMPS IN A HANDWIRED FASHION

10

- EPIPHONE VALVE JNR. - LEGACY VALVE EDITION 5 - HARLEY BENTON GA5
- CRATE PALOMINO V8
- VOX AC4TV

Many commercial tube guitar amps, especially those at the lower end, were built to a strict budget with low component quality and corners cut at every step. Many were poorly designed or wired (annoying hum or hiss, heater, and other voltages that are too high, thus shortening tubes' life) and lack even basic features. Often there is no tone control, no external speaker or headphone jack, no master volume, the list of grievances goes on and on.

Assuming you like your tube amp's looks and providing the cabinet, speaker, power, and output transformers are fine (although even they can be replaced by better ones), selling your amp's printed circuit board and rebuilding it in point-to-point fashion (hand wiring it) is a feasible option.

You can use superior components, a better layout, and optimized wiring that will not cause hum or hiss and thus end up with a more reliable and better-sounding amp.

The three case studies used in this chapter will illustrate the principles behind such rebuilds, so you can apply such methods to any amp.

REBUILT: EPIPHONE VALVE JNR. A.K.A. LEGACY VALVE EDITION 5

Legacy Valve Edition 5 seems identical to Epiphone Valve Junior, arguably the most modified budget-priced low power tube amp. Both brands come in combo and head versions. They are also the most Spartan of all modern amps, having just an on-off switch and volume control.

Further research uncovered that Harley Benton GA5 is an identical beast, except for the added "EQ" or tone control pot. More on that soon. Both Legacy's and Harley Benton's circuit boards have the etched words "Valve Junior" scraped off (marked "3" on the photo on the next page).

Looking inside the chassis (next page), the wire dressing is neat and straightforward, the printed circuit board is well laid out and can easily be removed for modifications and improvements.

Shielded cables were used between both the input jack and PCB (4) and the volume control pot and the PCB (5). The four silicon diodes provide HV rectification, followed by four 22mF caps (6) and associated resistors in the RC filtering circuitry. The fifth capacitor (7) is a low voltage type (4,700mF/16V), smoothing the pulsating output of the heater rectifier, a solid-state bridge (8).

Finally, notice a provision for two more output jacks on the smaller PCB (9). Since the output transformer has the unused (unwired) 8-ohm output wire, you can wire it up to one of those jacks, and the third one could be used as a footswitch jack. What you want that footswitch to do is up to you, but the two most obvious choices are some gain boost (bypassing the added tone stack) and "top boost" or bright voicing activation.

Moving onto the simple circuit, the first thing that one notices is two identical preamp stages. Due to their lowish 100k anode resistors and a cathode bias of around $1.7V_{DC}$, their gains are moderate. Notice the two 68k input resistors, paralleled due to the N/C (Normally Closed) changeover contact inside the input jack. Once a guitar is plugged in, they form a 2:1 voltage divider, so only half of the input voltage reaches the first grid. The same attenuation happens at the output of the 1st stage; notice another resistive voltage divider after the 22n coupling cap (1M+1M fixed resistors).

BELOW LEFT: The chassis is quite small, but there is still some room on top to install a filtering choke and/or a larger output transformer.

The messy bundle of wires behind the power transformer are various primary taps. Each tap is terminated by an in-line fuse holder socket.

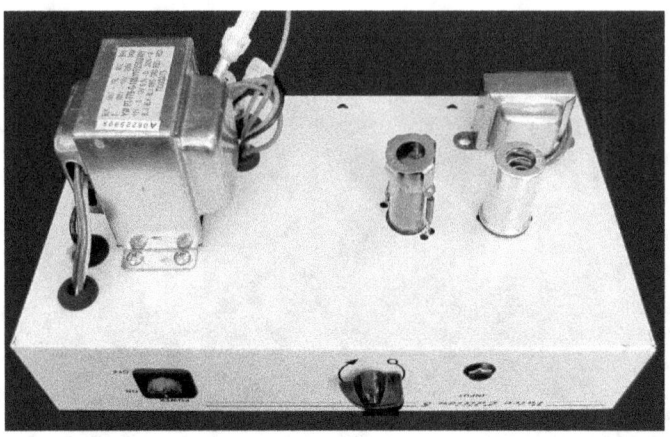

LEGACY VALVE EDITION 5 and EPIPHONE VALVE JNR.

- Controls: Volume
- 1x12AX7 + 1x6BQ5 power tube
- Solid state rectification
- PCB construction
- Output power: 5 Watts into 8Ω
- Jensen C8R speaker
- Dimensions: 335 (W) x 210 (D) x 430 (H)
- Weight: 8.3 kg (18.2 lb.)

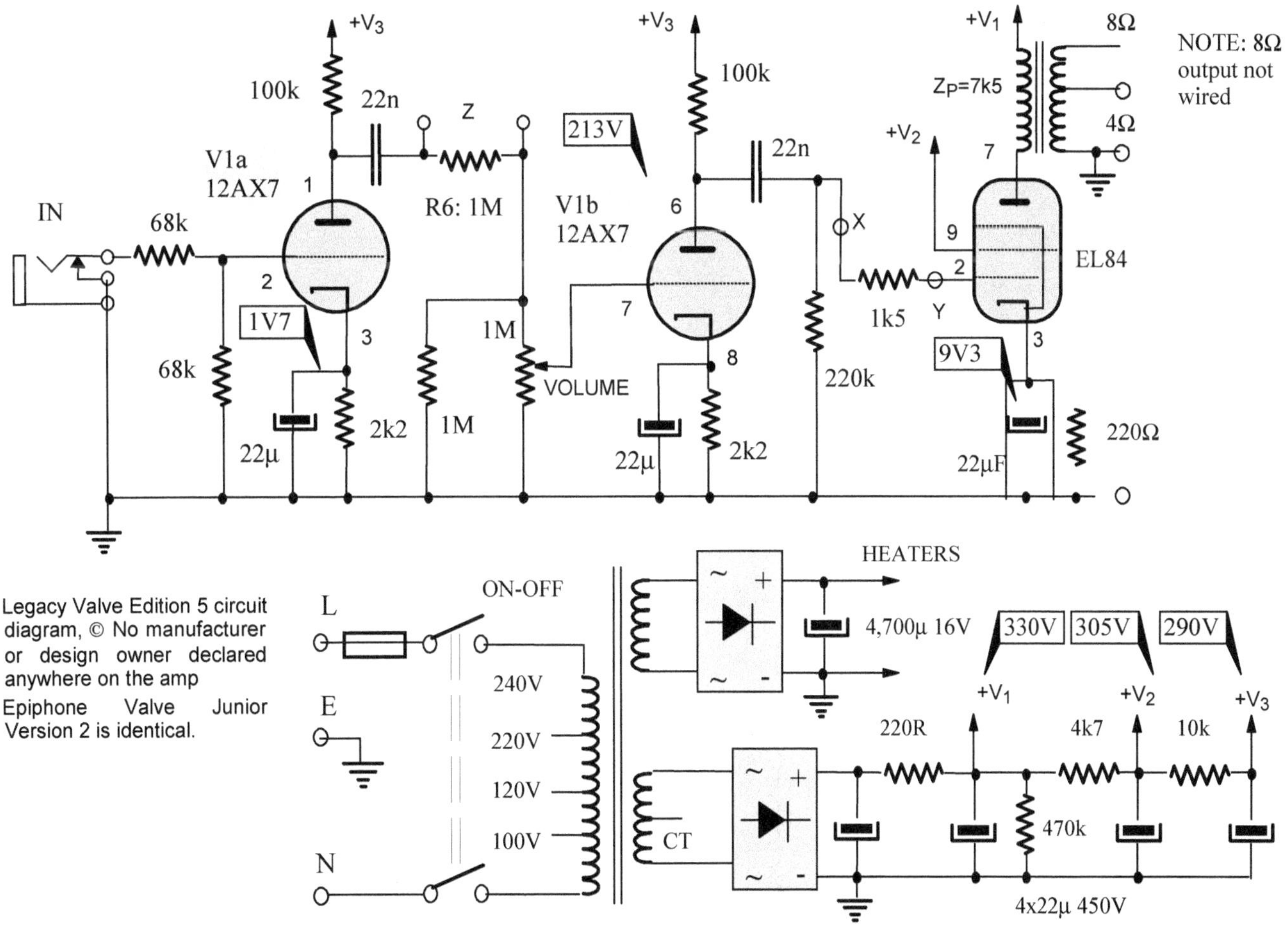

REBUILDING COMMERCIAL TUBE AMPS IN A HANDWIRED FASHION 183

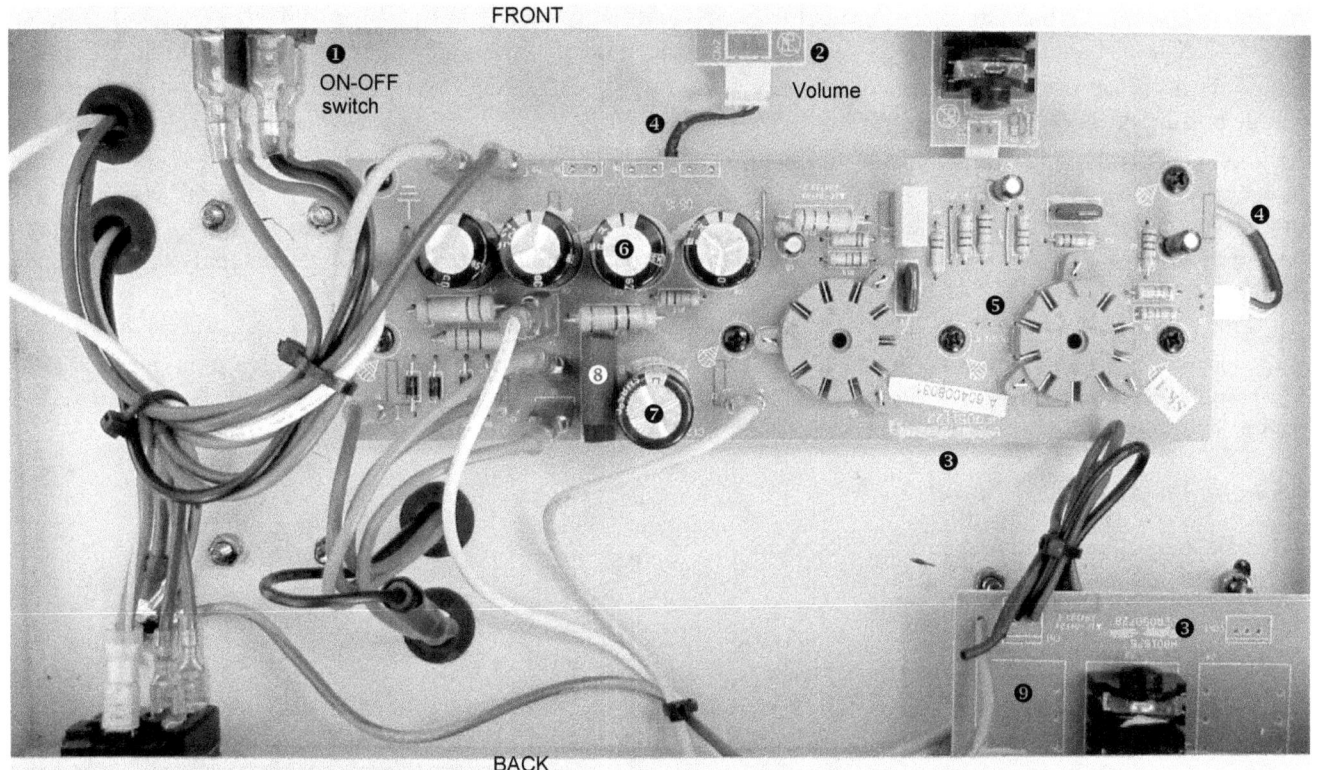

Legacy Valve Edition 5 circuit diagram, © No manufacturer or design owner declared anywhere on the amp
Epiphone Valve Junior Version 2 is identical.

All this gain lowering is acceptable because there are no gain-sapping tone controls. But, should you wish to add such controls, you need to boost gain by increasing the resistance of anode resistors and removing the two already mentioned voltage dividers. Notice the 1k5 grid stopper resistor at pin 2 of the output pentode (EL84) and its ends, marked X and Y. We will come back to those soon.

The anode currents of both stages are around 0.77mA, and, with the anode supply at $290V_{DC}$, the anodes are at around $213V_{DC}$. The HV supply of $330V_{DC}$ for the output stage is lowered down to $309V_{DC}$ due to the high DC voltage drop on the output transformer's primary winding (490Ω DCR). The cathode-biased EL84 runs an idle cathode current of 9.3/220 = 42 mA.

The good news is that the mains transformer features multiple primary taps for four different mains voltages, so if you happen to live in a 230 or 240V mains country and get your amp from the USA, you will not need to change the power transformer or install a step-down transformer. It also has a CT o the HV secondary winding, so you could install a tube rectifier with 6.3V heater such as EZ80 (6V4), providing the 6.3V winding can supply the additional heater current (0.6A).

Notice the space left on the PCB in parallel with R6, the 1M series resistor (marked "Z"). There are many things you can do here. You could permanently install a bright capacitor there or wire a switch to those two terminals that will switch such bright capacitor in and out. Alternatively, the switch could bypass (short circuit) the 1M R6 series resistor for a significant increase in gain!

Epiphone Valve Junior Version 1

The audio circuit was identical in Epiphone Valve Junior Version 1 and Version 2, but Version 1 had a simpler and inferior power supply, illustrated here. Notice that the high voltage for the output stage was taken directly from the first filtering capacitor (1), which is never a good idea (hum!), especially not in single-ended amps.s.

There is too much ripple (AC component) in that point, the DC voltage simply isn't filtered well enough for a single-ended amp and will introduce hum that cannot be mitigated except by adding another filtering stage in front of it. That is exactly what was done for Version 2 and our Legacy version.

Also, Version 1 used AC tube heating (2), Versions 2 and 3 used DC heating.

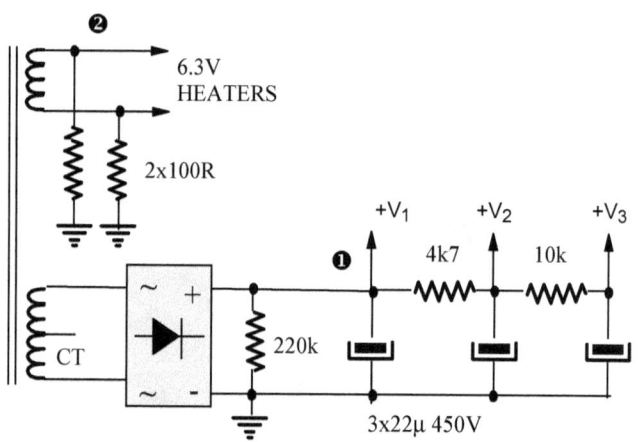

ABOVE: Epiphone Valve Junior Version 1 had an inferior power supply.

Epiphone Valve Junior Version 3

Version 3 audio circuit has undergone a few changes. First, the 68k grid resistor of the 1st stage is gone, replaced by a 1M resistor, thus increasing overall gain (3). The 2k2 cathode resistors of the two preamp stages were lowered to 1k5 (4). And finally, it seems that a different output transformer was used, this time with a lower 5kΩ primary impedance (5), while versions 1 and 2 used transformers with 7k5 primaries.

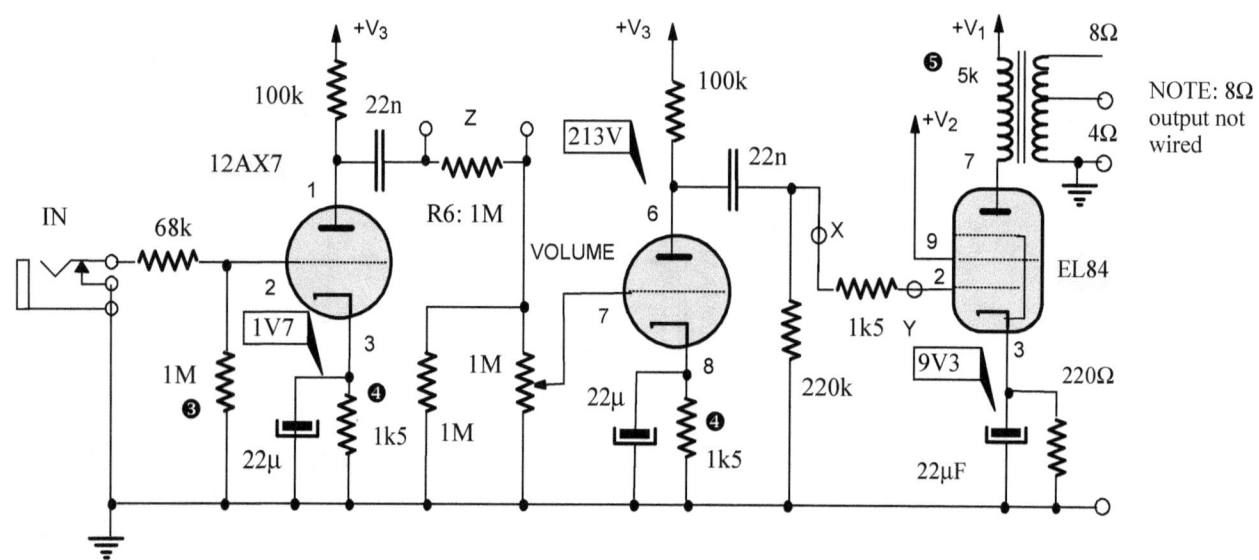

Epiphone Valve Junior Version 3 audio section

The output transformer

The output transformer uses EI48 laminations (16mm center leg width) with 15mm stack width, so $A = 1.6 \times 1.5 = 2.4 \text{ cm}^2$, resulting in the power rating of $P = A^2 = 2.4^2 = 5.7$ Watts. This is the lowest power rating of all small amps we've ever had on our test bench.

The primary's DCR was very high, almost 500 ohms, meaning that a very thin wire was used for the primary winding. The primary inductance was very high, 50.3H at 120Hz, dropping to 39H at 1kHz. However, the leakage inductance at 1kHz was also very high for such a tiny transformer, 166mH, resulting in a low quality factor of only 235.

The cabinet is quite well made and sturdy so that it can take a beefier speaker. Apart from speaker replacement, the priority on this amp would be to replace the output transformer with a larger and better-sounding alternative.

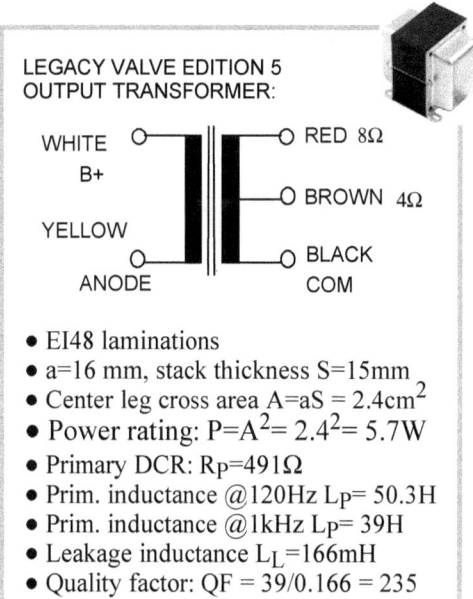

LEGACY VALVE EDITION 5
OUTPUT TRANSFORMER:

- EI48 laminations
- a=16 mm, stack thickness S=15mm
- Center leg cross area $A = aS = 2.4 \text{ cm}^2$
- Power rating: $P = A^2 = 2.4^2 = 5.7W$
- Primary DCR: $R_P = 491\Omega$
- Prim. inductance @120Hz $L_P = 50.3H$
- Prim. inductance @1kHz $L_P = 39H$
- Leakage inductance $L_L = 166mH$
- Quality factor: $QF = 39/0.166 = 235$

After a few simple modifications

The anode resistors were increased to 150k, the grid leak resistor of the 1st stage to 1M (so there is no 2:1 attenuation of the input signal). The 68k resistor should be on the other side of the 1M resistor, close to the tube's grid. A simple one-knob tone control works well (3), as does master volume. For that simple conversion, just replace the 220k resistor to GND with a 200-500k log potentiometer.

Since there is no power control or output attenuator of any kind, two simple mods were implemented. The PENTODE-TRIODE switch (sometimes called 1/2 power switch) changes the maximum output power and the voicing of the amp since triode mode sounds warmer, softer, and lusher.

The same applies to the NFB (negative feedback) switch (6). When engaged, it lowers the output power, but the sound is also cleaner - the distortion is reduced. These two switches give you four different voicings (2x2=4).

Due to the NFB signal coming into that point (2nd stage's cathode), the 22mF elco was replaced by a lower value, 2m2 (7)! The strength of the negative feedback can be regulated by changing the return resistor; typical values are in the 10k (very strong NFB, lower output power, and cleaner tone) to 47k (weaker NFB, louder sound output but a more distorted tone) range. The gain boost switch can be manually or foot-activated. It could be a pull-to-operate switch on a tone control or volume pot or a stand-alone toggle or push-button switch if manually operated.

The pentode/triode or full/ half power and NFB on/off switch are best added at the back of the chassis, while the gain boost switch and standby switch should go to the front, to the opposite sides, as illustrated on the next page.

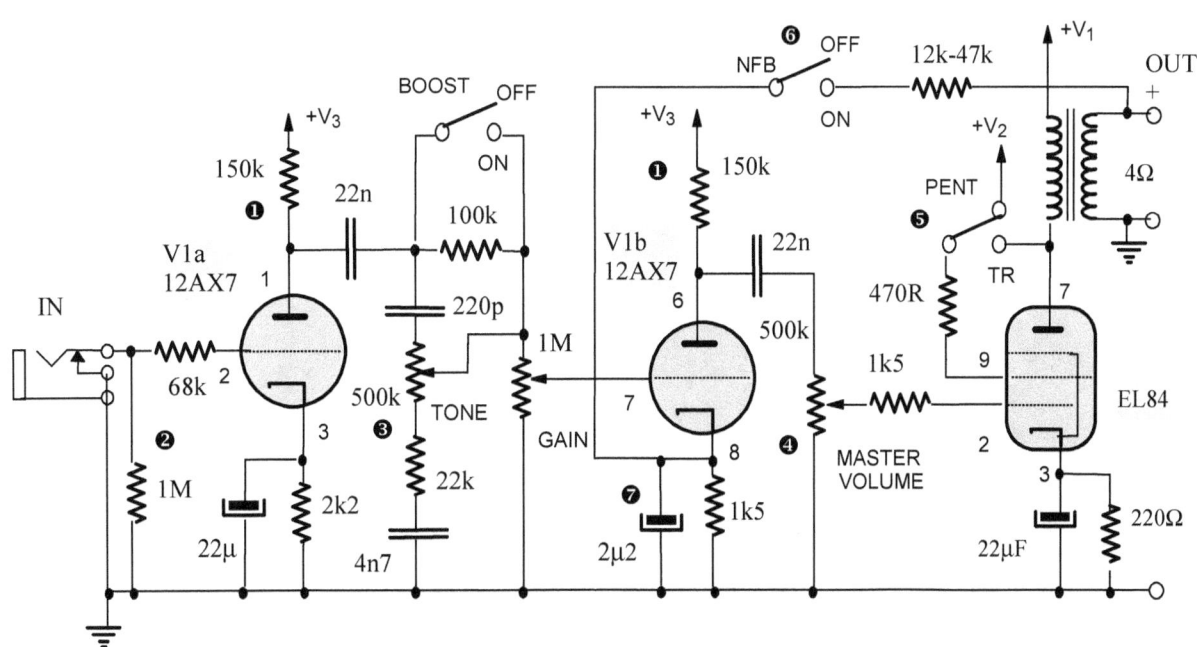

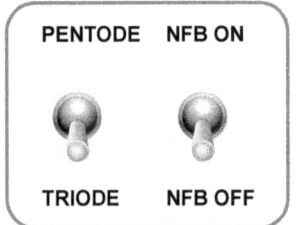 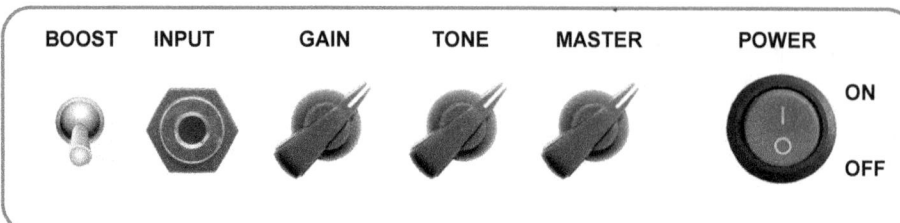

ABOVE: The added rear controls (LEFT) and a proposed layout of the front controls (RIGHT)

Implementing more intricate tone controls

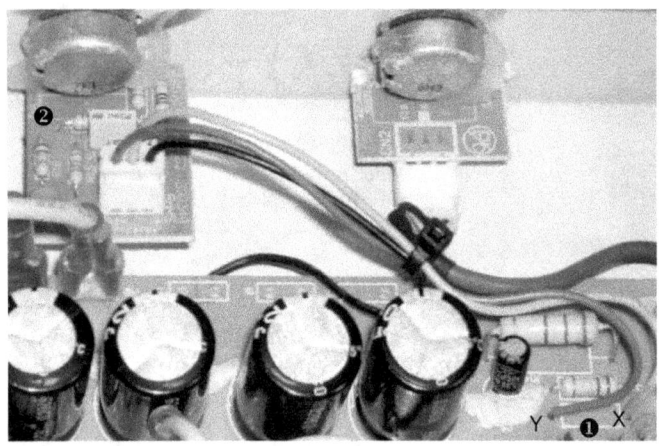

ABOVE: Harley Benton GA5 "EQ" PCB and wiring
RIGHT: The circuit diagram of its EQ control

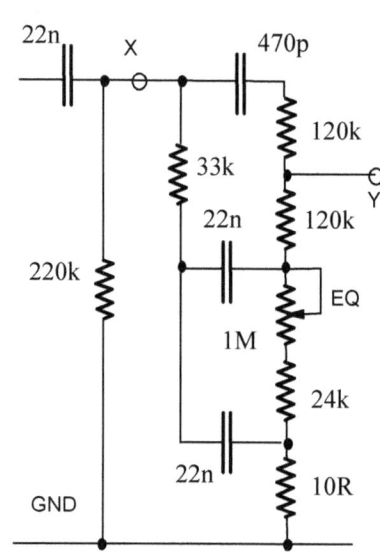

There is so much real estate at these amps' front fascia that you can add not just a simple one-knob tone control, but a 3-knob stack if you wish. In fact, that is exactly what was done in the case of the Harley Benton version. However, despite installing the whole stack, they only implemented a single control (effectively a BASS control pot). Fixed resistors replaced the midrange and treble potentiometers. The same main PCB was used, R15 (1k5 grid stopper resistor) was not installed (1), two wires between point X and Y were taken out to a small PCB (2) instead, with all tone control components mounted there.

This circuit (right) implements a full 3-way tone control stack. The fixed 220k resistor at the input of the tone stack can stay, or, if you want to add the MASTER VOLUME control, replace it with a 250 or 500kΩ log pot.

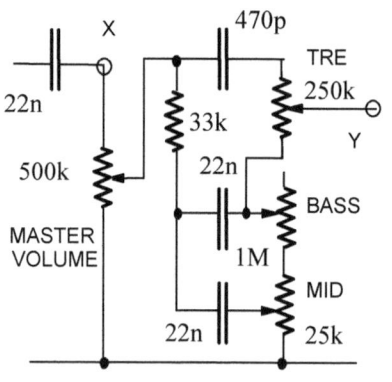

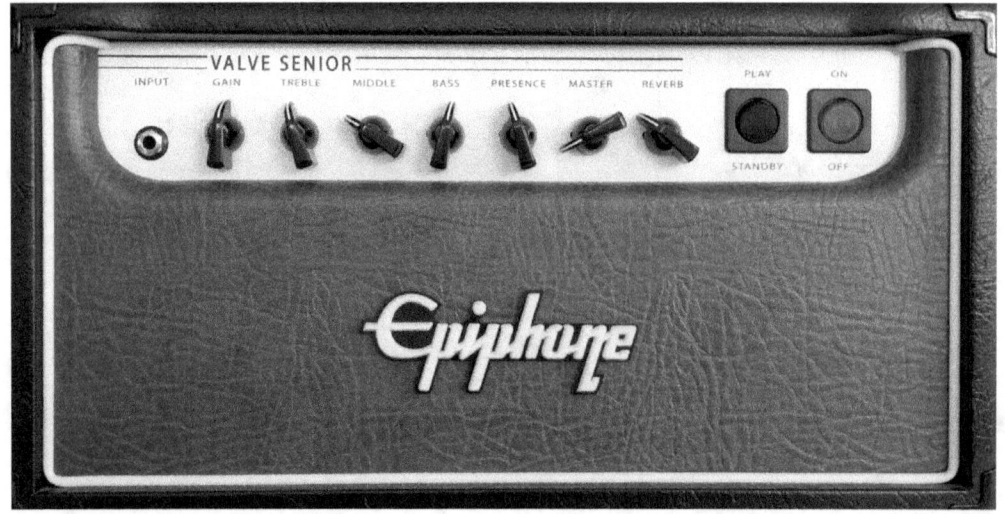

ABOVE: This simple mod gives you master volume control and three-way equalization.

LEFT: Epiphone Valve Senior has extensive controls, including presence and reverb, but is a totally different beast, a push-pull amp using two 6V6 in the output stage.

Bitmo™ mods

We didn't buy the Bitmo™ Trio™ kit (sells for US$49.95); instead, we did some research online and found a partial circuit diagram and color photos illustrating the installation of this kit and the first page of its instructions. This was enough to determine its topology and the wiring diagram, although we could not ascertain some component values although we could not ascertain some component values (R_X, C_X, C_Y and C_Z). Anyway, the goal of this exercise is not to disclose proprietary information and take away business from Mr Hutchinson (we wish him every success in his kit-selling business), but to illustrate his approach and way of thinking in devising these mods, purely as an educational exercise.

The kit is called Trio™ because it is comprised of three separate modifications. The instructions call the first two mods "Duo Tone Control with pull boost". That used to be a kit called "Duo™", to which the third modification was added, called "Voicing/gain switch".

The tone control is a standard CRC series filter between the output side of the 22n coupling capacitor and ground, 500p cap in series with a 1M "Tone" pot and 5n cap to GND. The two normally open switches SW1a and SW1b are part of the "Tone" pot and are activated by pulling its knob. The diagram shows the switches in the off (non-active) position. Pulling the knob would close both switches.

One contact (SW1b) adds another cap in parallel with the 500p cap. This increased capacitance means lower reactance and more treble (high) frequencies bled to GND, thus seemingly increasing bass frequencies.

The closure of the other contact (SW1a) short circuits the resistor R_X, which is in series with the 22µF cathode bypass cap. That removes the local NFB or "cathode degeneration", now the 1k5 cathode resistor is bypassed for most audio signals. The voltage gain of the 1st stage is increased by about 3dB, so a mild gain "boost" is achieved.

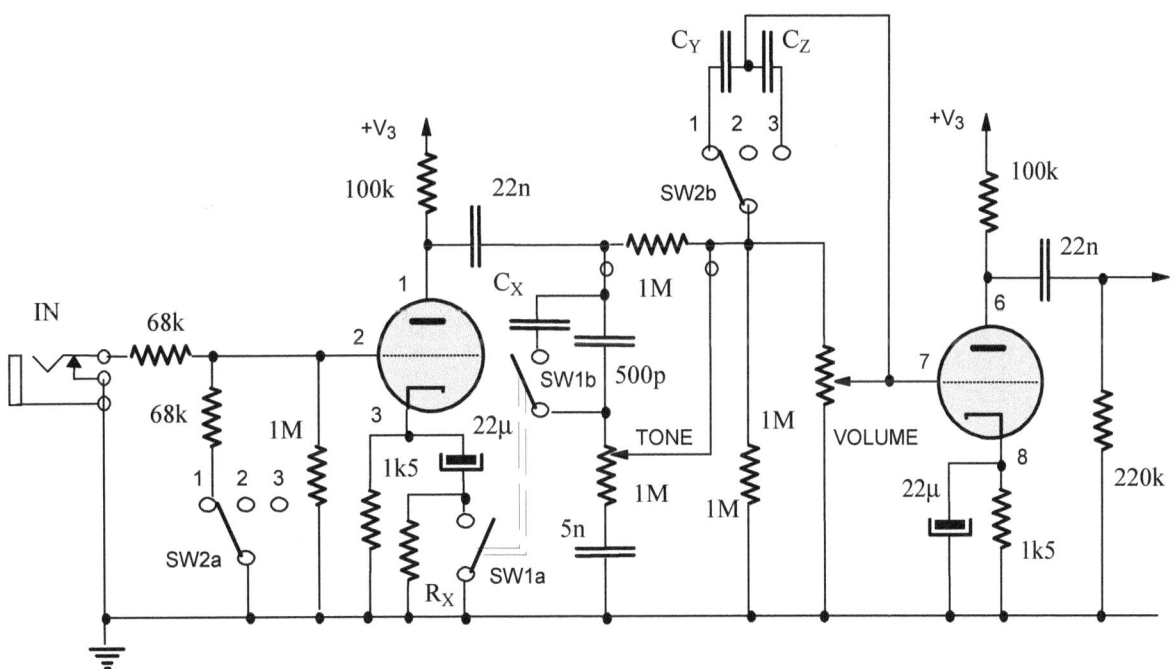

Epiphone Valve Junior Version 3 with Bitmo™ Trio™ mods © Bruce Hutcheon

The "Voicing/gain switch" marked as SW2 is a 3-position or double pole double throw (DPDT) switch with the middle "off" position. That off position is marked as position "2" on the diagram.

The way SW2a is wired differs slightly for Versions 1 and 2 of Epiphone Valve Junior; since they have a 68k grid resistor, version 3 already has a 1M resistor to GND. In position "1," the 68k resistor is in parallel with the 1M resistor, so the attenuation of the input voltage divider thus formed is almost the same as in versions 1 and 2, so that is the low gain position.

Capacitor C_Y is bypassing the upper portion of the volume control pot. Apparently, that mimics the bright channel on Fender's "black-face" amps. In positions "2" and "3", the 68k resistor is out of the circuit, so the gain is higher.

In position "2," there is no capacitive bypass of the upper portion of the volume control pot, and in position "3," Capacitor C_Z is bypassing the upper portion of the volume control pot, producing "more gain for a singing lead sound."

Rebuilding the amp

Instead of messing around with the stock PCB, we decided to install it into another combo cabinet and rebuild this amp in point-to-point or, rather (more accurately) terminal strip fashion. The guitar input jack was moved further towards the corner of the chassis, and two new holes were drilled on each side of the existing potentiometer hole, which now features the middle pot, Tone control (1) flanked by Gain (2) and Master Volume (3).

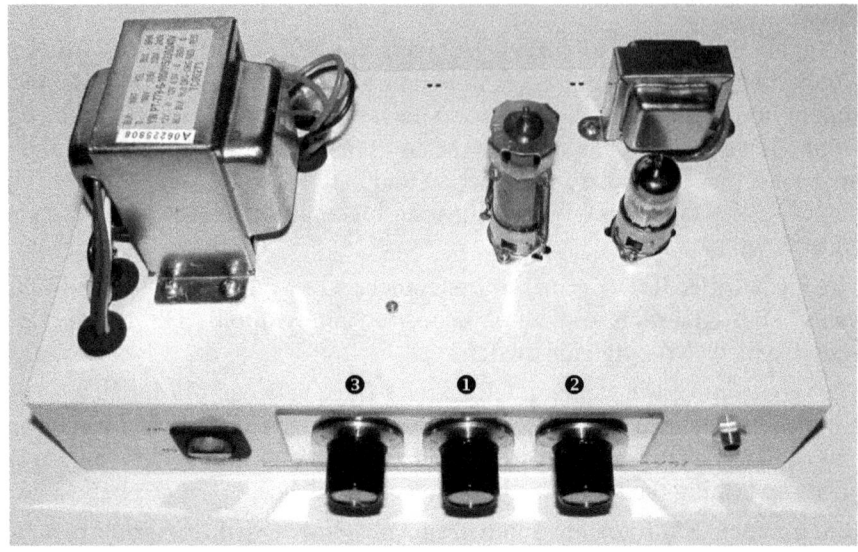

The Power on-off switch was retained, and the transformer and tube positions were not changed.

The old screen printing on the fascia was covered by a mirror finish gold Perspex, which now provides a cool reflection of the new large knobs.

The tone control worked really well, as did the triode-pentode switch. DC voltages in main points are given below for three different tubes, the ubiquitous 12AX7 (ECC83), the special dissimilar triode ECC823 (reversed 7247) by JJ and 12AT7 (ECC81). A wide variety of voicing options is thus available for a demanding guitarist.

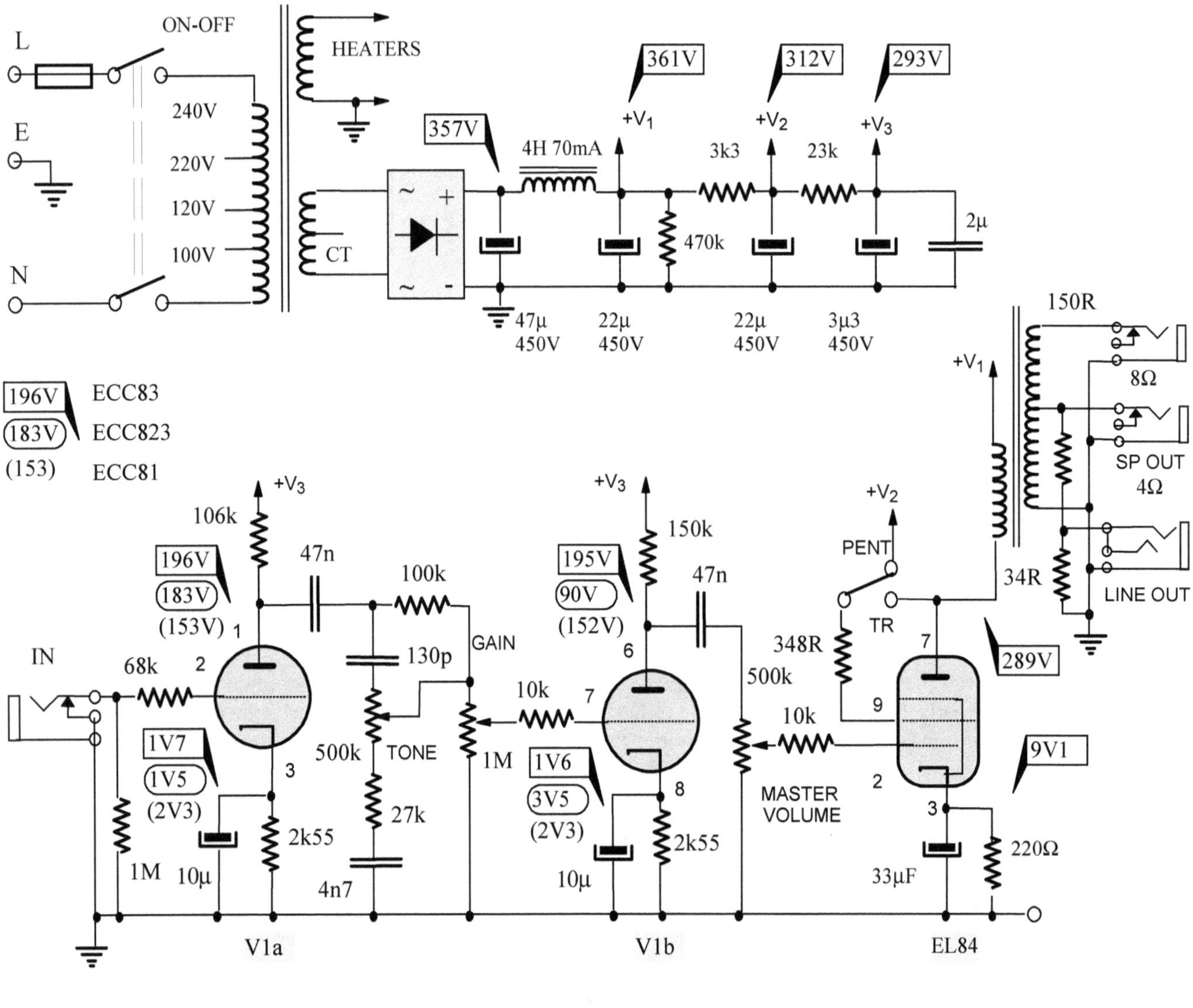

Two new phone jacks were added at the back, one mono for the 8Ω speaker output (1), which was not wired on the original amp, the other stereo for headphones output, but so designed that it can be used as a line output as well (2). The Triode-Pentode switch was also added next to it.

Moving onto the internal wiring (left), a small choke fits nicely between the IEC power inlet at the back and the On-Off switch at the front (3).

The power supply is on its own terminal strip, headed by a bridge rectifier (4), and followed by a filtering chain of four elcos and one film bypass capacitor (5) in parallel with the last elco, a trick that cleans the sound of the preamp stages and adds transparency to the tone.

Since the main terminal strip had only 12 pairs of lugs, about half-a-dozen components had to be wired in a true P-T-P fashion, two voltage divider resistors for the line/headphone output (6), the screen resistor of EL84 (7), its cathode resistor and bypass capacitor (8) and the grid stopper resistor of the input stage (9).

Since the mains transformer is on top of the chassis (the steel chassis acts as a magnetic screen) and the small choke is very far from the audio terminal strip, their EM radiation has no impact on the audio circuit.

Also, notice the neat and twisted heater wiring (7) and very short wire runs between the terminals and the three pots at the front, only 1-2 cm long. Thus, although the original amp's cable runs used shielded cables, there was no need for them here. The amp was absolutely quiet, something recording artists would genuinely appreciate.

RIGHT: The thin metal Legacy nameplate was binned and replaced by a golden bat logo (a few bucks on ebay), so the amp was christended "The Golden Bat"!

REBUILT: CRATE PALOMINO V8

The cabinet & chassis donor: Crate Palomino V8

This elegant looking combo amp was analyzed in Volume 1 of this book. originally, this was a hybrid design with IC-based preamp circuitry and tube driver and output stages (one 12AX7 and one EL84).

Unfortunately, the quality of sound did not match its good looks. It had a pronounced hiss and hum, so we sold its PCB and converted it to a hardwired (turret board construction) tube amp. The speaker seemed beefy enough for a small amp pushing and pulling 10-12 Watts from two 6V6 beam power tubes.

The existing hole where EL84 was protruding through the chassis (remember, the sockets were on a PCB) was of the right size for mounting an octal socket, and there was plenty of room at (3) for another. Notice the tube clamp had a provision for another output tube, so Crate obviously used the same chassis for slightly more powerful push-pull amps as well.

The clamp was too low for 6V6 tubes and due to the tight and snug fit of these tubes into their octal sockets, it wasn't needed any longer and was discarded.

The tiny mains transformer (1) had only a 115V primary, and the output transformer (2) was a single-ended design, so both had to be removed and replaced.

The amp only had *Gain*, *Tone* and *Volume* controls, which greatly simplified our design.

The topology

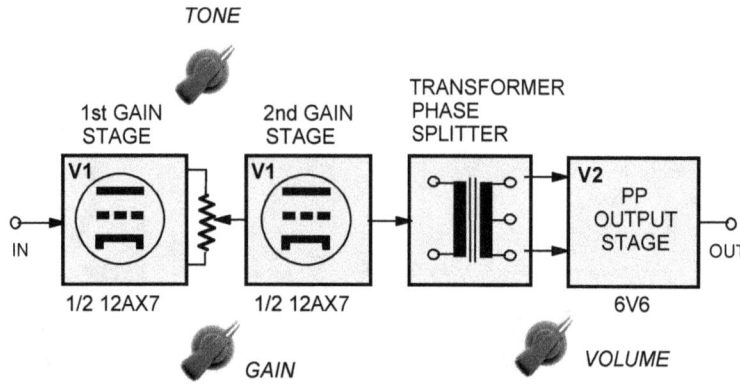

As always, the topology was dictated primarily by the controls of the host amplifier (assuming we wanted to retain all of them so all knobs would do what the front panel said they should do).

The *Gain* was be positioned after the first preamp stage and the *Volume* after the interstage transformer. This meant we needed a dual-gang volume control pot, the type used in stereo hi-fi amps.

The interstage transformer (phase splitter)

Instead of one tube dedicated to the task of splitting the output signal from a preamp into two out-of-phase signals to drive the two power tubes' grids, in this project, we used an interstage transformer. Some interstage transformers provide no voltage gain, having three identical windings, one primary and two secondaries. This Merit transformer has a turns ratio of 3:1, meaning each secondary winding has three times more turns than the primary winding. This also means the secondary voltage (each winding) is three times higher than the primary AC signal voltage. In other words, the transformer amplifies the voltage signal three times.

MERIT A2914 INTERSTAGE TRANSFORMER

- EI41, a=13 mm, stack thickness S=13mm
- Center leg cross section $A=aS = 1.7cm^2$
- Power rating: $P=A^2= 2.9W$
- Primary DCR: 485Ω
- Secondary DCR: 874Ω + 1,026Ω

REBUILDING COMMERCIAL TUBE AMPS IN A HANDWIRED FASHION

The instruction sheet, reproduced here, says that up to 10mA of DC current is allowed through its primary, so we can place it directly in the anode circuit of a preamp tube.

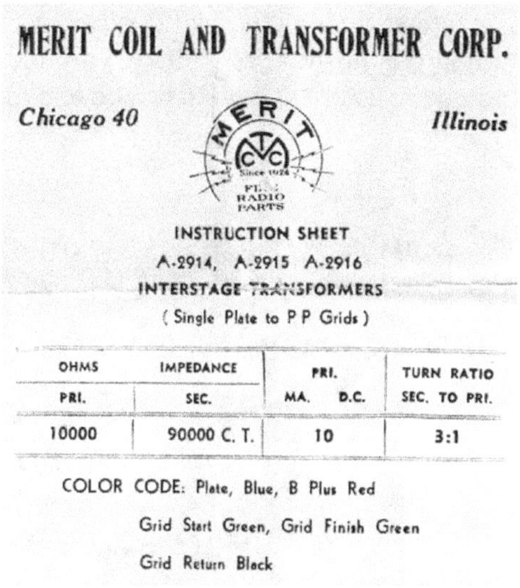

The circuit diagram

The input stage is fairly standard, as is the tone control circuitry that follows. The value of 2nd stage's cathode resistor had to be experimentally determined. In this case, the anode current is around 1.6mA, resulting in anode power dissipation of 272*0.0016 = 0.44 Watts. This means that it could be increased, but don't go over 1W/272V = 3.6mA otherwise, the 1 Watt anode dissipation of 12AX7 would be exceeded, shortening the tube's life.

The total cathode current of the power stage is 18V/270R = 67mA or 33mA per tube.

The standard terminal strip with 12 pairs of lugs could just fit transversally across the chassis depth (next page), so it was positioned at the end of the chassis, as far from the power supply as possible. The preamp tube is the exact location as before, a bit far from the input jack and the terminal board but close enough for the 22k grid stopper resistor to reach its grid at pin #2.

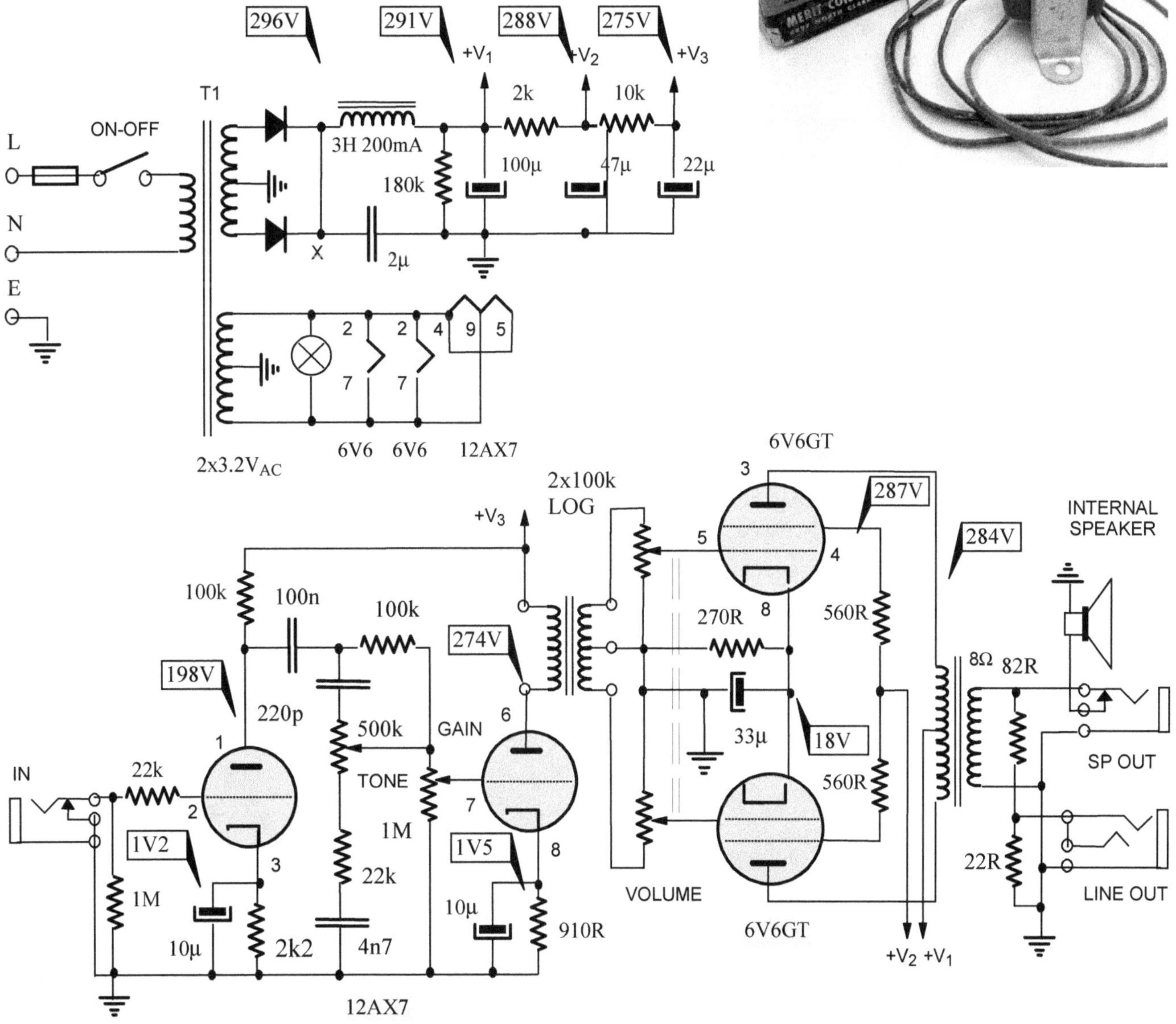

The interstage transformer (2) is very close to output tubes to minimize the wire runs between the two. The two large carbon composition resistors (3) are the 560R screen resistors, while the large 5W white resistor is the common cathode resistor (4), with its cathode bypass elco mounted on the power supply terminal board (5).

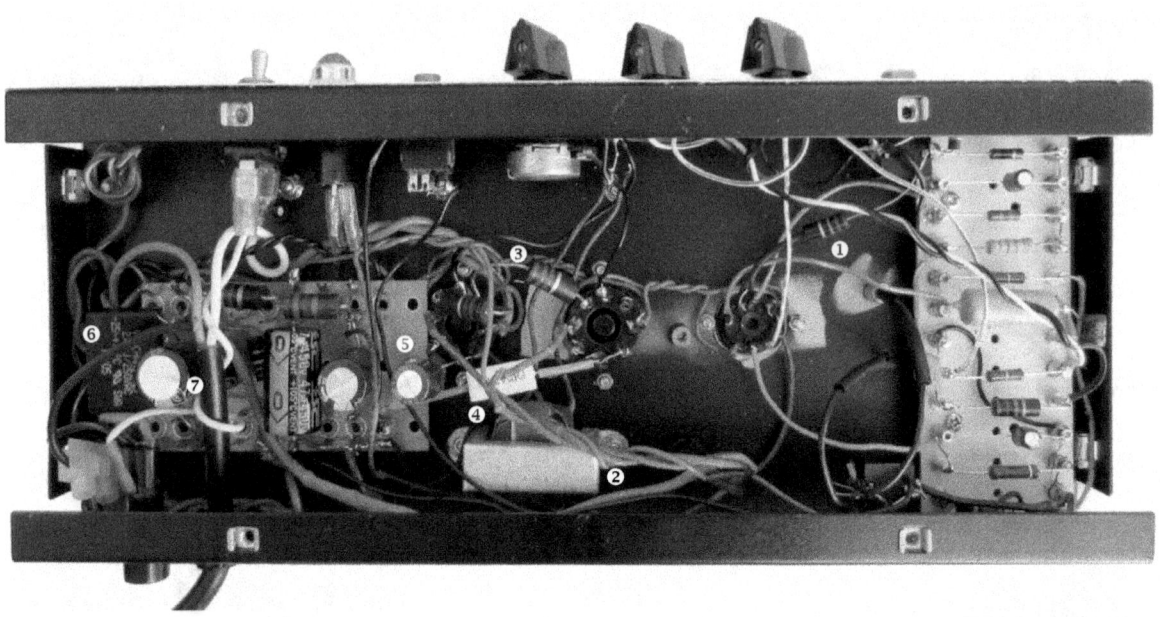

The DC voltage "tweaker" film capacitor (6) is black in color and cannot be seen very well on the photo; it is right next to the first filtering elco (7). The output transformer (8) and the filtering choke (9) were salvaged from vintage amps.

The power transformer was brand new, a bit too large for this 12 Watt amp, but it could physically fit, and we didn't have a smaller one anyway. The load on it is very low, so it runs as cool as a cucumber.

Trade trick: Fine-tuning the DC voltage with just a single film capacitor

Due to quite high AC voltage on the HV secondary winding, initially, we envisaged and built an LC filtered high voltage supply without any capacitance between point "X" and GND. The DC voltage with LC filtering is much lower than with CLC filtering, as discussed in the power supply chapter of this book.

However, the voltage V1 and all subsequent voltages were now too low. With a CLC filter and 22 to 47µF elcos all voltages would be too high. If that happens to you, don't despair.

Take a few high voltage film capacitors (rated at a minimum of 400V_{DC}, 630V_{DC} would be better), 0.22µF, 0.47µF, 0.68µF, 1µF and 2.2µF, and solder them one by one between point "X" and ground. The higher the capacitance, the higher the voltage rise in point "X" from the LC filter situation with no capacitance at all.

In other words, by changing the value of that capacitor, you can fine-tweak the DC voltage in point "X"!

The filter will operate as something "in-between" the LC and CLC filter. The mathematical description would be too complicated; the current and ripple voltage waveforms are also "neither here nor there," but the main thing is that this arrangement works well.

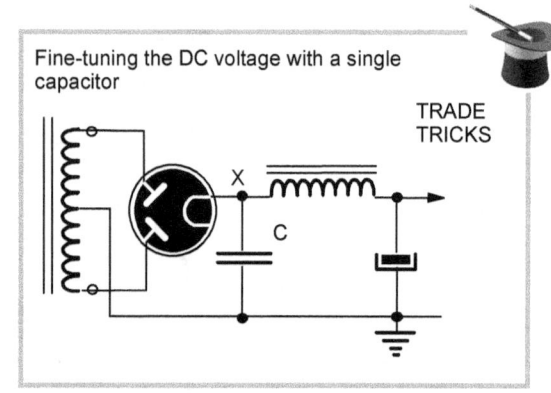

Fine-tuning the DC voltage with a single capacitor

TRADE TRICKS

REBUILT: VOX AC4TV

Apparently inspired by the 1961 vintage Vox AC4 amp and released in 2008, AC4TV is available as a head, an 8" and a 10" combo. The vintage AC4 used a tube rectifier and an EF86 pentode preamp stage. This amp has two 12AX7 triode pentode stages, so it's unclear what that "inspiration" was. Apart from the single-ended, cathode-biased EL84 pentode stage, the two amps are totally different. The maximum output power is 4 Watts.

Before: The inside view

There are quite a few problems with the small and crammed printed circuit board. Firstly, notice how close the cathode bypass capacitors (1) are to the heat-producing cathode resistor of the output stage and the output tube (2). Likewise, the power supply filtering caps (3) are too close to the hot power tube. Any heat above the ambient temperature will dry elcos out and severely shorten their life.

If you study the photo showing the rear view of the amp with the removed back panel and the "chassis" carrying the PCB, you will notice that the two tubes are mounted horizontally.

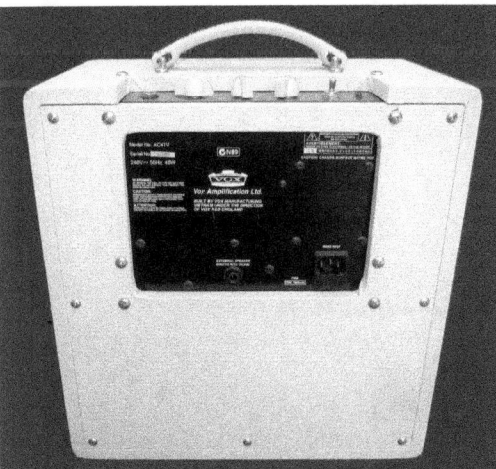
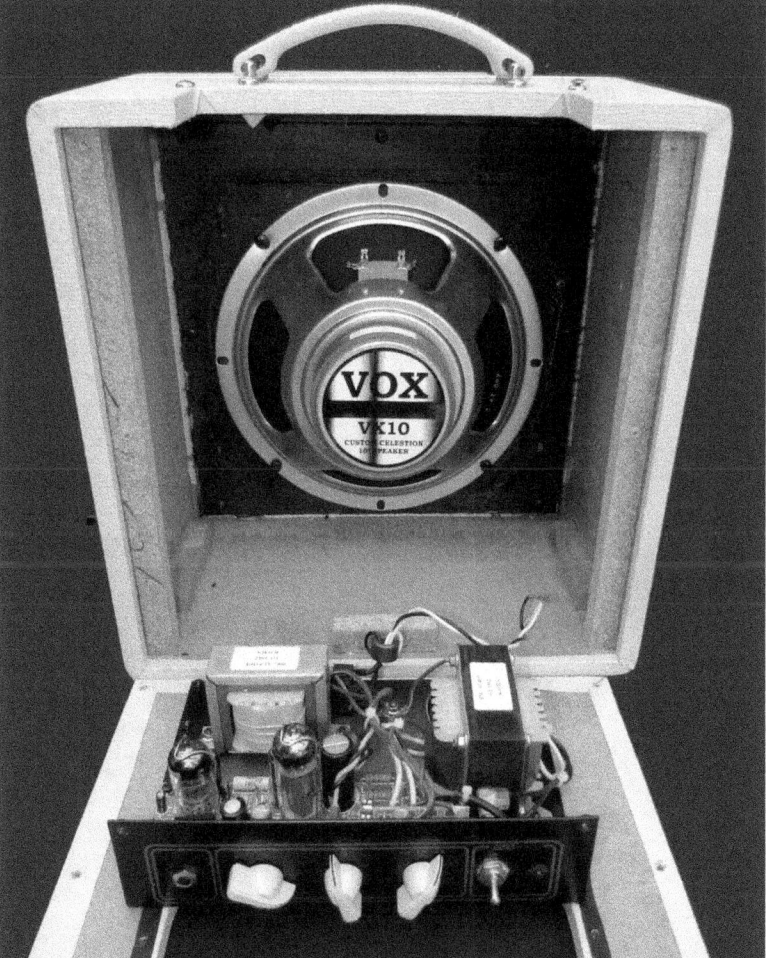

ABOVE: The outside and internal views of Vox AC4TV

RIGHT: A close-up of its printed circuit board

That is never a good idea. Unless you really have no other option (due to small or awkward internal space and topology), tubes should be mounted vertically, either pointing upward (as on hi-fi amps) or hanging downwards, as on many guitar amps, but never horizontally.

The control, screen, and suppressor grids of power pentodes are delicate structures or helixes wound using thin wires that get hot and stretch in operation. When horizontally mounted, this thermal dilatation combined with gravitational force can cause internal shorts, which may damage or even destroy the tube. It seems that whoever designed this amplifier either lacked some basic knowledge about vacuum tubes or simply did not care about the reliability aspects of this amp, for instance, how long the tubes and elcos last.

There are two internal fuses (4), so replacing them is a major and time-consuming exercise if one blows during practice or performance since the rear panel needs to be unscrewed and removed. An externally accessible fuse would be a better option.

To get the output stage to distort at lower volume levels, VOX included a 3-way switchable passive attenuator (6), which limits the output power at 4 Watts (attenuator out of the circuit), 1 Watt, and 1/4 Watt. The output tube still works at the attenuated power levels, but instead of all its output driving the speaker, the attenuator diverts some of that power into heat dissipated on the resistor mounted on a small PCB. That further increases the temperature levels inside this tiny enclosure.

Circuit diagram before

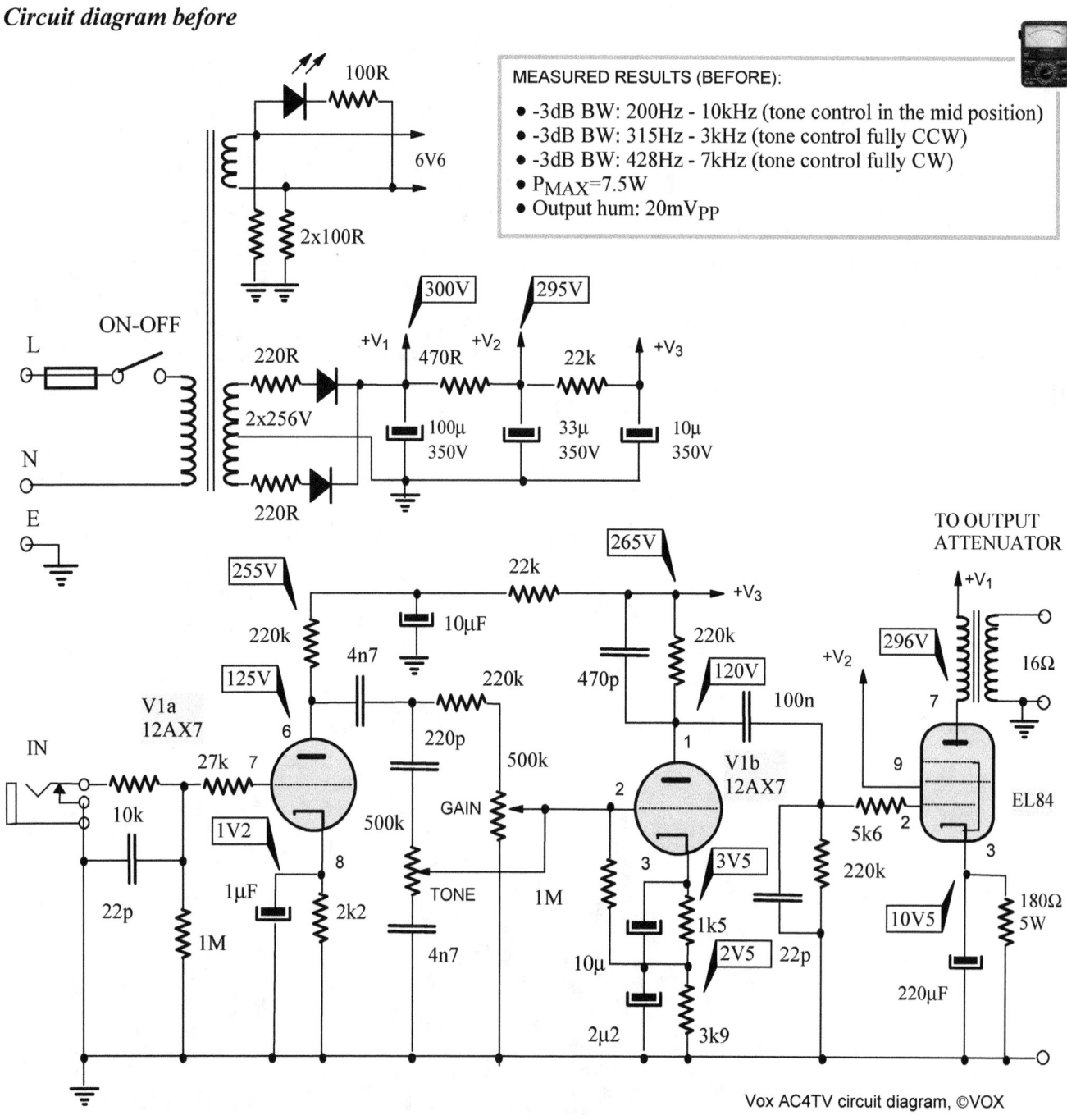

MEASURED RESULTS (BEFORE):
- -3dB BW: 200Hz - 10kHz (tone control in the mid position)
- -3dB BW: 315Hz - 3kHz (tone control fully CCW)
- -3dB BW: 428Hz - 7kHz (tone control fully CW)
- P_{MAX}=7.5W
- Output hum: 20mV_{PP}

Vox AC4TV circuit diagram, ©VOX

Notice three frequency compensation capacitors, 22p at the input, 470p across the anode resistor of the 2nd stage, and 22p shunting the grid resistor of the power tube. Despite all that, the amp sounded too harsh and shrill.

The hum (AC ripple) at the speaker terminals was 20mV$_{PP}$ (peak-to-peak), way too high, so the amp wasn't suitable for any recording or performance. Even during practice the hum was annoying.

The cathode DC current (the sum of anode and screen grid currents) can be calculated from the measured cathode voltage of 10.5V and the 180Ω resistor, so using Ohm's Law, I_K=10.5/180 = 58.3mA. The total power dissipation (cathode + screen) is P=I_KxV_{AK}= 0.583x(296-10.5) = 16.6 Watts.

The datasheet for EL84 says P_{AMAX}=12 W and P_{SMAX}=2 W or 14 Watts together, so VOX designers have exceeded this maximum dissipation by almost 3 Watts or 19% (16.6/14 = 1.186)! This is unforgivable. In its stock form, this amp will chew through power tubes like a starving beaver in a Canadian forest.

The designer tried to emulate the voltage sag of a tube rectifier by inserting a 220Ω resistor in series with each HV rectifier diode. While this does achieve a similar sag of the high voltage at higher power levels, it cannot emulate the sound of a tube rectifier. Remember, in single-ended amps, the power supply, including the rectifier, is in the signal path and has a major impact on the voicing of an amp!

Solid-state diodes, especially the cheap & cheerful 1N4007 or similar garden variety used by most manufacturers, make those amps (including this one) sound harsh and brittle. The slightly more expensive fast and soft recovery epitaxial diodes sound better, but even they cannot match the sonics of tube rectifiers.

The output transformer

With its lamination stack rated at 32 Watts, the output transformer is quite large for such a low power amp, meaning plenty of clean bass and low distortion should be possible.

The 16Ω speaker is reflected back into the anode circuit as 8.7kΩ impedance, which is very high for the EL84 tube, normally paired with 5kΩ transformers. Again, a higher load will reduce the maximum output power and lower the distortion in the output stage.

A very high-quality factor means low HF attenuation, extending the possible high-frequency limit of the amp. The upper -3dB frequency of the whole amp was only 10kHz, but the limitation was not in its output transformer!

VOX AC4TV OUTPUT TRANSFORMER:

- EI66 laminations
- a=22 mm, stack thickness S=26mm
- Center leg cross section A=aS = 5.7cm^2
- Power rating: P=A^2= 5.7^2= 32W
- Voltage ratio: 23.3
- Primary impedance with 16Ω speaker: 8.7kΩ
- Primary DCR: R_P=132Ω
- Primary inductance @1kHz L_P= 8.5H
- Leakage inductance @1kHz L_L=6.9mH
- Quality factor: QF = 1,232

Rebiasing the output stage

The original quiescent point Q1 was positioned above the parabola of maximum dissipation, so we lowered the anode current slightly. The published graph shows grid bias in Q1 as -9.7V, but in our case, it was -10.5V. These discrepancies are normal.

The published graphs are for specific screen voltages, in this case for V_S=300V, but our V_S was slightly lower. Secondly, tube parameters vary, and the actual tube used had somewhat different characteristics than the bogey or average tube used to publish the graphs.

Increasing the cathode resistance from 180Ω to 220Ω fixed the problem.

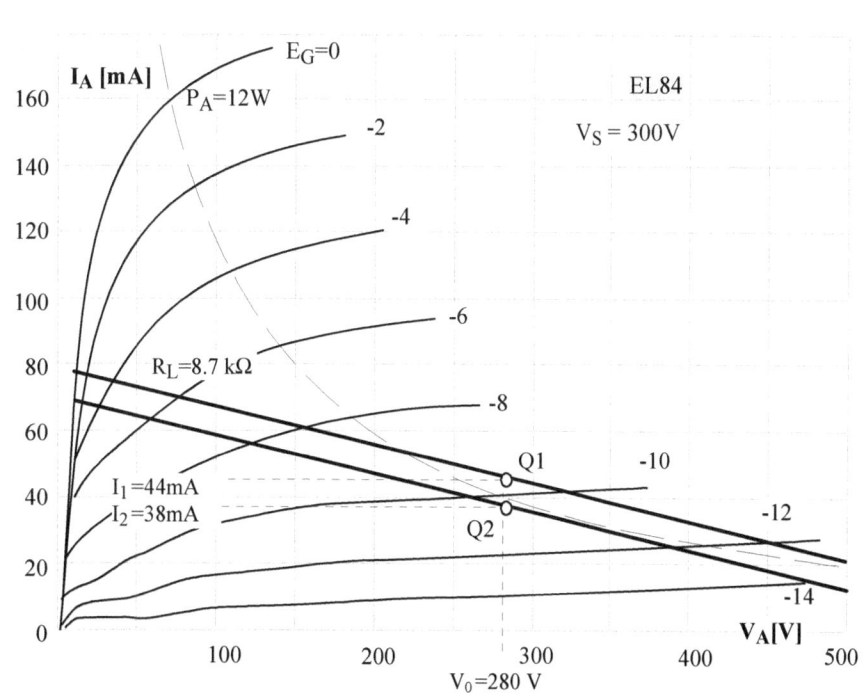

Rewiring the amp in a point-to-point fashion

The first goal of the rewiring exercise was to give the amp some air, to spread out the previously cramped insides. The power (1) and the output transformer (2) were retained but mounted on the back timber cover and not on the steel chassis. That eliminated their mutual coupling through the magnetic chassis and reduced the output hum from 20mV$_{PP}$ to 2mV$_{PP}$!

The maximum output voltage remained the same, around 11V, which on a 16Ω speaker resulted in maximum power of $P=V^2/R = 11^2/16 = 7.5$ Watts.

The three tubes were mounted on an aluminium L-shaped profile which also acts as a shield between the two transformers and the components mounted on an RS (Radio Spares) terminal strip, which, if you search for it on their website, is for some strange reason called a "group panel."

Master volume control was added where the attenuator was (4), on the front panel, and the attenuator was moved to the back (5). Since there was no room left on the terminal board, the power supply filtering elcos were glued directly to the chassis (6).

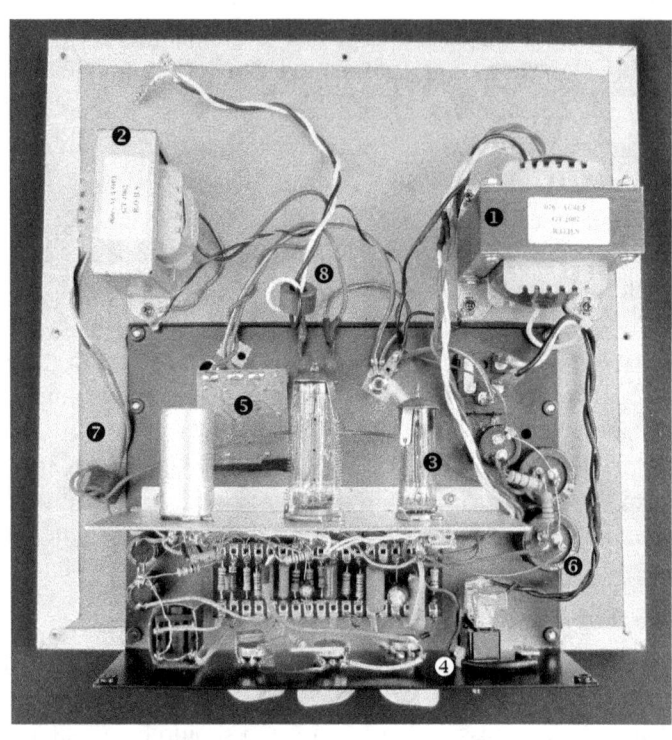

ABOVE: The close-up of the terminal strip, pots, tube sockets and the associated wiring
LEFT: The back cover of the rebuilt Vox AC4TV

The 22nF coupling capacitor between the gain potentiometer and the grid of the second triode, marked on the original VOX diagram, was missing on our amp, so we returned it in the rewired version.

Two ferrite O-rings were added to reduce RF interference and hum, one in the line between the output tube's anode and the output transformer's primary (7), the other in the secondary's output to the speaker (8). One or two loops of the wire through the ferrite rings are sufficient.

Installing a tube rectifier

Since the high voltage secondary of the stock power transformer had a center tap, replacing silicon diodes with a rectifier tube was an option. That would soften the sound and bring back some of the magic of the original, vintage AC4 amp, which used a 6V4 (EZ80) tube rectifier.

We added 6X4 (EZ90), a similar indirectly heated rectifier tube (3). While 6V4 uses a Noval (9-pin) socket, 6X4 needs a miniature 7-pin socket.

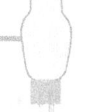

TUBE PROFILE: 6X4 (EZ90)

- Indirectly-heated dual rectifier
- Heater: 6.3V/600mA
- V_{MAX}= 350V, I_{DCMAX}= 70mA
- V_{HKMAX}= 450V
- Voltage drop @70mA: 22V
- Peak inverse AC voltage: 1,250V

Two important conditions that must be met before you install a tube rectifier

Since the amp's anode supply voltage is around $300V_{DC}$ and the maximum heater-cathode voltage of the 6X4 rectifier tube is $450V_{DC}$, it could be heated from the same heater circuit as the rest of the tubes; it did not require its separate heater secondary winding.

Of course, you cannot simply add another tube's heater to the existing power transformer; the additional heater current draw (600mA in this case) could overload it. Luckily, this amp's heater winding happily supplied the additional current since it had some reserve capacity.

Circuit diagram after modifications

The lower frequency limit of the stock amp was too high, in the region of 200-428 Hz, so we increased the coupling capacitances (1) and (2), and the lower -3dB limit dropped from 200 to 92Hz.

With the removal of two bypass capacitors, 470p across the anode resistor of the 2nd stage, and 22p across the grid leak resistor of the power stage, the upper-frequency limit increased from 10 to 22 kHz (with tone control in the mid position). The amp acquired an added sparkle; it sounded more lively yet less harsh! Perhaps the tube rectifier softened the sound.

Since we performed all the modifications together, we will never be sure of the specific contributions of each change. Only when you change one thing at a time can you link each mod with a sonic change (if any)!

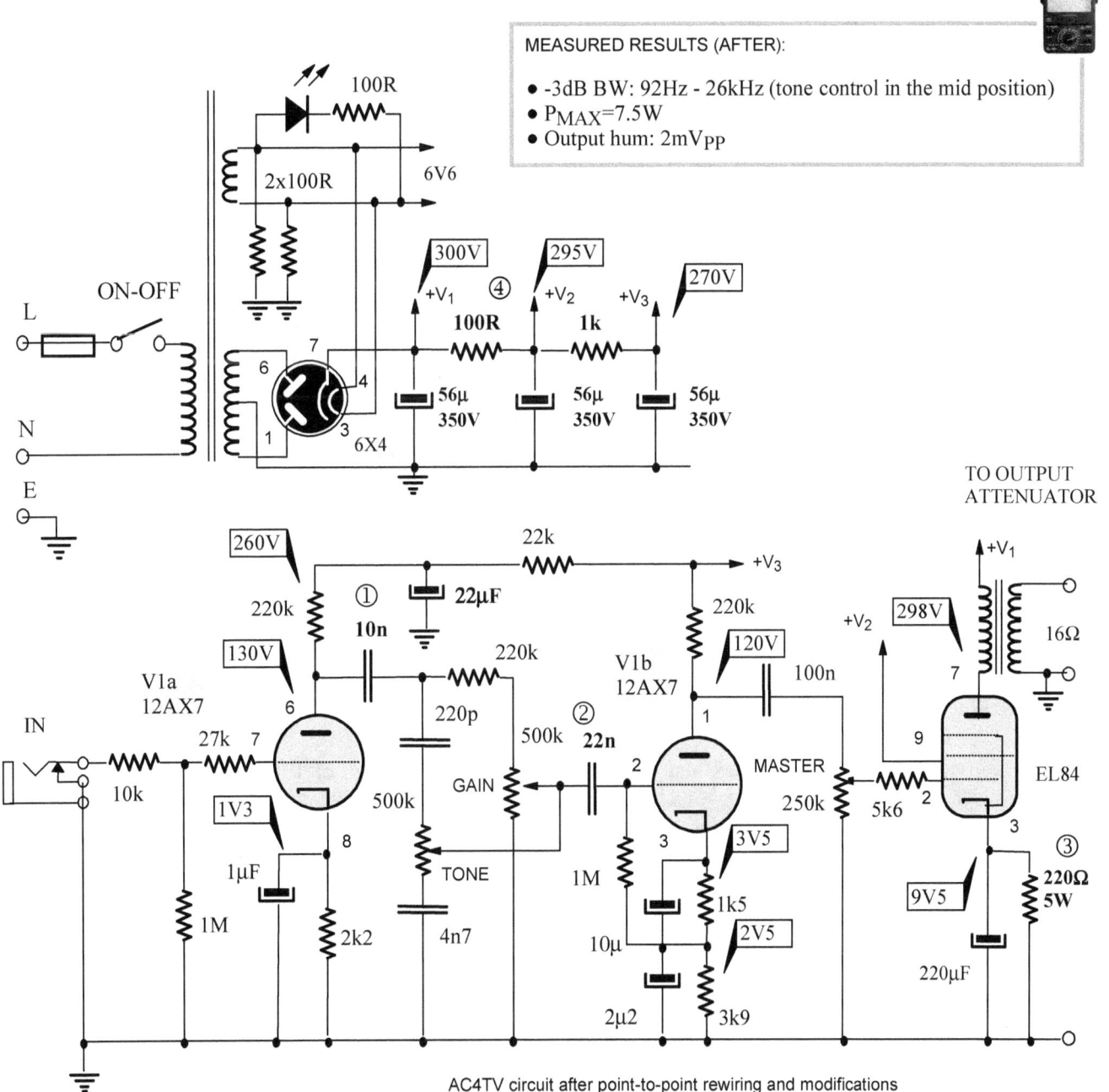

AC4TV circuit after point-to-point rewiring and modifications

As already mentioned, the power tube's cathode resistance (3) was increased from 180 to 220Ω, which reduced the cathode current to 9.5/220 = 43.2 mA. The anode and screen combined power dissipation was now P=(297-9.5)*0.0432 = 12.4 Watts, below the 14 Watts maximum.

The two series resistors in the CRC HV power supply filtering were adjusted to achieve the same DC supply voltages of 300 and 296 Volts (4).

The output attenuator

As always, to understand the operation of a switched circuit, eliminate the switch and draw as many simplified circuits as there are switch positions. In this case, there is no need to draw the circuit when the switch is in the "5W" position; the input to the attenuator is simply switched to its output, the RC components aren't doing anything.

The circuits for the other two power levels are identical, except that the value of the resistor in parallel with the output in the voltage divider is 15Ω in the "1 Watt" position and 5.6Ω in the "1/4 Watt" position.

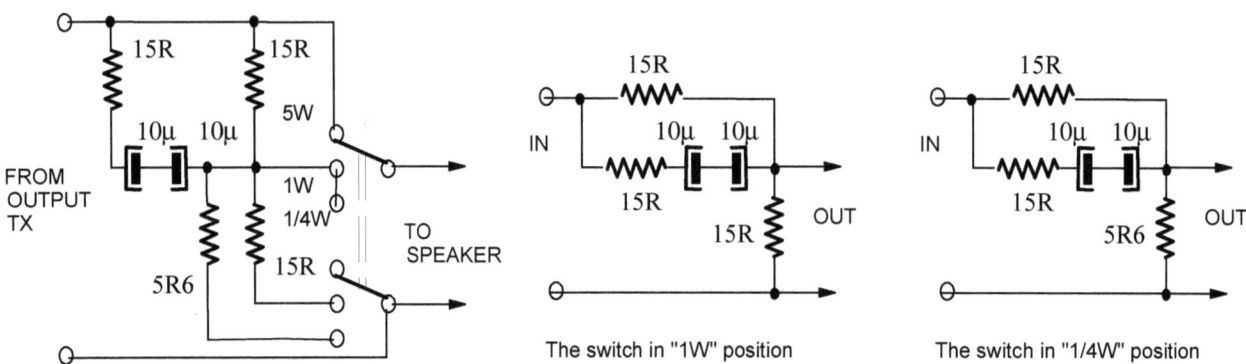

The switch in "1W" position   The switch in "1/4W" position

Notice two identical series capacitors connected back-to-back. Their total capacitance is half or 5mF. Without these caps, the attenuator would be a simple resistive voltage divider, but their presence in one of the two 15Ω branches means the impedance of that branch will change with the signal's frequency. In other words, this is a frequency-compensated voltage divider.

To understand how their impedance will change and what difference such a change will have on the attenuation of the circuit, let's choose two frequencies, one in the bass and one in the treble region, say 200Hz and 4,000 Hz. Then we will calculate the reactance of the capacitors and the attenuation of the attenuator in 1W and 1/4W positions.

At f=200 Hz, $X_C=1/\omega C = 1/(2\pi fC) = 159$ Ω, or approx. 10X higher than the value of the resistor. Since the phase angle between the voltage drop on the capacitor and the resistor is 90 degrees, we cannot simply algebraically add those two impedances up; we have to use Pythagora's Theorem and calculate the hypotenuse of the right angle triangle they form: $Z= \sqrt{(R^2+X_C^2)} = \sqrt{(15^2+159^2)} = 159.7$ Ω.

Since X_C is so much higher than R, we could have just assumed that the presence of 10X higher reactance will not change the overall impedance of that branch. Indeed, a change from 15Ω to 15.9Ω is negligible.

At f=4,000 Hz, $X_C=1/\omega C = 1/(2\pi fC) = 31.8$ Ω, and $Z= \sqrt{(R^2+X_C^2)} = \sqrt{(15^2+31.8^2)} = 35$ Ω.

So, at 200Hz, the two series branches of 15R and 15R form one equivalent branch of 7R5, so in the "1W" position, the voltage divider has an attenuation of A= 15/(15+7.5) = 0.67, and in the "1/4W" position the attenuation is A= 5.6/(5.6+7.5) = 0.43!

At 4kHz the two series branches of 15R and 35R form one equivalent branch of 10R5, so in "1W" position the voltage divider has attenuation of A= 15/(15+10.5) = 0.59 and in the "1/4W" position the attenuation is A= 5.6/(5.6+10.5) = 0.35! Thus, we conclude that the presence of two capacitors reduces the attenuation of high frequencies, but only slightly.

Since the full power position (no attenuation) is marked 5W and power is $P=V^2/R$ and since R is the constant load impedance (16Ω), the ratio of two power levels depends on the square of their voltages: $P_{OUT}/P_{MAX}= (V_{OUT}/V_{MAX})^2$ and since attenuation $A=V_{OUT}/V_{MAX}$, we have $P_{OUT}=A^2*P_{MAX}$!

At 200Hz the output power is approx. $P_{OUT}=A^2*P_{MAX} = 0.67^2*5 = 0.45*5 = 2.25W$ in "1W" position and $0.43^2*5 = 0.185*5 = 0.93W$ in "1/4W" position. So, in neither position the actual power output matches the levels declared by VOX.

DIY PROJECTS: QUIRKY & UNUSUAL DESIGNS

11

- THE NEW YORKER: EVERYTHING OLD IS NEW AGAIN
- GOLDEN GECKO: A CHAMELEON AMP THAT CAN USE SIX DIFFERENT OUTPUT TUBES
- DOUBLE TROUBLE: BLENDED SE OUTPUT STAGES
- LOW VOLTAGE TUBE GUITAR AMPLIFIERS

If the idea of building a tube amp inside a gutted state amp's cabinet does not light your fire, your choice of a suitable cabinet is limited only by your imagination. To get your creative juices flowing, three eclectic (could be translated as "weird" or "quirky") examples are covered in this chapter.

"The New Yorker" is a vintage Heathkit hi-fi amp, modified for guitar use and housed inside an old-fashioned hi-fi speaker cabinet.
"Golden Gecko" is a universal tube head (many different Octal power tubes can be plugged in without any adjustments) built inside a large metal ashtray as its chassis.
"Double Trouble" is much more than an amp, almost a work of art, a dual-channel amp built inside a timber wall shelf.
Finally, we designed and built a couple of low-voltage tube guitar amplifiers.

NEW YORKER: EVERYTHING OLD IS NEW AGAIN

The speaker box

A pair of these dark teak hi-fi speaker boxes were spotted on a verge during council pickup days and ended up in our stash, where they sat pretty for a few years. Eventually, while looking for projects to include in this book, an idea was born. Why don't we use it as a "housing" or a cabinet to accommodate a small PA or hi-fi amp that will be used as an example of a conversion to a guitar amp?

While the boxes are unmarked (no brand name), the speaker inside is a 6.5" Australian-made ROLA, so the boxes were probably locally made and sold with one of many radio-turntable consoles popular at the time. The stamp on the speaker says 65/8, which could be the 8th week of 1965.

The DC resistance of the speaker's coil is 8.6Ω, indicating that its nominal impedance is most likely 16Ω. Using a digital LCR meter on R (resistance) function, the readout was 11.4Ω. That "resistance" is actually a misleading name, the LCR meter sends continuous 1 kHz tone pulses through the speaker (you can hear them during measurements) and thus measures an overall impedance at that particular frequency, not the DC resistance.

Judging by the size of its magnet, the speaker should be good for at least 15 Watts. There is plenty of room inside the enclosure, so even a 12" speaker can be installed.

ABOVE RIGHT: Teak finish and dark cloth make the vintage speaker box look classy. The leather handle on top was just placed there temporarily to see what it would like as a guitar combo. The etched aluminum panel is there for esthetic purposes only; it is simply glued to the front baffle.

FAR RIGHT: Measuring DCR of the speaker coil with a multimeter (8.6Ω).

RIGHT: Measuring the impedance of the speaker's coil at 1 kHz using a digital LCR meter (11.1Ω).

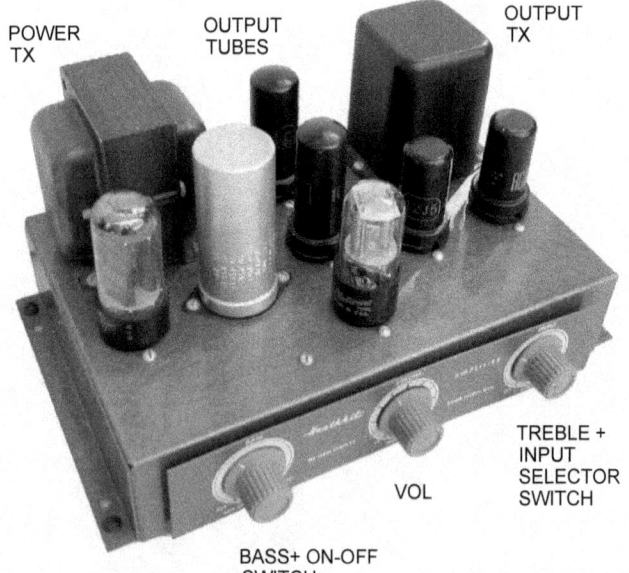

POWER TX — OUTPUT TUBES — OUTPUT TX — TREBLE + INPUT SELECTOR SWITCH — VOL — BASS+ ON-OFF SWITCH

The amp: Heathkit A7-C

This petite cutie dates back to 1955, when it sold as a kit for US$17.50. Typical of its vintage, it is an all-tube design, and the construction is point-to-point, or, in guitar circles, "hand-wired."

A maximum of six Watts of output power was claimed by Heathkit.

Two RCA inputs (1) are at the back (next page) and will not be used once we convert this baby to a guitar amp. Leave them physically in place but disconnect and remove the two shielded cables (6). The guitar input jack will be in position (2), very close to the channel selection switch (3). That way no shielded cable will need to be used.

DIY PROJECTS: QUIRKY & UNUSUAL DESIGNS

However, the switch is mounted at the back of the treble potentiometer, which needs to be rotated fully clockwise to click into one switch position and fully CCW to select the other input. This is a very awkward arrangement; installing another switch at (4) would be better.

As for the outputs, you can replace the existing terminal strip with three separate phone sockets for 4, 8, and 15Ω speakers (5), or just one speaker socket and an impedance selector switch.

Since that would require drilling of the chassis, wiring of the switch, and significant time investment, we kept the terminal strip and only wired the 8Ω output permanently to the internal speaker.

Most of the original film caps had high dielectric leakage (losses) and had to be replaced. A few were still fine (of a different brand), so it seems Heathkit used various brands and component quality levels.

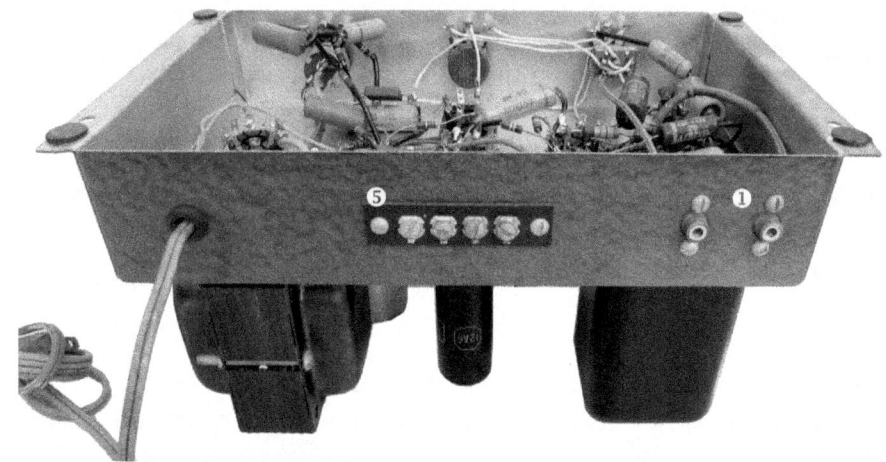

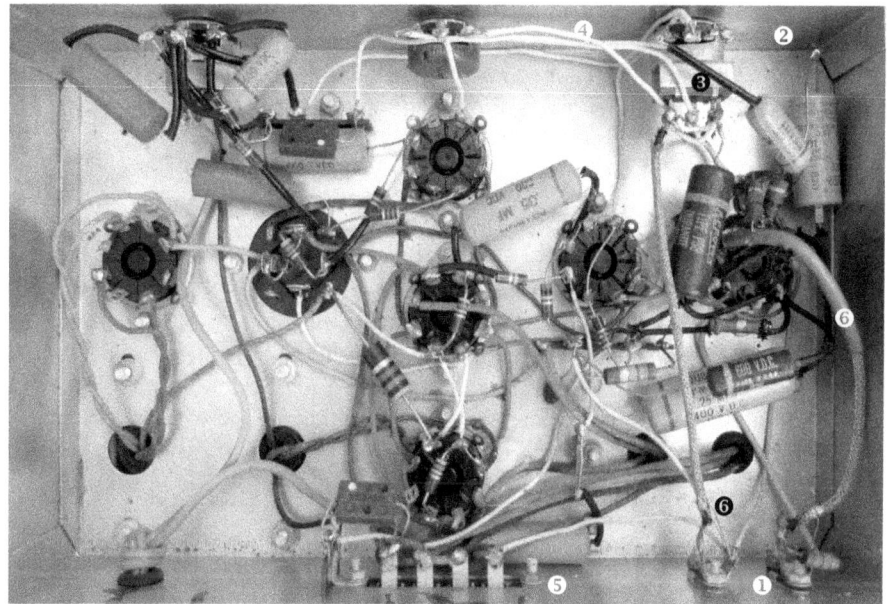

ABOVE: The under the chassis view of Heathkit 7-C in its original condition, just as it was on the day some keen DIY enthusiast wired it up in 1955!

The original topology

To Heathkit's designer's credit, all tubes in this monoblock are octal and use 12.6V heating. Sweet and simple. Input 2 is a line level, so a sensitivity of around $1V_{RMS}$. After the volume control, the signal is amplified by a single triode (12J5) and tonally shaped by a two-knob EQ circuit. The gain loss in the tone control section is compensated for by another common triode stage, this time using one-half of a 12SL7 duo triode. Finally, the other 12SL7 triode acts as a split-load phase inverter and drives the grids of the 12A6 output pentodes.

Input 1 is for a phono cartridge or a microphone. Phono equalization filter comprising one resistor and two capacitors must be installed for a turntable input and must be removed for a microphone or a guitar input. The input terminating resistor of 27k also needs to be removed for guitar input. The two inputs are at the back, together with three speaker outputs, impedances of 4, 8, and 15Ω.

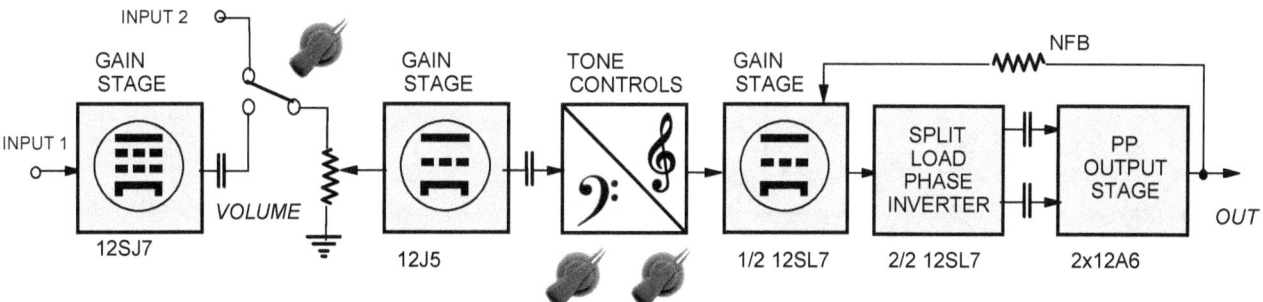

The output tubes: 12A6

12A6 octal beam tetrodes were not common even in the 1950s, and today they are almost unheard of. With 7.5 Watts anode and 2.5 Watts screen dissipation, they aren't much weaker than 6BQ5 or 6V6. Their pinout is identical to that of 6V6, so providing the amplifier's power supply (primarily the heater and HV secondary windings) can supply the additional current draw of 6V6, they can be substituted instead of 12A6. Of course, 6V6's heaters would need to be rewired in series.

TUBE PROFILE: 12A6

- Indirectly-heated beam tetrode
- Octal socket, heater: 12.6V, 150mA
- V_{AMAX}=250V, P_{AMAX}=7.5W
- V_{SMAX}=250V, P_{SMAX}=1.5W
- gm=3 mA/V, µ=210, r_I=70kΩ
- SE pentode (7k5 load): P_{OUT}= 3.4 W @ 5% THD

The preamp pentode: 12SJ7

A 12.6V heater version of the more common 6SJ7 high gain octal pentode, dating back to 1939. The S indicates a "single-ended" body or case, meaning without the top cap. The 12J7GT is a glass version ("G") with the control grid connected to the top cap (thus no "S") to ensure as much separation and isolation between the grid and anode circuits and hence low hum. The tube was designed for low frequency (audio) use, so its anode curves are relatively linear.

TUBE PROFILE: 12SJ7, 12SJ7GT

- Indirectly-heated sharp-cutoff pentode
- Octal socket, 12.6V, 150mA heater
- V_{AMAX} = V_{SMAX} = 300 V_{DC}
- P_{AMAX}= 2.5 W, P_{SMAX}= 0.7 W
- TYPICAL OPERATION PENTODE:
- V_A=250V, V_S=100V, V_G=-3V
- I_{A0}=3 mA, I_S=0.8 mA
- gm =1.65 mA/V, µ=1,650, r_I=1MΩ

The preamp triode: 12J5

An obscure low-to-medium mju single triode. However, a decent plate dissipation of 2.5 Watts and a relatively low internal impedance of 7-8kΩ at higher anode currents make it a good choice to drive tone stacks, as Heathkit is doing here. This means the loss of voltage gain is minimized compared to triodes with very high internal impedance, such as 12AX7, which are the worst choice for driving such low impedance loads!

TUBE PROFILE: 12J5

- Indirectly-heated general purpose triode
- Octal socket, heater: 12.6V, 150mA
- V_{AMAX}=300V, P_{AMAX}=2.5W
- gm=2.6 mA/V, µ=20, r_I=7.7kΩ

The preamp duo-triode: 12SL7

12SL7 and its 6.3V heater sister 6SL7 may be relatively unknown in tube guitar amps but are pretty common in high fidelity circles. 12SL7 has similar static parameters to 12AT7, a more modern Noval duo-triode, a high mju of around 70, but a lower anode power rating, only 1Watt, compared to 12AT7's 2.5 Watts.

TUBE PROFILE: 12SL7

- Indirectly-heated duo- triode
- Octal socket, 12.6V, 150mA heater
- V_{AMAX}=300V, P_{AMAX}=1W
- Static parameters (at V_A=250V, V_G=-2V and I_A=2mA): gm=1.6 mA/V, µ=70, r_I=44kΩ

The original circuit diagram

For the sake of consistency, tidiness and clarity, most circuit diagrams in this book have been redrawn. It saves valuable space as well, since many original diagrams are very large and spread out. Also, some contain detailed information of no interest to us here, such as terminal numbers and colors of interconnecting cables.

There is also an issue of copyright in drawings, getting permissions, etc., while nobody can prevent you from drawing your own circuit diagram of an amp.

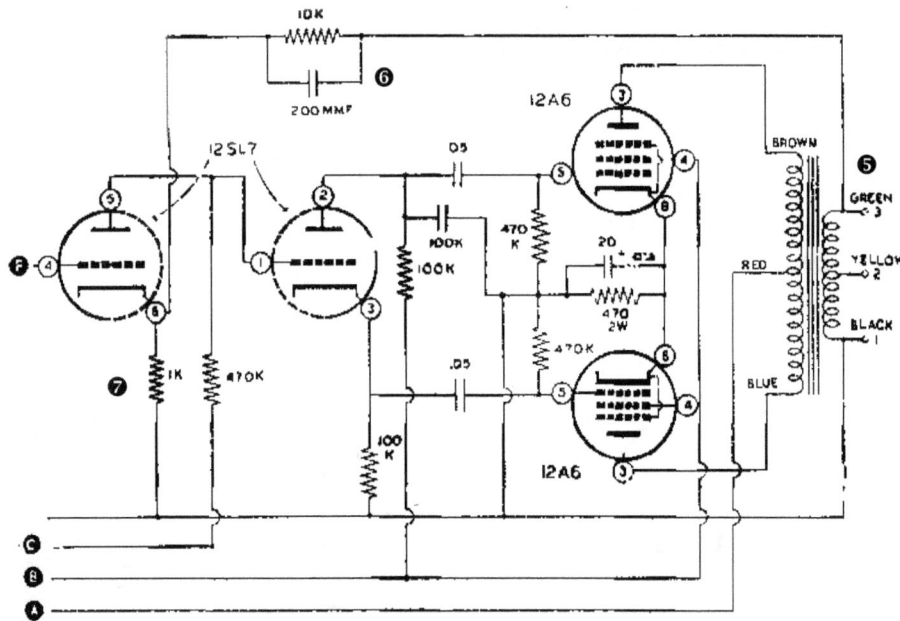

ABOVE: The driver, phase splitter and output stage of Heathkit A7-c amplifier

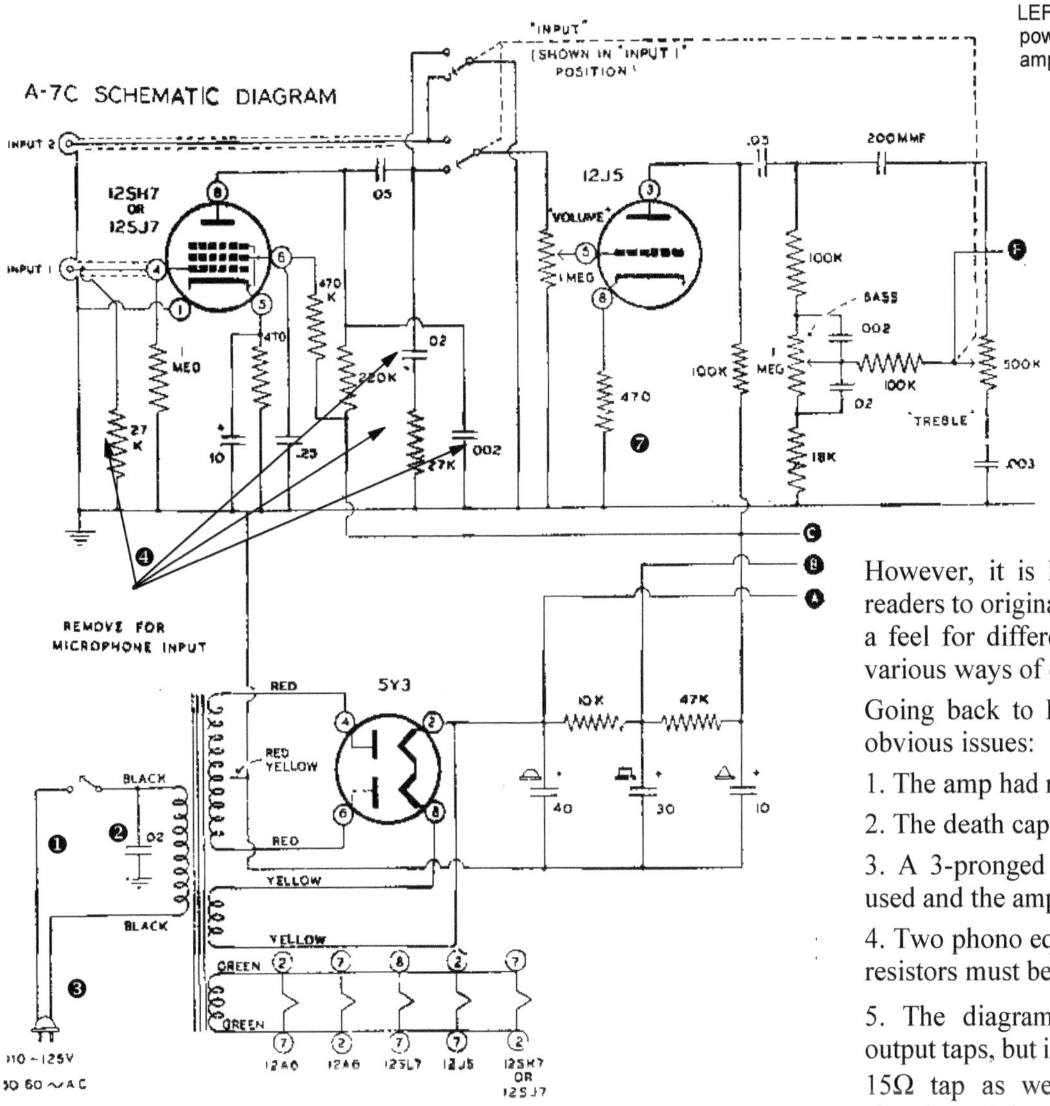

LEFT: The preamp stages and power supply of Heathkit A7-c amplifier

However, it is helpful to expose the readers to original diagrams so they get a feel for different symbols used and various ways of drawing circuits.

Going back to Heathkit A7-C, a few obvious issues:

1. The amp had no fuse

2. The death cap has to be removed

3. A 3-pronged mains cable must be used and the amp's chassis earthed.

4. Two phono equalization caps & two resistors must be removed.

5. The diagram shows 8Ω and 4Ω output taps, but in reality the amp had a 15Ω tap as well, from which NFB signal was taken.

6. The NFB components should be removed to increase the output power and avoid a sound that is too clean.

7. Both triode gain stages have un-bypassed cathode resistors, which further reduces both gain and distortion.

The initial topology

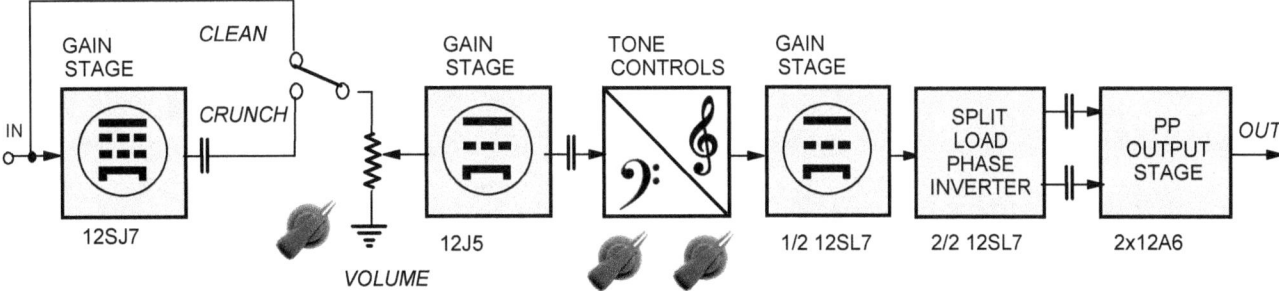

Initially, we assumed that the two triode gain stages would be enough to amplify a guitar pickup signal to maximum power levels, especially after removing global negative feedback and bypassing the cathode resistors in both those stages, which significantly increased the input sensitivity. Thus, the thinking went, we should make the pentode input stage switchable and remove it from the signal path in the Clean mode.

However, the Clean mode provided a great clean vintage sound but not much loudness, but at low volumes, the Crunch mode sounded even better. Also, there was a huge jump in loudness when Crunch was engaged. As the old military adage goes, no plan can survive the first contact with the enemy, or, as guitar amp makers would say, the proof of a design is in the tone. So, we changed the two stages around.

The final topology

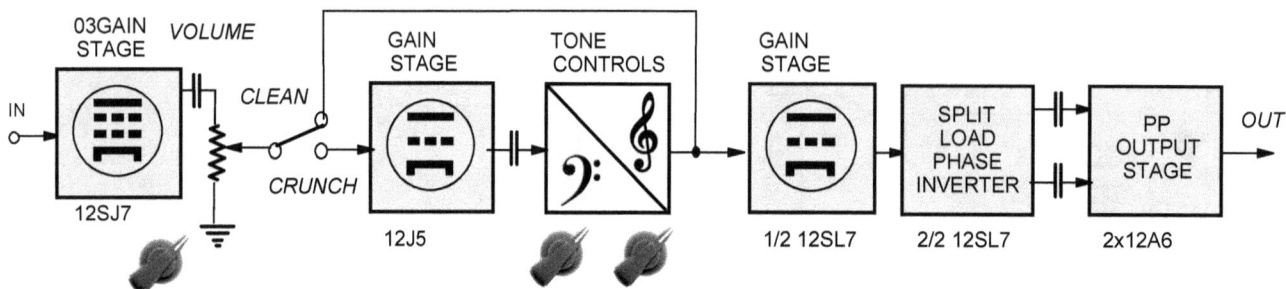

This solved the issue of a huge jump in loudness between the two modes. The high gain pentode stage is now active in both modes, the 12J5 triode stage is bypassed in *Clean* mode, as are, for good measure, the tone controls.

However, if you look at the circuit diagram, you will notice that the wiper (slider) of the Treble pot and all associated RC components are still connected to the grid (input) of the 12SL7 stage, and indeed, they do reduce gain somewhat. So, alternatively, use a DPDT switch, so its second contact can completely disconnect the Tone Controls from the 12SL7 stage, as per the arrangement below.

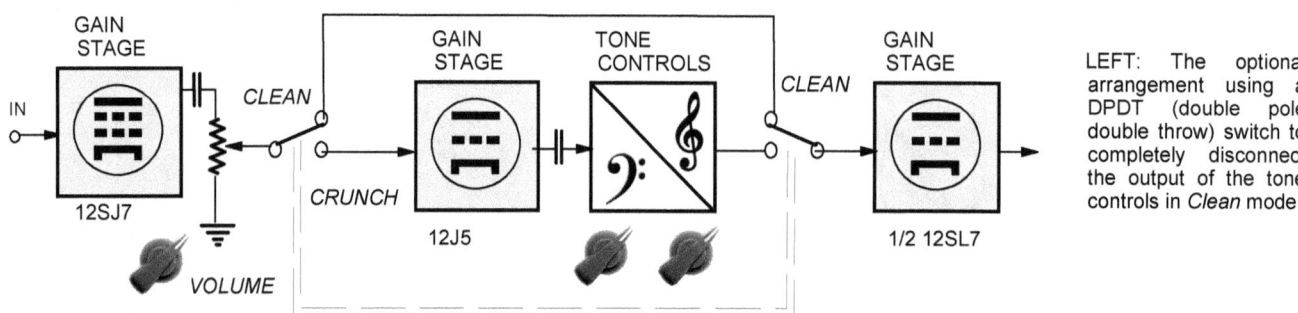

LEFT: The optional arrangement using a DPDT (double pole double throw) switch to completely disconnect the output of the tone controls in *Clean* mode

Circuit diagram after mods

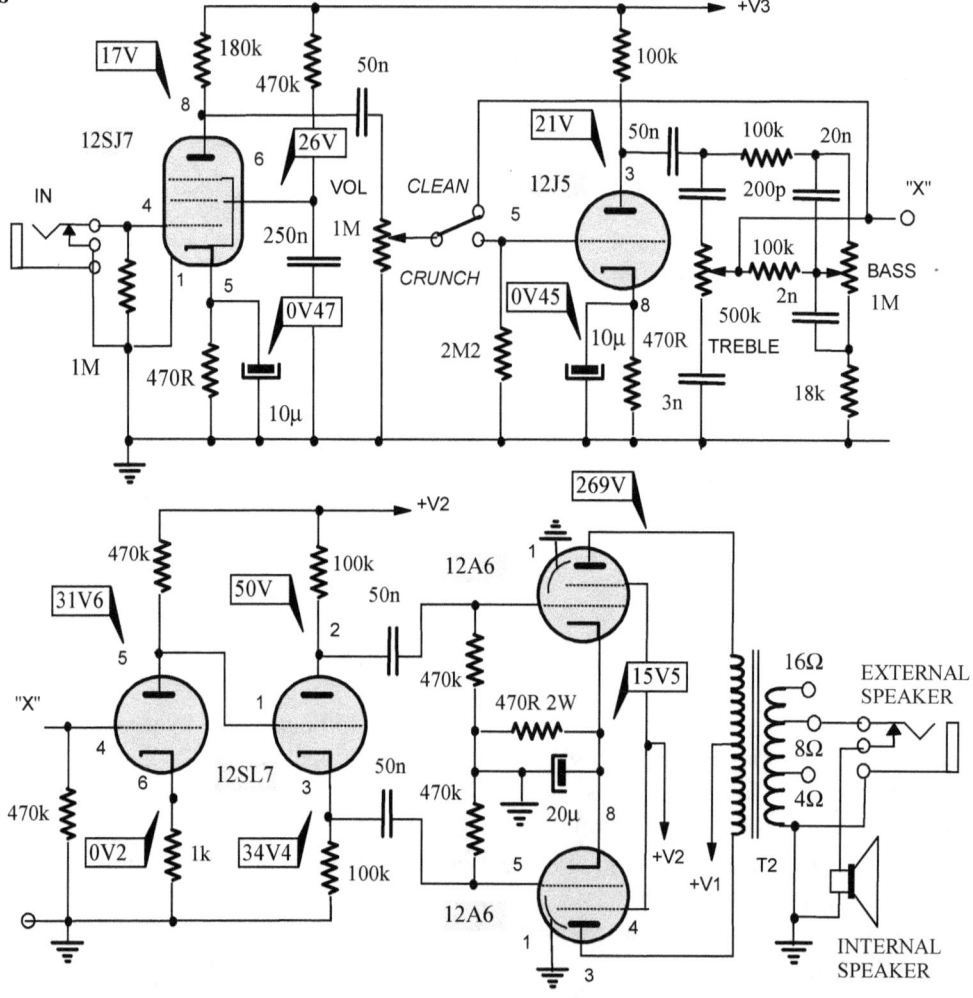

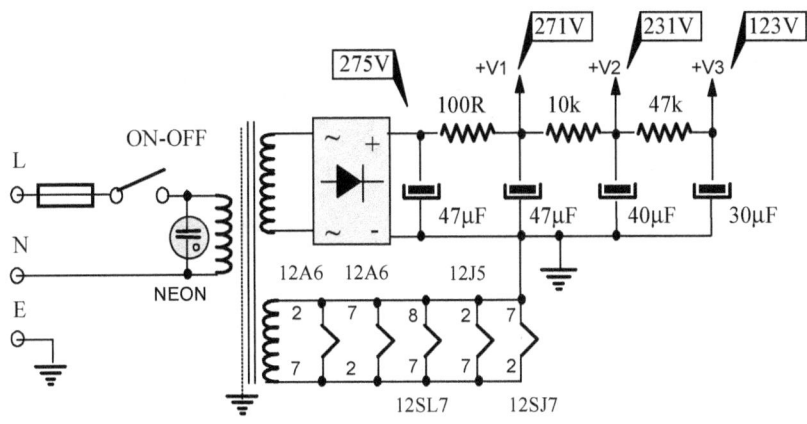

GOLDEN GECKO: A CHAMELEON AMP THAT CAN USE SIX DIFFERENT OUTPUT TUBES

Every designer has their favorite power tube(s), and in the hi-fi realm, 7027 and 7027A are definitely on our list of desirables. While the distortion of the ubiquitous 6L6 in SE triode mode is up to 10%, triode connected 7027A in the SE output stage easily achieves 5 Watts at only 2.8% distortion. As a pentode, two 7027A in push-pull can produce 50 Watts at only 1.5% harmonic distortion and 76 Watts at only 2% THD!

For the same heater power, 7027A has a 35 Watt anode power rating, higher than 30 Watts of 6L6 varieties, meaning it is also a more efficient tube. After all, 7027 was developed 20 years after the granddad 6L6 (which dates back to 1936!), so you'd expect the tube technology to progress somewhat over those two decades.

Design goals

Although 7027 and 7027A are such nice and clean sounding tubes, ideal for bass and keyboard amps, that does not mean that when pushed, they will not distort. We wanted to name this amp "Dr Jekyll and Mr Hyde," after the famous novella by the Scottish author Robert Louis Stevenson, first published in 1886. It describes a case of a split or multiple personality disorder, which would transform the respected, decent, and "clean" Dr Jekyll into a nasty murderer, Mr. Hyde. However, we bought a cool golden gecko emblem on eBay for a dollar, so we had to find a place for it somewhere.

Our amp will also have a split personality, a clean, articulate, and warm amp but with a dirty and crunchy flip side. That was one of the most challenging goals to achieve in one amp. Design goal #2 was for a player to be able to use as many octal power tubes as possible simply by swapping them around without any rewiring or adjustments.

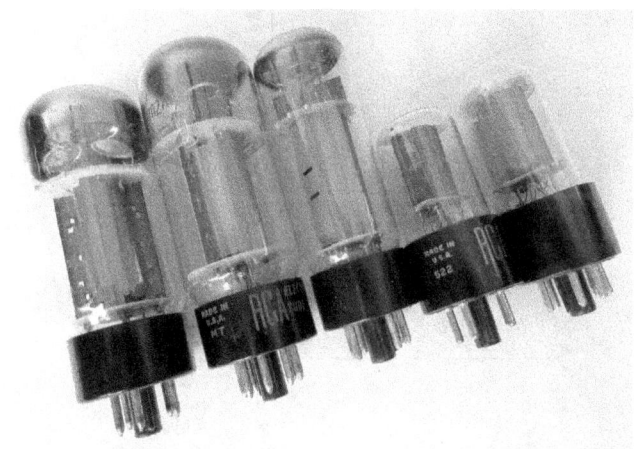

While 6L6 (far left) and 7027A (2nd from left) have an identically sized anode, the EL34 (middle) has a somewhat lower anode dissipation rating. All three dwarf the 6V6 (1st from the right), especially the slightly smaller and lower-powered 6F6 (2nd from the right). All are USA-made NOS tubes, most by RCA, except the Chinese-made 6V6. Notice the added cooling fins on both sides of the anode on the Chinese tube.

6V6, 7027A, 6L6 and EL34 in the same amp!

EL34 and 6L6 are pin-compatible, providing pin 1 on the tube socket is connected to pin 8, since EL34 does not have an internally connected suppressor grid, as 6L6 has (beam forming plates internally tied to its cathode). Luckily, 7027 and 6L6 are also pin-compatible beam power tubes, providing that in amplifiers designed for 6L6, nothing is connected to pins 1 and 6! Pin 1 is a screen grid in 7027A (also connected to pin 4, as in 6L6), and pin six is connected to pin 5, the control grid.

Sometimes designers of point-to-point wired equipment use such unused lugs on tube sockets as tie-points to save on terminal lugs. This is a risky practice and should be avoided. However, we are designing and building our amp from scratch, so we don't have to worry about these issues. The amp will be fully compatible with both tubes, so both can be used, 6L6 for slightly lower output power and higher distortion.

SIDE-BY-SIDE	EL34	6L6	7027A	6V6
Heater volts / amps	6.3/1.5	6.3/0.9	6.3/0.9	6.3/0.45
Max anode voltage [V]	800	500	600	315
Max screen voltage [V]	800	450	500	285
Anode dissipation [Watts]	25	30	35	12
Screen dissipation [Watts]	8.0	5.0	5.0	2.0

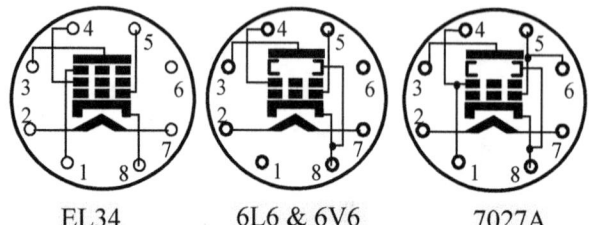

Pin 1 in EL34 pentode is its suppressor grid which needs to be externally connected to its cathode (the most common option in commercial amps) but can also be strapped to the anode or screen grid. Providing we join socket pins #1 and #4 to strap EL34's G3 (suppressor grid) to its screen grid, EL34 will also be compatible with our design. These pins are already internally connected in 7027 and will not affect 6L6 and 6V6, where pin#1 isn't used.

In some designs, EL34s operate at anode voltages of up to 800V, way above the allowed maximum of 6L6. Swapping 6L6 for EL34 would destroy the 6L6 tubes. Again, we will not design our amp for such high voltages, so the higher anode and screen ratings of EL34 will be irrelevant.

However, notice a much higher heater draw of EL34, 1.5A versus 0.9A for 6L6 and 7027A, so our heater winding must be capable of delivering a minimum of 2 Amps.

Since 6V6 is pin-compatible with 6L6, it can also be used in The Golden Gecko amp. Usually, the 6L6-optimized bias would not be suitable for 6V6, and its much lower maximum anode and screen voltages would be exceeded in a 6L6 amp. However, in this case, due to the very low anode and screen voltages (in the 230VDC ballpark), 6V6 can be substituted, and the output power approximately halved compared to the other three tubes.

Circuit diagram

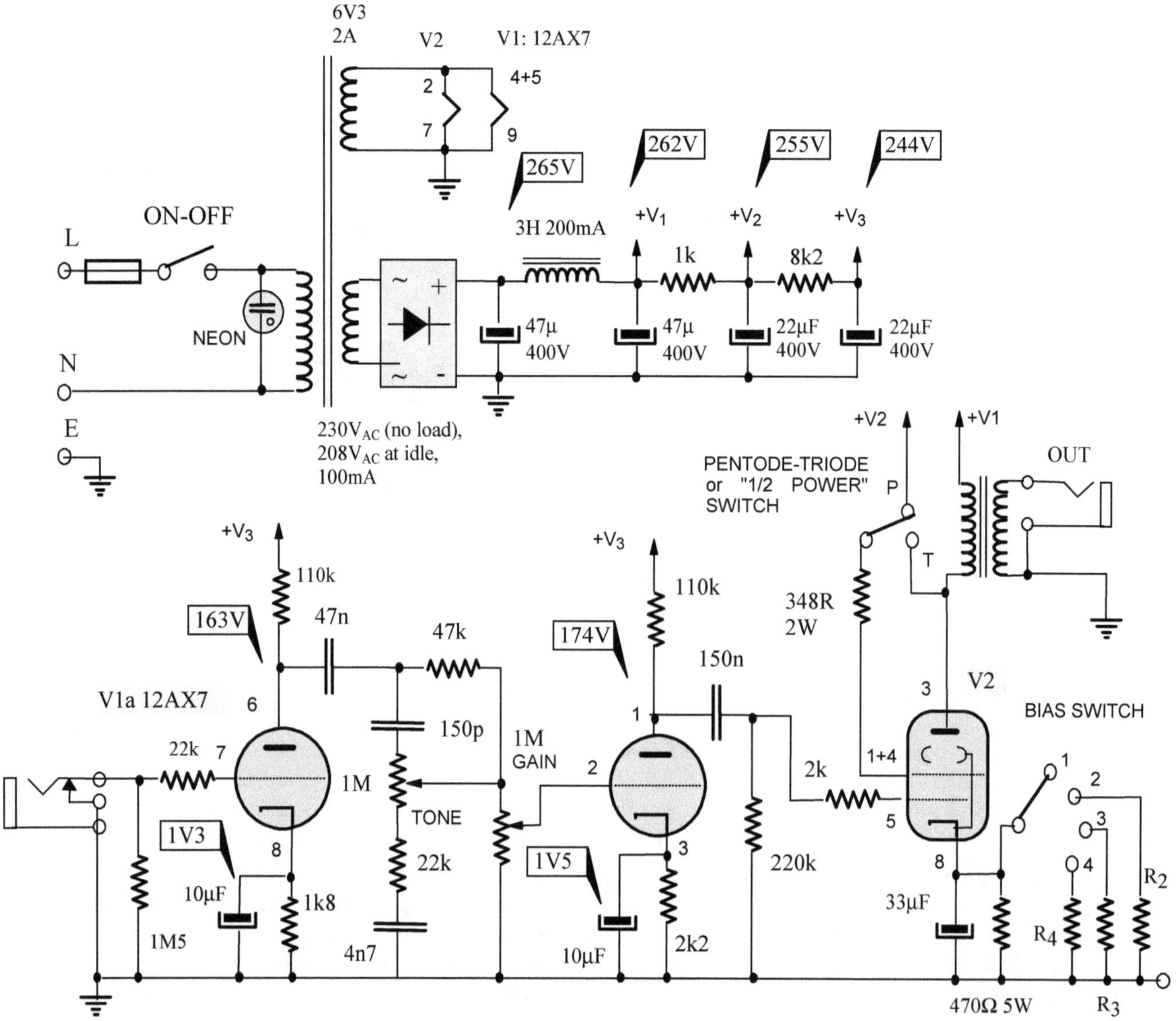

Calculating the values of cathode resistors for four different bias regimes

A switch-selectable cathode resistance of the output stage (4-positions) would give the guitar player 4x4=16 different settings in the pentode and another 16 settings in the triode mode. In short, it would turn this relatively simple little amp into 32 differently voiced amplifiers. Perhaps we should have called it The Chameleon, but Golden Gecko is close enough.

In the first instance, let's choose four relatively equally "spaced" values of 470, 420, 380, and 330 ohms. One resistor will stay permanently connected, which is the one with the largest resistance of 470Ω in switch position #1. In positions 2, 3, and 4, resistors R_2, R_3 and R_4 would be added in parallel.

So we know that R_1=470Ω, now we need to calculate the values of R_2, R_3 and R_4. Looking at R_2 first, we know that their parallel combination $R_P=(R_1*R_2)/(R_1+R_2)$. By some algebra "magic" we can express R_2 from that equation as $R_2=(R_1*R_P)/(R_1-R_P)$, and since we want R_P=420Ω in position #2, we get R_2= (470*420)/(470-420) =3,948Ω, so we will use a standard value 3k9 resistor.

By the same token, $R_3=(R_1*R_P)/(R_1-R_P)$, and since their parallel combination R_P is now 380Ω (switch position #3), we get R_3=(470*380)/(470-380)=1,984Ω, so we should use a 2k resistor.

Finally, in position #4, we want the total cathode resistance of R_P=330Ω, so R_4=(470*330)/(470-330)=1,108, so we should try to find 1k1 or a standard 1k2 resistor.

We didn't want to rummage for hours through our resistor stash to try to find the precise values just calculated, so for R_2, we used a 3k3 resistor, a special quality 2k5 resistor with gold leads for R_3, and two of those 2k5 resistors in parallel (total of 1k25) for R4. These resistance jumps don't have to be precise anyway since the cathode bias jumps they determine don't have to be precise either. The main thing is that we have some means of biasing the three different output tubes "hotter" or "colder"!

The analysis of the operating conditions - pentode mode

Once the amp was finished, we took dozens of measurements of various DC voltages around the power stage. Despite all that tabulated data, a few things became apparent once we drew the quiescent operating points Q1 and Q4 on 6L6 anode characteristics.

The range of anode currents was minimal, from 42mA in Q1 to 60mA in Q4. The power tube was grossly underutilized. The DC load-lines A-Q-B are very far from the curve of maximum power dissipation, meaning the output power levels will be far from their maximum attainable levels of 7-10 Watts pentode mode and 3-5 Watts in the triode mode.

Finally, notice that Q1 wasn't positioned in the middle of the DC load line; the Q1-A1 swing was almost double the Q1-B1 swing. This means that one side of the signal waveform would be cut off, and the distortion at higher power levels would be significantly increased.

All these were a direct consequence of our very low V_0 of around 230V. However, that may not be a bad thing. On the bright side, these were the positives:

1. The output stage would start (noticeably) distorting at much lower output volumes.

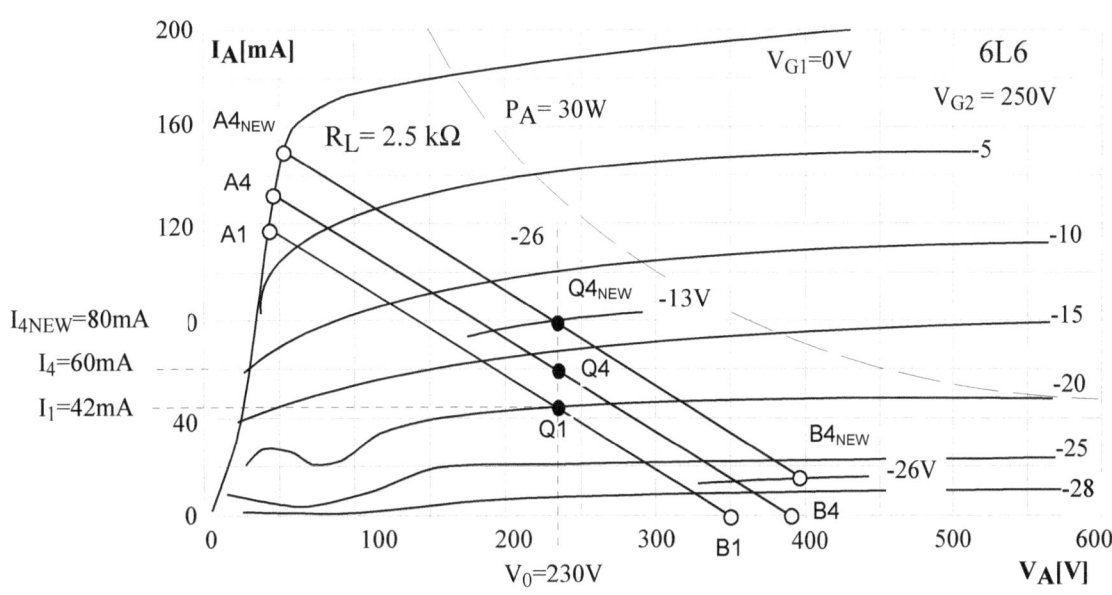

2. It would be easy to get any desired distortion (preamp or power stage) at low volumes, ideal for recording situations and miked live performances.

3. With the idle anode power dissipation under 10 Watts at all times, the three larger power tubes (rated from 25-35 Watts) would be cruising through their effortless and long life in this amp and would last a very, very long time. Even a 12 Watt 6V6 would be fine.

4. The HV secondary winding's voltage dropped from 230V at no load to 208V at idle. Such a small power transformer inevitably has poor regulation, resulting in high voltage droop at higher power levels. Although a solid-state rectification is used, the amp will inherently possess the spongy feel and compression of vintage amps with tube rectifiers such as 5Y3, which is a desirable aspect to many a guitar player.

Revising the switched biasing and new values for R2-R4

The power transformer's HV winding can supply up to 100mA continuously, and we are sitting at 40-60mA. It is thus possible to change the biasing resistors to bring points Q2, Q3, and Q4 higher. For clarity, only a revised point $Q4_{NEW}$ is indicated at the idle anode current of around 80 mA, a very hot bias situation. That would make the two halves of the anode voltage swing almost identical ($A4_{NEW}$-$Q4_{NEW}$ and $Q4_{NEW}$-$B4_{NEW}$). The output power will also be significantly increased.

The only caveat is that the maximum power dissipation of 6V6 tubes (12 Watts) will be significantly exceeded (P_0=0.08*230=18.4 Watts) so you'd not be able to use the 6V6 lovelies (for long) in that situation.

You can calculate the required values of R_2, R_3 and R_4 for your own chosen values of the operating points, assuming you will keep Q1 and its 470Ω cathode resistor. For instance, in $Q4_{NEW}$ the bias is -13V so the cathode voltage needs to be +13V and R_K= 13/0.08 = 162.5 Ω! Finally, R_4 in parallel with R_1 must be $R_2 = (R_1*R_K)/(R_1-R_K) =$ (470*163)/(470-163) = 250 Ω.

The analysis of 6L6 operating conditions - triode mode

The operating conditions with 6L6 output tube and bias switch in position 1 (R_K=470Ω) show a bias of -19.8V (-20V used for clarity) and idle I_A=48mA. Cathode current was measured at 42 mA, but such discrepancies are normal; these are generic triode curves, and those of the particular tube used were slightly different.

Nevertheless, with a maximum signal centered around -20V bias (a peak of 20V), we can estimate the 20mA to 94mA anode current swing and 125V to 300V anode voltage swing, ΔV = 300-12 = 175V, ΔI = 94-20 = 74 mA, so the maximum signal power output is $P_{OUT}= \Delta V \Delta I/8 = 175*0.074/8 = 1.6W$

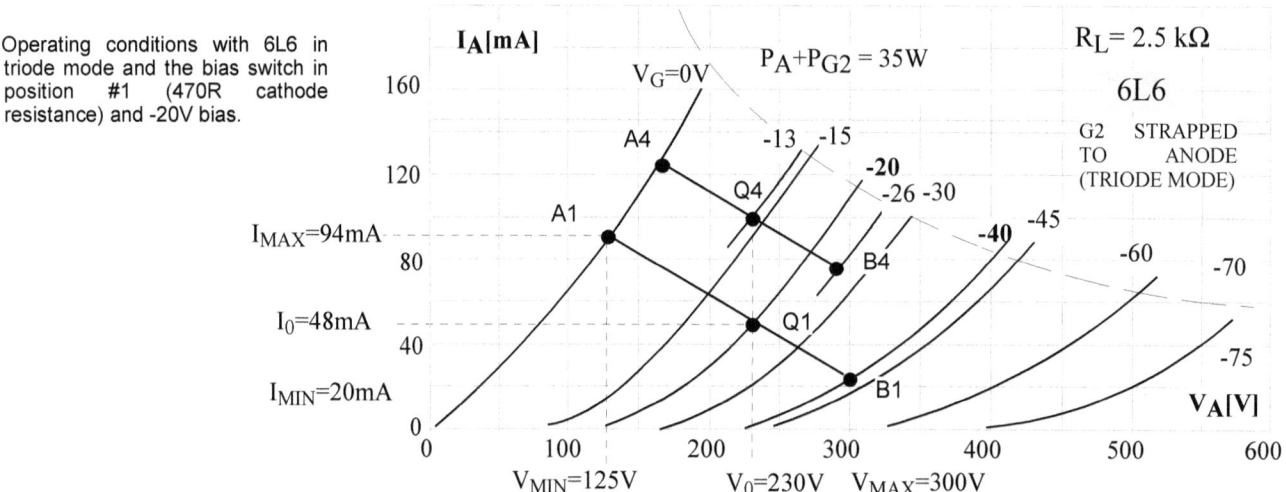

Operating conditions with 6L6 in triode mode and the bias switch in position #1 (470R cathode resistance) and -20V bias.

DC bias and cathode currents table (pentode mode)

Once we'd changed the three switchable cathode resistors to their new values of 1,063Ω, 557Ω, and 325Ω (the actual measured values), we measured the bias voltage on the cathode for our four octal power tubes, and for good measure, we added 6K6 as well.

For 6L6 we got the range of idle cathode currents of 41.7mA to 66.5mA. In all cases the 7027 and EL34 draw only a few mA less than 6L6, while 6V6 draw about 15% less current. As envisaged, 6F6 current draws were very close to those of 6V6, only 1.5mA less with 465Ω cathode resistance increasing to 8.5% less with 253Ω R_K and almost 11% less with the hottest bias using R_K of 191Ω!

Metal ashtray makes a cheap yet strong chassis

Steel communal ashtrays are sold in Australia for around AU$30, but with a bit of shopping around can be had for AU$20 (US$14 at the time of writing). Made of 1.6mm steel and powder coated in a light gray gloss finish, they measure 304 x 176 x 90 mm.

The ashtray makes an ideal chassis; the size is just right for a single-ended head or even for a small push-pull amp. It is quite deep, 90 mm, so the power transformer and choke or the output transformer can be mounted inside the chassis. Notice that it has "lips" on three sides, so mounting it on some kind of base is easy. We chose a kitchen drawer front from Ikea. By some lucky coincidence, it's of identical width (176mm) as the steel chassis.

Topology

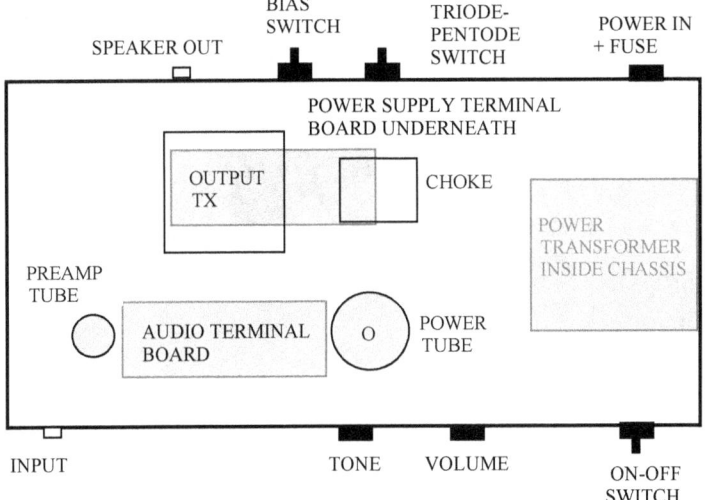

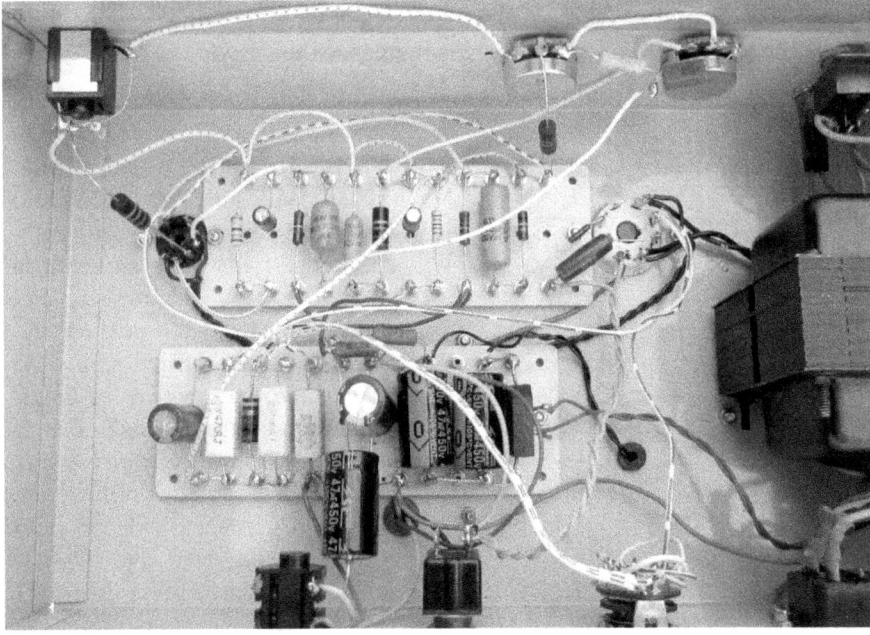

TUBE	V_K [V]	I_K [mA]
R_K=465Ω		
6L6	19.4	41.7
7027A	18.6	40.0
EL34	17.3	37.2
6V6	15.6	33.5
6K6	14.9	32.0
R_K=323Ω	465Ω ||	1,063Ω
6L6	16.6	51.4
7027A	15.8	48.9
EL34	15.2	47.0
6V6	13.3	41.2
6K6	12.4	38.4
R_K=253Ω	465Ω ||	557Ω
6L6	14.7	58.1
7027A	14.2	56.1
EL34	13.8	54.5
6V6	11.8	46.6
6K6	10.8	42.7
R_K=191Ω	465Ω ||	325Ω
6L6	12.7	66.5
7027A	12.2	63.9
EL34	12.2	63.9
6V6	10.0	52.4
6K6	9.0	47.1

ABOVE: The bias table
LEFT: The chosen topology (top view)
BELOW LEFT: The internal view of the chassis and wiring

The power transformer is under the chassis, the choke and output transformer on top. The triode-pentode switch and the bias switch are at the back, together with the speaker output jack. There is plenty of room to add a headphone or line-out socket as well.

The audio section and the power supply will be built on identical tag boards with 11 pairs of terminals. The cathode capacitor and resistors of the output stage are on the power supply board.

Make sure you insulate the exposed lugs and terminals carrying the mains voltage, in this case, $240V_{AC}$, such as the IEC power inlet's terminals. While you are testing and measuring the energized amp, it is very easy to forget the presence of such voltage and accidentally touch those terminals while your focus and attention are elsewhere.

Sonic and playing impressions

Just as we predicted, our guitar player friends loved the versatility of this small amp and the variety of tones it can produce, from clean and creamy triode warmth to the sharp bite of the overdriven pentode. If you have time to kill on a rainy (or snowy) weekend, this baby will keep you amused for hours!

1) Guitar input
2) Output transformer
3) HV power supply filtering choke
4) 4-position "Bias switch"
5) "Triode-pentode" switch
6) Speaker jack

DOUBLE TROUBLE: BLENDED SE OUTPUT STAGES

Normally, single-ended output stages with 6V6 or 6BQ5 work with smallish output transformers of 4-6kΩ primary impedance. Since their primary DC current is under 50 mA in the idle state, their primaries could be wound with a thin wire, thus making it possible to fit many turns on a relatively tiny bobbin. This achieves decent primary inductance levels of 10H or even more, which improves their bass response.

However, we have two power tubes in parallel here, so the idle primary current is at a 75-80mA level, requiring the use of a thicker primary wire. To fit the same number of turns as before, a much larger winding window is needed, meaning the EI laminations used must be larger and their stack thicker. This increases the size and the cost of the finished output transformer.

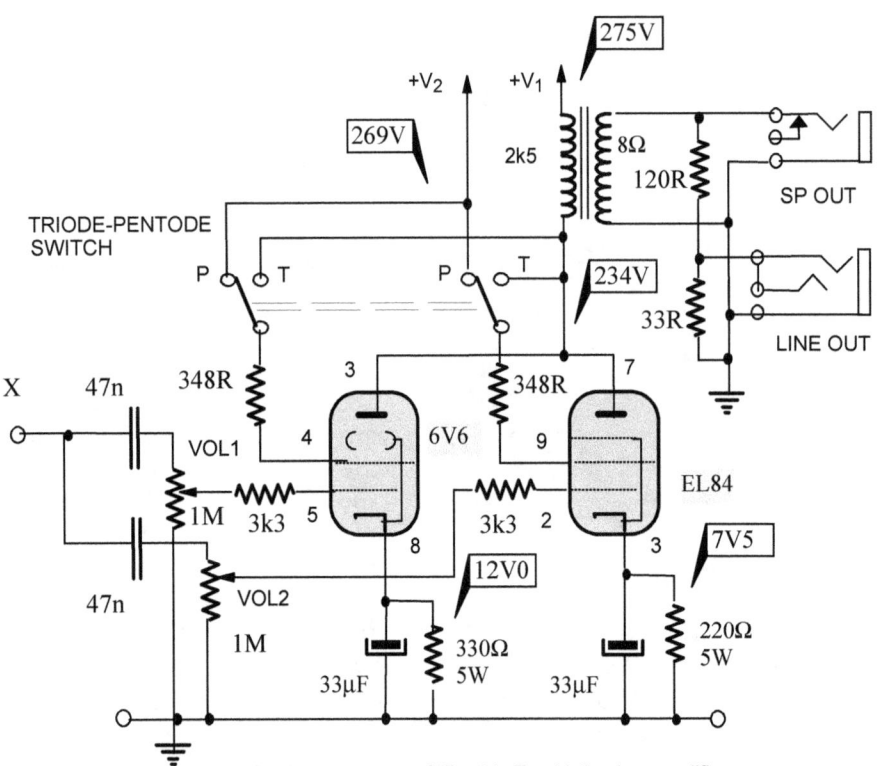

Dual output stage of "Double Trouble" guitar amplifier

Primary impedances of 2.5-3.5kΩ are the most common in hi-fi, for SE tubes such as 6L6, EL34, 2A3, 300B and KT88, so there are plenty of currently produced transformers to choose from. Most are sectionalized and quite large, and thus expensive. Plus, hi-fi transformers sound too clean and even sterile in guitar amps since they don't distort (much)! Therefore, don't waste your hard-earned money on hi-end (read: overpriced) transformers; even the smallest and worst-performing hi-fi (or rather "lo-fi") transformer is perfectly OK for a guitar amp.

We ended up with only 275V anode supply, and due to very high voltage drop across the output transformer's primary (41V!) only 234V at the anodes of output tubes.

With 36mA of cathode and 35mA anode current, 6V6 dissipates (234-12)*0.0035 = 7.8 Watts, which is very conservative.

The cathode current of the Noval tube, EL84, is almost identical, 7.5/220=35mA, of which the anode current is 33mA. Its anode dissipation is (234-7.5)*33= 7.5 Watts, only 7.5/12 = 0.625 or 62.5% of its maximum.

With a different choke and lower DC primary resistance output transformer, 260-270V should be obtainable at the anodes. That would increase the anode dissipation of power tubes to 9-10 Watts each, which would be ideal. The maximum output power would increase as well.

The junk box tube: 12FQ8

A vintage Wurlitzer tube organ had almost thirty 12FQ8 tubes, half-a-dozen 12AX7 lovelies and a pair of 7868 power tubes in a push-pull output stage. The 12FQ8 duo triodes were used in frequency divider and percussion circuits, a purpose they were originally designed for. Although each triode had two independent plates (anodes), in the Wurlitzer circuit they were paralleled.

Each of the anodes is rated at 0.5Watts, enough for a guitar preamp stage. By connecting both anodes in parallel we'd get a 1 Watt rating (just as that of 12AX7) and a halved internal resistance (from 76kΩ to 38kΩ). The transconductance would double to 2.5mA/V, and although mju would remain around 95, the stage's voltage gain would increase due to the lower internal resistance of such a triode.

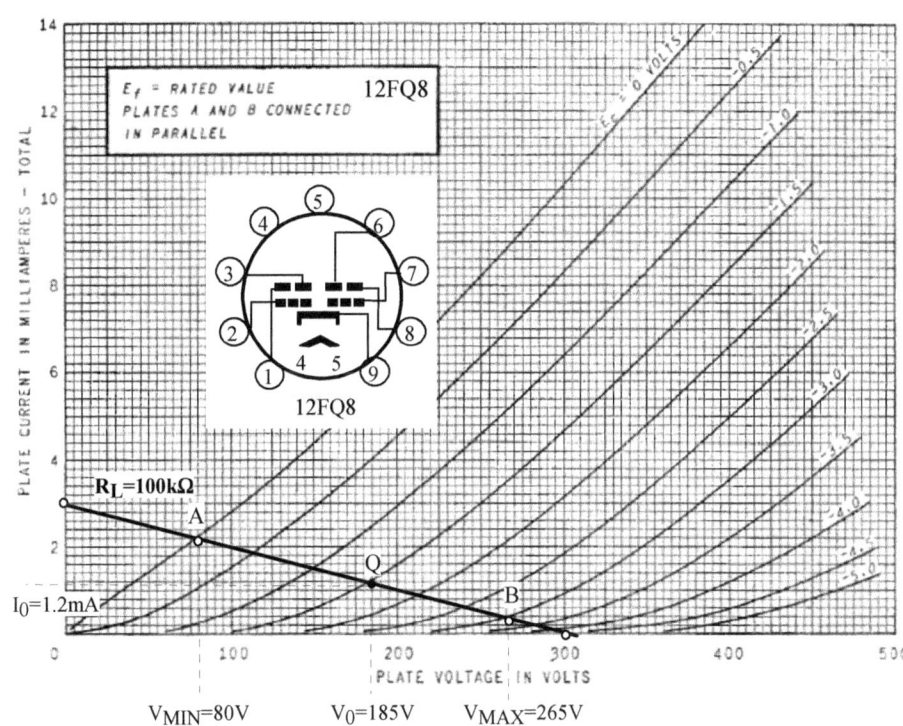

TUBE PROFILE: 12FQ8

- Duo-triode
- Noval socket
- 12.6V, 150mA heater
- V_{AMAX}=330V
- P_{AMAX}=0.5W (each anode)
- Typical operation (one triode with its two anodes paralleled):
- V_A=250V, V_{G1}= -1.5V, I_A=1.5mA
- gm=1.25 mA/V, μ=95, r_I=76kΩ

LEFT: The anode characteristics of one triode within 12FQ8 tube, with both its anodes strapped together.

A DC loadline of 100kΩ is drawn for gain estimation purposes.

12AX7 has a relatively low transconductance of around 1.4 mA/V (in a typical operating point used in guitar amps), so 12FQ8 should sound more dynamic and responsive.

Assuming a 300V HV supply and a 100k anode resistance, we position the Q point at -1.5V bias and get V_0=185V and I_0=1.2mA (diagram below). The anode voltage would swing from 80V to 265V, a total of 185V, which when divided by the grid signal swing of 3V would mean a voltage gain of 185/3 = 62 times. This is higher than a typical gain of a 12AX7 stage under the same conditions. The 2nd harmonic distortion coefficient can be estimated as $D_2 = (\Delta V_+ - \Delta V_-)/2\Delta V *100$ [%] = (105-80)/2*185 = 6.8%, quite high, but this is meant to be a higher distortion head, so the more fuzz, the merrier!

Since these tubes are dirt cheap, one could even parallel both triodes and all four plates, resulting in one 2 Watt tube with high gain, high transconductance (5mA/V in the same operating point as in our tube profile box), and very low plate resistance of only 19kΩ! That's what we will do in this project.

ABOVE: 12FQ8 tubes look good and sound even better!

Other possible 12FQ8 applications

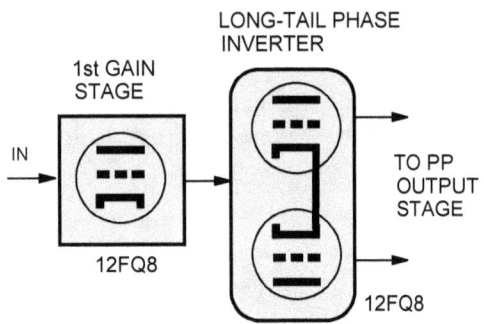

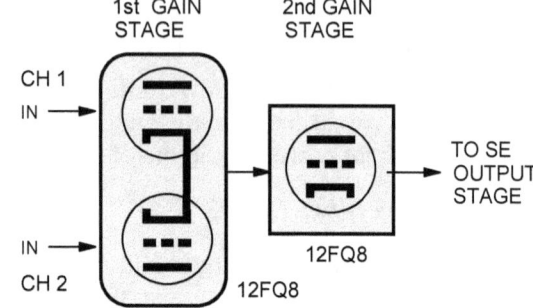

RIGHT: The two halves of one 12FQ8 lend themselves well to a 2-channel input stage.
LEFT: Two paralleled triodes in the first stage, followed by a long-tail phase inverter

Apart from the obvious topology, with one paralleled 12FQ8 tube followed by another, as in our case here, the two halves offer a possibility of two channels, which can have different gains and be voiced differently. The cathode resistor and bypass cap must be common due to the single cathode shared between the two triodes.

The common cathode limits other uses to amplification stages where two triode's cathodes would be connected anyway, such as the long-tail phase splitter (cathode-coupled phase inverter).

The preamp stages with 12FQ8

Due to the low internal resistance of two paralleled triodes in each stage, the gains are higher than with 12AX7 triodes. However, keep in mind that you will need 12.6V heater supply for 12FQ8. If your power transformer has two 6.3V heater windings, you are in business. If it hasn't, you can always build this project using 12AX7 tubes.

Since our tubes were salvaged from a Wurlitzer organ and were tested at the lower acceptable limit (just over minimum), the currents through both stages are quite low, 0.9mA for the 1st and 0.85mA for the 2nd stage.

We ended up with too much gain but left the amp in that state to get some early breakup and distortion.

For more clean headroom, reduce the 2nd stage's anode resistor to 56k or even 47k, remove its cathode bypass capacitor and increase the 100k resistor (3) to 330k or 470k.

You could even use a 500k Gain pot instead of 1M. That would cut the input signal to the 2nd stage down due to the voltage divider action. Now we have 100k-1M, so most signal at the 1st stage's output reaches the 2nd stage. With a 470k-500k combination, that signal would be halved.

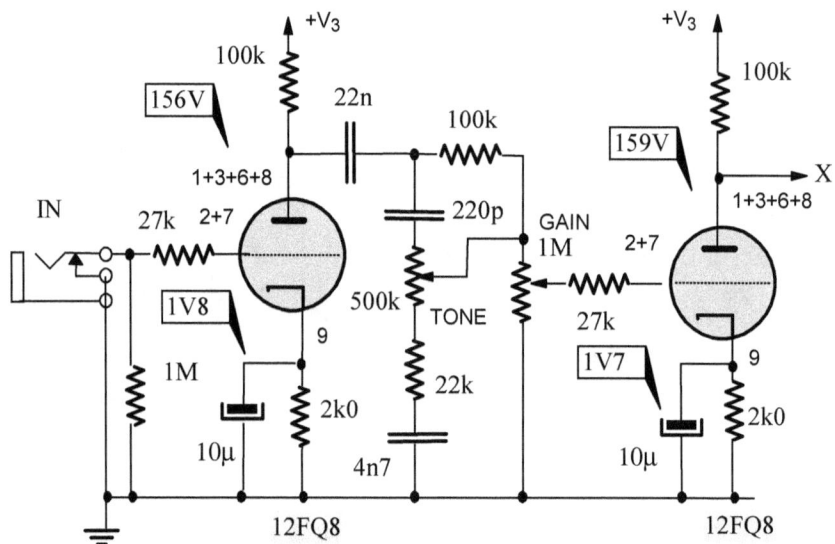

ABOVE: The high gain high transconductance version with two 12FQ8 tubes
BELOW: Alternative version with one 12AX7 tube

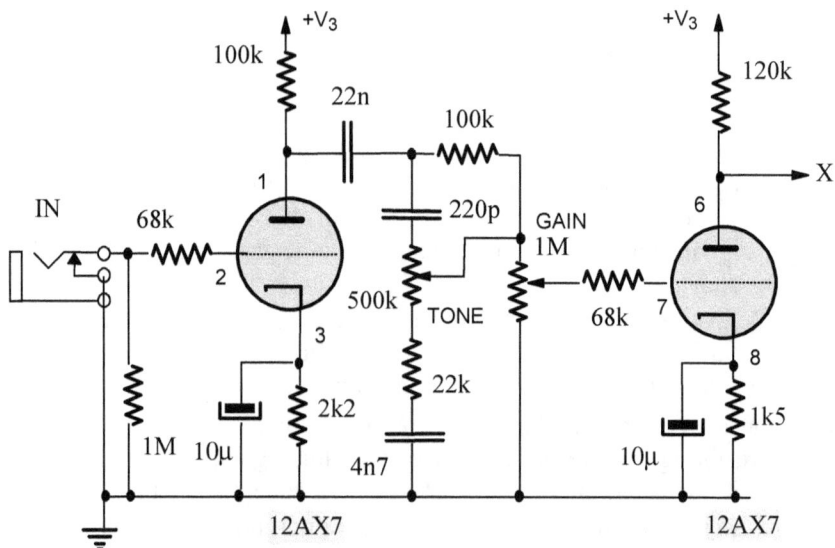

DIY PROJECTS: QUIRKY & UNUSUAL DESIGNS

The power supply

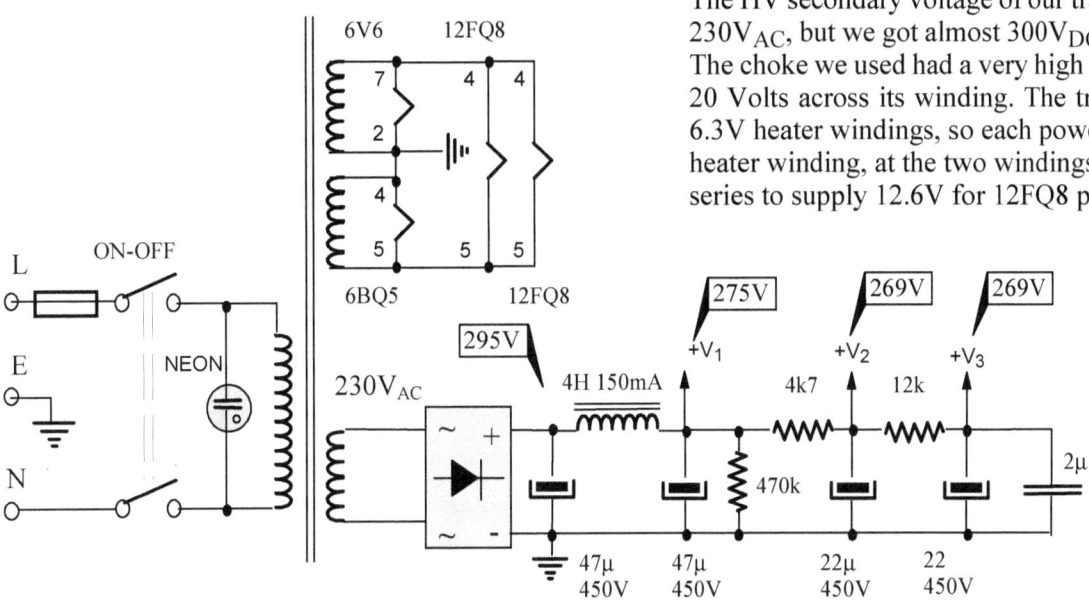

The HV secondary voltage of our transformer was only $230V_{AC}$, but we got almost $300V_{DC}$ after rectification. The choke we used had a very high DCR, so it dropped 20 Volts across its winding. The transformer had two 6.3V heater windings, so each power tubes has its own heater winding, at the two windings were connected in series to supply 12.6V for 12FQ8 preamp tubes.

The controls and finished amp

The chassis is quite small, and to keep things sweet & simple, we'll only have Gain and Tone controls in the preamp stages and two individual Master Volume controls, one for the Octal, the other for the Noval tube.

The funky metal feet were salvaged from a vintage Heathkit tube amplifier and resprayed in metallic silver to match the silver handles, front/back fascia, and transformer covers. The Double Trouble looks much more impressive in real life than here on a B&W photo, like a true work of art that it is!

BELOW: A bright red circular wall shelf served as the cabinet for this project, with satin chrome handles and metal feet salvaged from a vintage Heathkit hi-fi amplifier. The front acrylic cover was laser cut for AU$20, and the front and rear panels were cut and engraved for AU$30, so the total cost of the cabinet was AU$80 (around US$50.-).

The amp's chassis was the rectangular metal ashtray, just like the one used in the Golden Gecko amp.

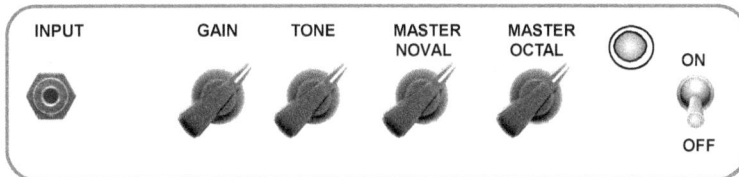

Front panel controls

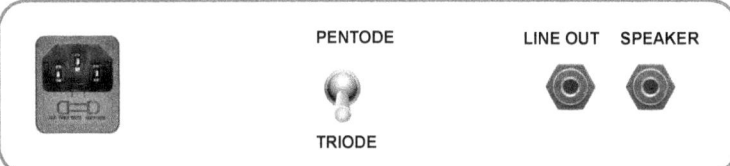

Rear panel controls

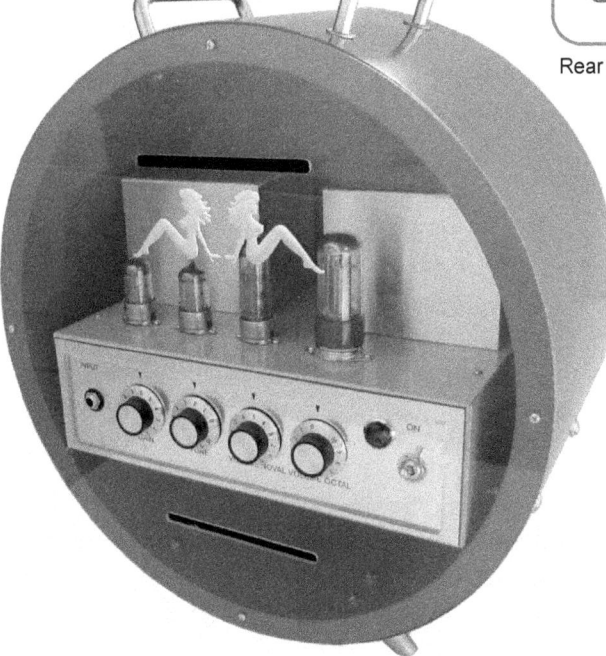

Sonic impressions

Our resident guitarist liked the "Dr. Jekyll and Mr. Hyde" split personality of this amp so much that he spent more time testing it out than any other amp in this book.

As envisaged, the Noval channel with EL84 sounded louder and brighter. It had more gain and oomph. The Octal 6V6 was creamier and smoother, perfect for those jazz and bluesy tones. Blending the two outputs produced quite an unusual tonal mixture, "an iron fist in a velvet glove" would be a fitting description.

LOW VOLTAGE TUBE GUITAR AMPLIFIERS

When semiconductors became commercially available in the late 1950s, designers and equipment makers welcomed them with open arms. Transistors were much smaller and lighter than vacuum tubes and produced less heat (did not require heaters). More importantly, being low voltage high current amplifying devices, transistors did not need output transformers and worked at low DC voltages.

High voltages inside tube amps scare away many prospective DIY amp builders. Also, power transformers with both high voltage secondary and heater windings are quite expensive. What if we designed a guitar amp that used a cheap and widely available mains transformer (with low voltage secondaries), or, even better, one that will use an external AC or DC power pack?

Vacuum tubes weren't designed to operate at low DC anode voltages, so the first question is how tubes would work at low DC voltages, if at all? If it is possible, what tube types would be the best choice for such an operating regime? And finally, what kind of output power levels can such low voltage amps produce, and how would they sound?

Low voltage single-ended amp

This section looks at two relatively simple low voltage guitar amps, one single-ended, the other push-pull. As always, we start with the simpler one, the single-ended design.

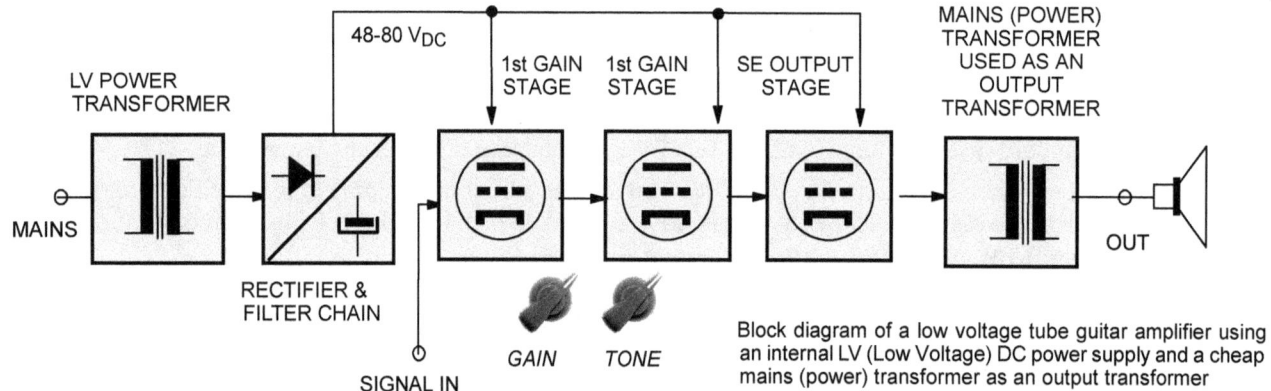

Block diagram of a low voltage tube guitar amplifier using an internal LV (Low Voltage) DC power supply and a cheap mains (power) transformer as an output transformer

The output tubes

PL508 and EL508 beam power tubes were used as vertical sweep tubes in early European color TV receivers. Although in production for only a few years, since unused by commercial amplifier manufacturers, the NOS stocks are still plentiful, even by desirable tube makers such as Mullard, Philips, Valvo, and Telefunken.

The sonic signature is detailed, open and transparent, without false sweetness or syrupy tubey coloration. Above all, PL508 produces a great bass. Is that because in its original service the tube worked at a mains (bass) sweep frequency of 50 or 60Hz, but this tube *loves* the bass frequencies.

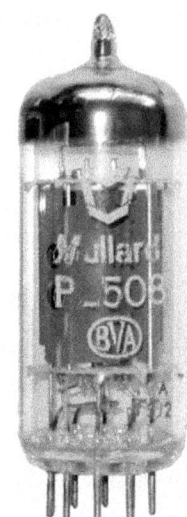

TUBE PROFILE: EL508 (6KW6) & PL508 (17KW6)

- Indirectly-heated beam power tube
- Magnoval socket
- Heater: 6.3V, 1.2A (EL508), 17V, 0.3A (PL508)
- $V_{AMAX}=400V$, $V_{G2MAX}=275V$
- $I_{K2MAX}=90mA$, $P_{AMAX}: 12W$, $P_{G2MAX}=3W$

Circuit diagram

Remember, we needed 17V for the PL508 heater and 6.3V for the 12AXT heater in series (same current, 300mA), a total of 23.3V. The power transformer had one 24V secondary (two 12V secondaries can be used, connected in series), perfect. When rectified, that produced 32V on the first filtering elco. The amplifier worked well with such a low anode voltage, but its output power was minuscule (0.1-0.2 Watts). So, to get more power, we changed the rectifier into a voltage doubler by replacing two of the four diodes with elcos, and got +60V.

The 1st stage has a voltage gain of 18, the 2nd amplifies 12 times and the power stage amplifies 20 times in pentode and 5 times in triode mode.

With a 0.1V AC signal at the input, we got 1.8V on pin 6, 1.8V on the control grid input of the power tube, 36V on its anode, and 1.2V$_{RMS}$ on the 8-ohm load (0.18 Watts of output power) in pentode mode.

In both preamp stages, triodes work in the space charge regime and are biased using contact bias. A small grid current flows into the control grid; electrons flow in the opposite direction, from the CG through the grid leak resistor (1M here) into GND. This current flow creates a small DC voltage drop across the resistor and biases the tube by placing CG at a slight negative potential (around half-a-Volt here). In this low voltage working regime, the gain of the 2nd stage is not constant.

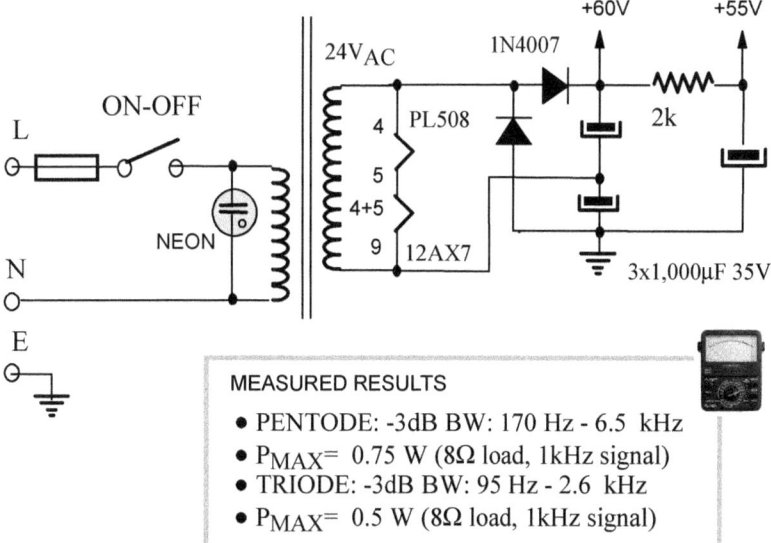

MEASURED RESULTS
- PENTODE: -3dB BW: 170 Hz - 6.5 kHz
- P_{MAX} = 0.75 W (8Ω load, 1kHz signal)
- TRIODE: -3dB BW: 95 Hz - 2.6 kHz
- P_{MAX} = 0.5 W (8Ω load, 1kHz signal)

The position of the Volume control potentiometer changes the DC bias on pin 2 and, therefore, the voltage gain and distortion of the whole stage. Finally, the resistance of the power tube's cathode resistor needs to be determined (and therefore its bias voltage). No calculations or models can help us here; it has to be done experimentally.

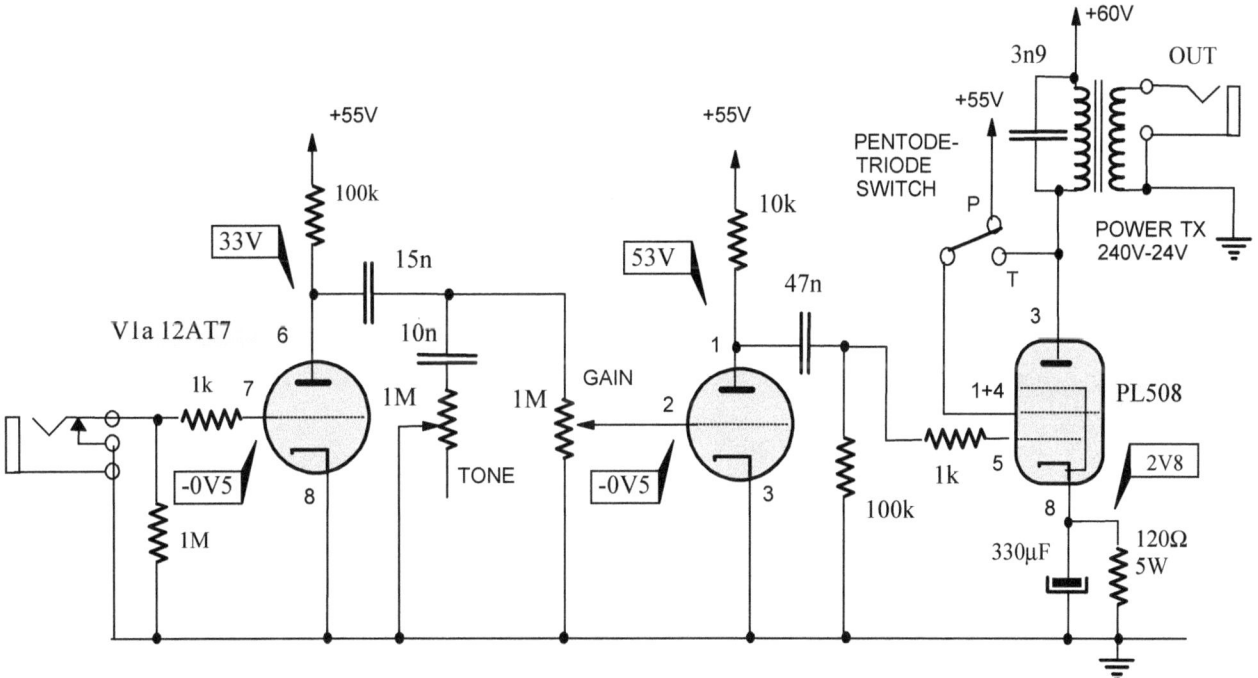

The output transformer

Once we got the audio circuit to work, the attention shifted to the output transformer. After trying transformers with various voltage ratios 240-6V, 240-12V, and 240-18V, the 240-24V voltage ratio (10:1) was the most suitable for the PL508 tube.

In this regime, the internal impedance of the PL508 was unknown (it was much lower than when operating on the usual 300-350V!), so the 100:1 impedance ratio seemed optimal, indicating that the primary impedance of the output transformer should be under 1k in low voltage tube amps.

Since the mains transformers were not designed to be used in a single-ended mode, with 25-60mA flowing through their primary winding and without any air gap, early saturation of their magnetic core was likely. To achieve higher output power levels, the magnetic core should be as large as possible.

A succession of various transformer trials, starting with A=4cm^2 (cross-sectional area of the EI magnetic stack), followed by A=6cm^2 and A=10cm^2 resulted in the maximum output signal voltage of 1.7, 2.2 and 2.5V respectively. These increases don't seem significant, but when expressed as output power, they become 0.4W, 0.6W and 0.8W (8Ω load), so an increase of 10/4=2.5 or 250% in transformer size results in a power jump of 2:1 or 200%!

Plotting the transfer characteristics of the output tube at low anode & screen voltages

Tube catalogs and data sheets don't contain any data or characteristics for low voltage tube operation, so no matter what power tube you decide to use, you need to do a little experiment to determine the required bias of the output stage. A transfer curve of the tube at the DC voltages in your circuit needs to be plotted, +60V on the anode and +55V on the screen grid in our case.

You'll need the heater, anode, and screen grid power supplies, a 9V battery, 1k to 5k LIN potentiometer, and two multimeters. One will measure the grid voltage (bias), on "DC Volts," 1-10V range, the other will be connected as mA-meter, in series with tube's anode (it can also be placed in the cathode circuit if you prefer). Start at 0V on the CG and record the reading of the mA-mater, in this case, 58mA, then increase the grid voltage in -1V steps by adjusting the potentiometer and record the anode currents. Once you have your table with data, you can draw the TX curve if you wish, as illustrated.

With a zero bias, the maximum anode current is only 58mA, so at such low anode/screen voltages, the tube is safe from overheating and destruction. The cutoff is just below -7V, so we can choose the quiescent point at any point around the middle of the curve to ensure the maximum possible input signal swing, which is 7V peak-to-peak or around $7/2.8 = 2.5$ V_{RMS}!

We chose -3V bias, and since the anode current is around 25mA (plus say 1mA of screen current) a cathode resistor of $R_K = 3/0.026 = 115\Omega$ is needed to bring the cathode up to +3V potential. 120Ω is the standard value to be used.

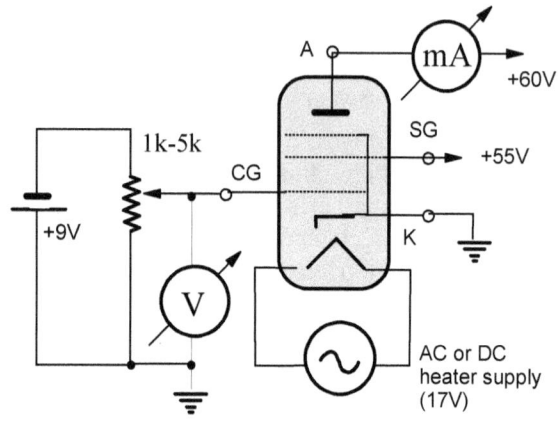

ABOVE: The test circuit for plotting transfer curve (I_A versus V_G) of an output tube
BELOW: Transfer characteristics of PL508 pentode with +60V anode and +55V screen grid voltages

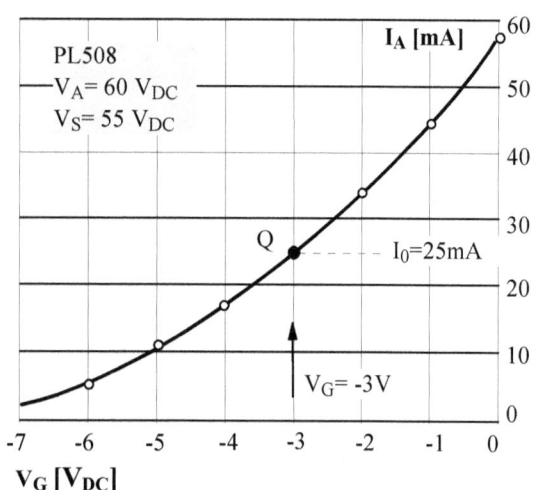

Low voltage push-pull amp

Having chosen a split-load phase inverter, a single triode stage at the input would not have enough gain, so a pentode is needed. Instead of using two separate preamp tubes, a pentode, and a triode, a single pentode-triode tube simplifies the design and construction.

ECF80 was released in 1954 to work as a frequency changer (mixer) in vintage black & white TV receivers. Although it wasn't designed for audio use, it is very similar to the American 7199 tube, which was.

Unfortunately, while the triode pin numbers are identical, the pinouts of the pentode sections are different, so the two types are not interchangeable without rewiring. However, since you'd be building this amp from scratch, you have an option of using 7199 tubes instead if you wish.

TUBE PROFILE: ECF80 (6BL8)
- Indirectly-heated triode-pentode
- Noval socket, heater: 6.3V, 430 mA
- MAX. VALUES PENTODE:
- V_A=250V, V_{G2}=175V, V_{KH}=100V
- P_A=1.7W, P_{G2}=0.5W, I_K=14mA
- TYP. OPERATION PENTODE:
- V_A=V_{G2}=170V, V_{G1}=-2V, I_A=10 mA,
- I_{G2}=2.8 mA, gm=6.2 mA/V, μ_{G2G1}= 47
- MAXIMUM VALUES TRIODE:
- V_A=250V, P_A=1.5W, I_K=14mA, V_{HK}=100V
- TYP. OPERATION TRIODE:
- V_A=105V, V_G=-2V, I_0=12 mA
- gm=5 mA/V, μ= 20, r_I = 4.0 kΩ

Heater connection

Since $24V_{AC}$ is available for heaters, two PL508 heaters are paralleled, pulling a total of 600 mA, to which ECF80 heater is added in series. However, its heater only draws 430 mA, so 600-430=170mA needs to be shunted through a parallel resistor.

At 6.3V across the resistor and ECF heater and 170mA current draw, using Ohm's Law, the required shunt resistance is R_S=6.3/0.170 = 37 Ω. A standard value of 39 Ω would be fine, underheating the preamp tube slightly and reducing noise as an additional benefit.

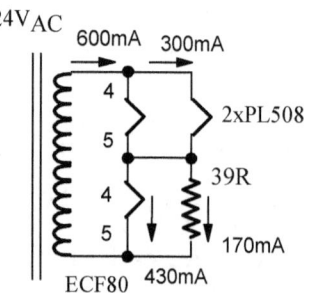

DIY PROJECTS: QUIRKY & UNUSUAL DESIGNS

The conceptual block diagram

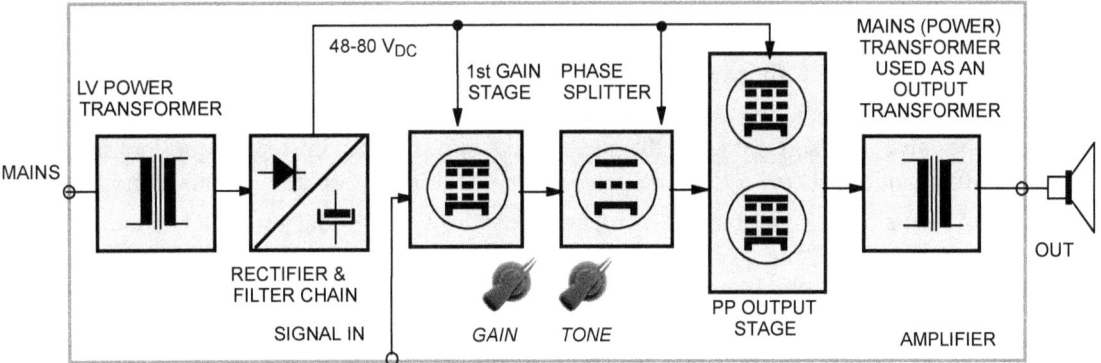

Block diagram of a low voltage tube guitar amplifier using an internal power supply and a cheap power transformer as an output transformer

Audio section

Standard cathode biasing was used this time, and more clean headroom was achieved compared to the SE version. Except for the very low anode voltage of the 1st stage and low cathode levels of the output stage (only 1.7V), the circuit is pretty standard. With standard RC biasing of the screen grid, the input pentode stage was very sensitive to changes in the anode and screen grid voltages.

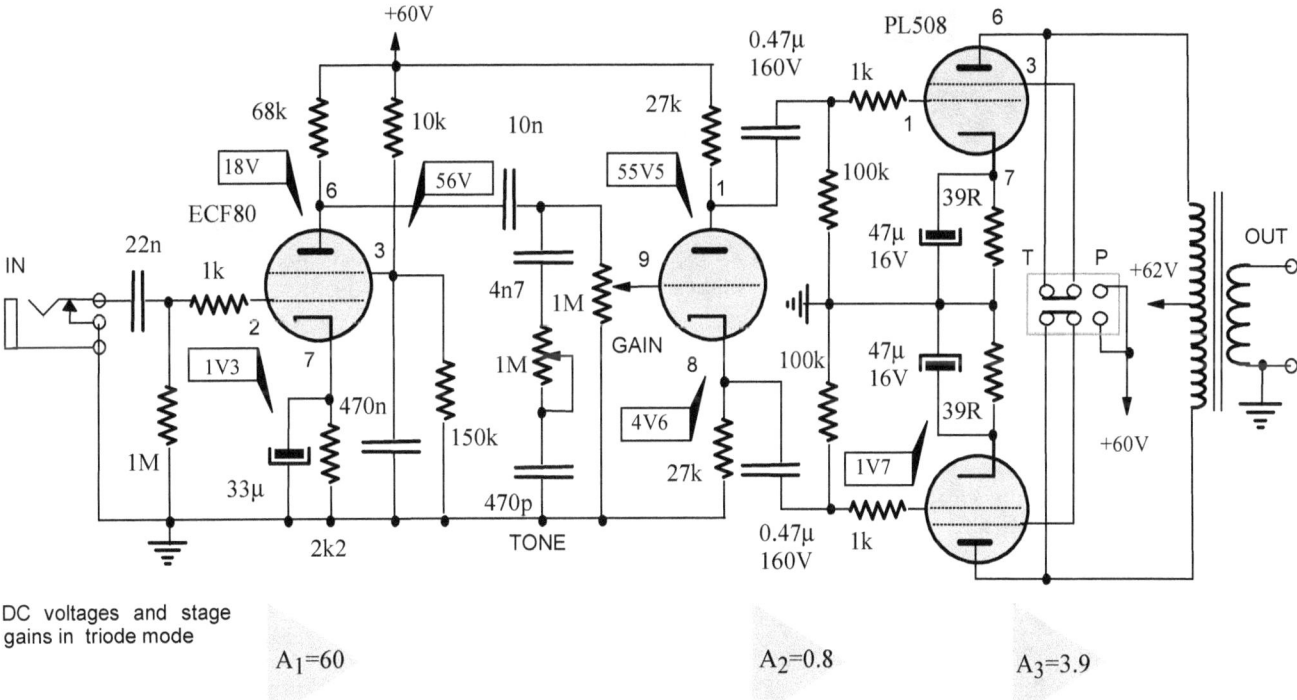

DC voltages and stage gains in triode mode

$A_1 = 60$ $A_2 = 0.8$ $A_3 = 3.9$

Since neither was regulated (a similar power supply as for the SE version), switching any larger loads on the mains circuit (heaters, air-conds, washing machines, etc.) would change operating conditions, output power, and tone of the amp. Adding a 150k resistor in parallel with the 470nF screen capacitor created a resistive voltage divider (10k-150k) and stabilized the screen DC voltage at around 56V.

Just as with the SE version, the difference in maximum output power levels between triode and pentode modes was not large, 0.75 W versus 0.9 Watts.

However, the frequency range widened significantly. In the pentode regime, the -3dB bandwidth was much wider, almost hi-fi like 59 Hz - 45 kHz, narrowing somewhat in triode mode to 15 Hz - 12 kHz. Notice how the BW shifted towards the bass (low frequency) region in triode mode; this is normal, since the load impedance is the same in both modes (the same output transformer), so it is impossible for it to be optimal in both modes.

MEASURED RESULTS
- PENTODE: -3dB BW: 59 Hz - 45 kHz
- $P_{MAX} = 0.9$ W (8Ω load, 1kHz signal)
- TRIODE: -3dB BW: 15 Hz - 12 kHz
- $P_{MAX} = 0.75$ W (8Ω load, 1kHz signal)

Devise your own logo (symbol) and a unique and memorable name for your amps

Once you've built your boutique amp, don't leave it nameless. The Fender, Marshall, and Traynor's mainstream amp makers simply named their products after themselves, using their surnames. However, unless you have a melodic-sounding surname, I would advise you to give it a nicer, more appealing name.

Also, if you heavily modify an amp, why keep its name (especially if it's a less desirable or obscure one)? It is not the same amp anymore, so give yourself some credit and make your creation into a unique-looking and sounding machine. Give it a fitting name that will reflect the personality of both the amp and its maker/modifier-you!

While thinking about it, consider a logo or symbol you could use for your "brand" of built or modified amps. A few years back, I bought a few identical wall clocks on a clearance sale. Since its base represents an electron's orbit inside an atom, "Atomium" would be a fitting brand name ...

LEFT-TO-RIGHT: A metamorphosis of an aluminium wall clock into an atomium logo. One clock back-plate makes four corner logos.

Chinese factories are churning out various logos (badges) or decals for cars, such as the "V6" and "V8" stick-on emblems pictured below. There are also lizards, snakes, spiders, wolves, bulls, and other similar animal-inspired decals, should you be more inclined towards fauna.

Vintage car badges such as the "Fifth Avenue" and the "New Yorker" (in two sizes) can often be bought for a few dollars, either online or from wrecking yards. They are not stick-on; some have integral pins, others can be drilled and bolted onto your amp's metalwork.

Another option are metal house numbers sold by home depots and hardware stores, such as the 6 and 8 digits pictured on the right. They are quite large, so they are suitable only for bigger combo amps, but they look very smart in brushed chrome (satin) finish.

CONVERTING VINTAGE TUBE GEAR INTO GUITAR AMPS

12

- TUBE SIGNAL TRACERS AS LEARNING & CONVERSION PLATFORMS
- TUBE AUDIO GENERATORS
- TUBE PUBLIC ADDRESS (PA) AMPLIFIERS AS CONVERSION PLATFORMS
- AMATEUR TUBE TRANSCEIVERS
- SONY SSA-464 AMPLIFIER SPEAKER SYSTEM

While the idea of converting or modifying vintage PA amps into guitar amps has been around for decades now, many vintage test instruments contain (among other circuits) a whole tube amplifier. These are usually of a low power variety (a few Watts) but are almost always much cheaper to buy than by now overpriced PA amps. Tube signal tracers and audio generators are two examples of such instruments.

This approach is especially recommended for those just starting on their building & modifying journeys. If you damage or destroy these cheap learning platforms, it won't be a significant loss. So buy a few pieces of such tube gear, study their design and construction, measure their DC and AC (signal) voltages, modify them, take them apart and learn from them all you can. Then move to the more expensive tube amps.

TUBE SIGNAL TRACERS AS LEARNING & CONVERSION PLATFORMS

EICO 147A signal tracer

In the 1950s and 60s, oscilloscopes were expensive, so signal tracers found their place in amateur workshops. As troubleshooting tools, they weren't that popular even then and are obsolete now. However, tube signal tracers have an educational value as learning platforms.

It does not matter what brand it is (Heathkit, Eico, Knight, PACO); a signal tracer is a complete mono single-ended audio amplifier. You get the mains and output transformers, two amplification stages, a single-ended output stage, tube-rectifier, and speaker. There is even a voltage indicator circuit built around a magic eye tube so that you can learn about them too.

Instead of building your first tube amp from scratch and paying 5-10 times more for parts only, get one of these or any low-grade vintage amplifier. It does not have to be a hi-fi or even a stereo amplifier; tube public address (PA) amps are also suitable. Study its operation, measure its voltages, currents, impedances, and amplification factors (stage gains). Observe the waveforms on an oscilloscope.

If you do something daft and it all goes up in smoke, better to destroy a $30 signal tracer or a $60 PA amplifier than to blow up a valuable guitar amp.

EICO MODEL 147 A SIGNAL TRACER

C1	cap., disc., .005 mfd (5K or 5000 mmf) ±10%
C2,3,4	cap., disc., .025 mfd (25K or 2500 mmf) ±10%
C5	cap., elec., 2 x 20/450 V - 2 x 10/350 V
C6	cap., molded, .25 mfd - 400 V
CR1,2	rectifier, 1N48
R1	res., 10MΩ (brown, black, blue, silver) 1/2W, ±10%
R2,8,20	res., 470KΩ (yellow, violet, yellow, silver) 1/2W, ±10%
R3	res., 100KΩ (brown, black, yellow, silver) 1/2W, ±10%
R4,7	res., 220KΩ (red, red, yellow, silver) 1/2W, ±10%
R5	pot., 500KΩ audio taper
R6	res., 2.2KΩ (red, red, red, silver) 1/2W, ±10%
R9	res., 330Ω (orange, orange, brown, silver) 1W, ±10%
R10,16,17,19	res., 1MΩ (brown, black, green, silver) 1/2W, ±10%
R11	res., 47Ω (yellow, violet, black, silver) 1/2W, ±10%
R12	res., 560KΩ (green, blue, yellow, silver) 1/2W, ±10%
R13	res., 1KΩ, 5W, ±10%
R14	res., 47KΩ (yellow, violet, orange, silver) 1/2W, ±10%
R15	res., 68KΩ (blue, grey, orange, silver) 1/2W, ±10%
R18	pot., 50KΩ, linear

ABOVE and LEFT: The component list and circuit diagram from Eico 147A manual.

BELOW: The internal view of Eico 147A signal tracer. Mains transformer (1), output transformer (2), magic eye tube (3).

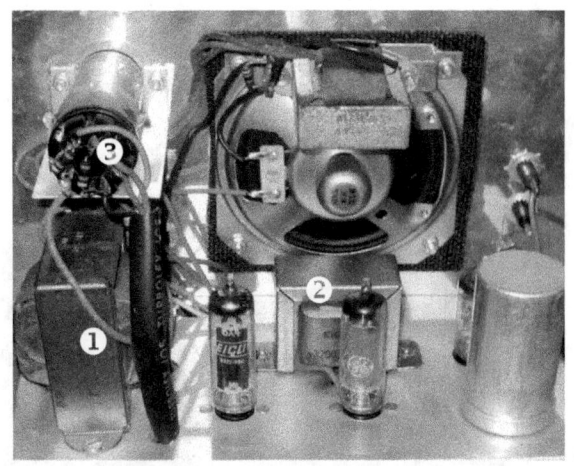

CONVERTING VINTAGE TUBE GEAR INTO GUITAR AMPS

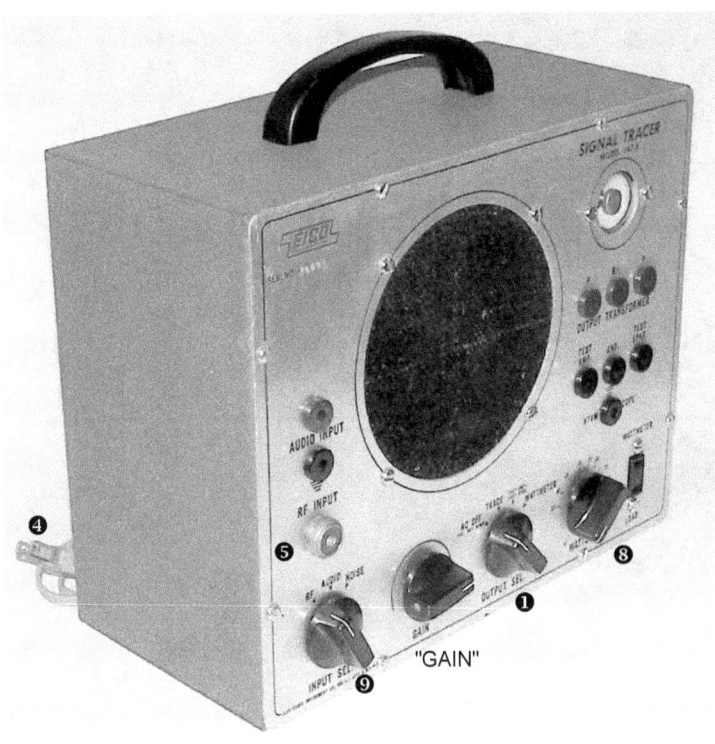

Starting with the power supply, one contact of the 4-position switch S2 (1) is used as an on-off switch (2). If you are retaining that switch, fine, if not, install a separate power toggle switch.

T3, R18 rheostat, the rectifier and filtering circuit can be removed (3), as should C4. Replace the two-pin plug and cord with a modern power cord and ground the metal chassis (4). The J3 input (5) needs to be converted to a 1/4" guitar jack. J2 and J1 (banana post sockets marked "audio input") will not be used.

You could connect them to the cathode of the output tube as test points to measure bias (DC voltage on the cathode resistor).

The first stage uses grid leak biasing through 10MΩ R1 resistor. Install a cathode resistor and bypass cap instead (6). Install bypass caps in the cathodes of the 2nd and output stages (7).

Instead of 220k (R7) you can install a 250 or 500k master volume pot (8) in the physical location of R18 rheostat.

The 3-position S1 switch (9) that was used as an input selector can be used for voicing, for instance, as DARK-NORMAL-BRIGHT selector at the input of the amp. However, it only has one wafer and a simple contact; it only switches two terminals together at a time. You may need to ditch that switch and replace it with one with the necessary contact arrangement. As an exercise, draw the input RC filtering circuit and the switch contacts to get the DARK-NORMAL-BRIGHT arrangement.

There are seven banana sockets below the magic eye. As you can see from the schematics, they give you access to most points in the output stage. One or two should be converted to a speaker output jack, but most will not be needed anymore. You can leave them in place as they are, or remove them all and blank that rectangular space with a nice logo or custom decorative plate. The only limit is your imagination. Incidentally, the magic eye (proper name: cathode ray tube) can stay; it looks cool glowing green.

TUBE AUDIO GENERATORS

EICO 377 audio generator

While tube signal tracers are an obvious choice for conversion into a small single-ended guitar amp, they are by no means the only vintage test instruments that can find a new lease of life in one of those quirky adaptations.

The sturdy metal case of Eico 377 looks cool, and since there are only a few terminals and controls, all will be used in the conversion to a guitar amp, and all will be functional.

An audio generator also works at audio frequencies; it consists of an oscillator and one or more amplification stages. Although most don't have an output transformer (some do!), the output stage (usually a cathode follower) can easily be converted into an anode-loaded output stage by adding a small SE transformer.

Apart from changing the old elcos, the power supply does not need modifications (diagram on the next page). There is even a filtering choke for better filtration of the AC ripple on the high voltage DC line, something most guitar amps don't have!

The 6SJ7 pentode works as an oscillator (1) whose frequency is determined by the RC network to the left (2).

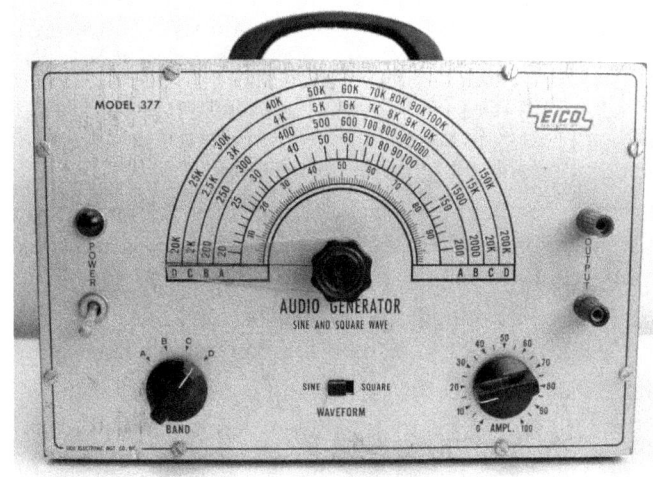

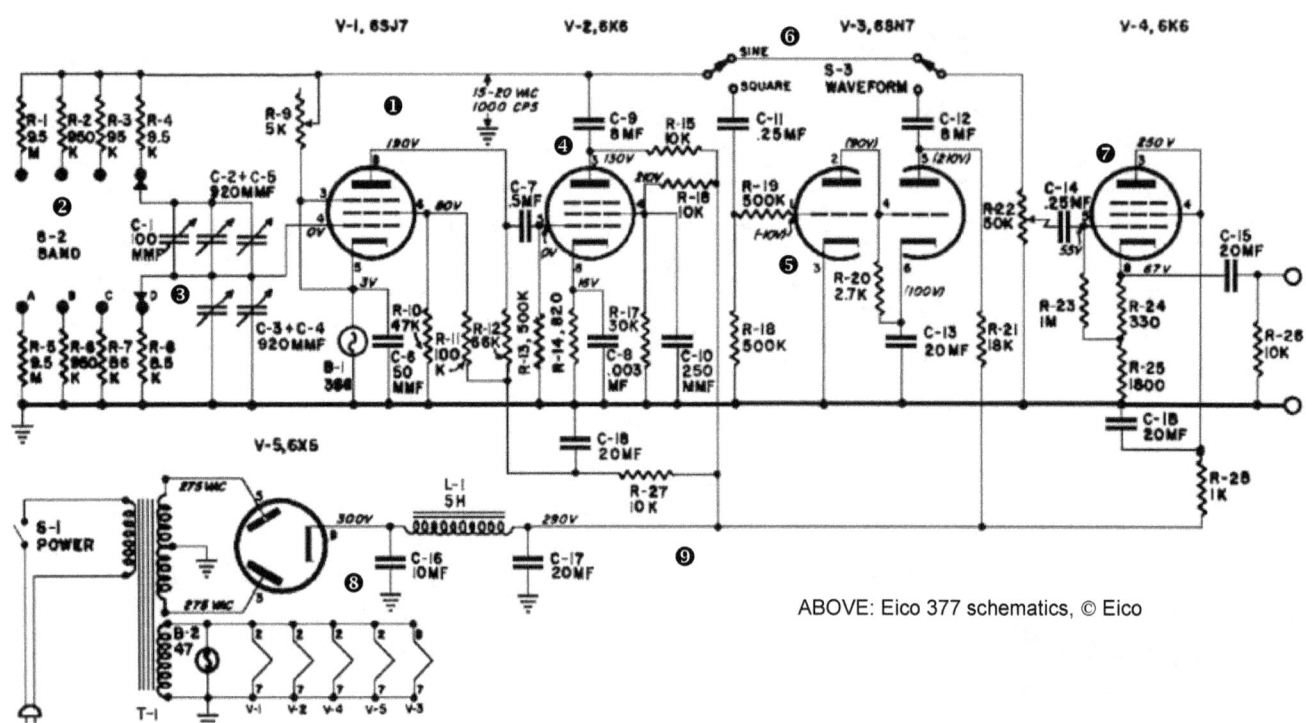

ABOVE: Eico 377 schematics, © Eico

The 4-position switch ("BAND") determines the range, while the variable capacitors (3) fine-tune the desired frequency within each range. This whole shebang is not needed anymore, so that stage will be converted into a conventional amplification stage. The "BAND" switch can be retained with some of the resistors reused to change the gain or voicing of that stage. Alternatively, convert it into a phase-shift tremolo oscillator.

V2, the first 6K6 power pentode, works in a standard voltage amplification stage (4) with a 10k anode resistor and 820W cathode resistor. This is a power tube and can be utilized better; we'll come back to it in a moment.

Conversion options

The venerable 6SN7 duo-triode, revered amongst the audiophiles for its magic sound, works as a voltage clamp or limiter only when "SQUARE" is selected by the "WAVEFORM" switch (6). This stage will be converted to two low gain stages. Even if each achieves a gain of only 10, two stages in cascade (one after the other) would have a gain of 10x10=100, which is much higher than any 12AX7 single triode stage can ever achieve!

Finally, the output stage is another 6K6 pentode working as a cathode follower (7). You have two options. Either use that 6K6 and keep V2 idling in its socket as a spare or connect both in parallel to double the amp's output power!

If you are ambitious, use both in a push-pull arrangement, use the 1st stage as a high gain input stage as before, the first 6SN7 triode as the second stage, and the 2nd 6SN7 triode as a cathodyne (split-load) phase splitter.

The 1st filtering elco is only 10μF, the 2nd is 20μF, so ditch both, make both 33 or even 47μF. You will need to add another RC filtering stage afterward (9) to decouple the input stages from the power stage. A 1-10k resistor in series (the value depending on how much you wish to lower the anode supply voltage for the 1st two stages, currently 230V) and a 22μF elco is all that is needed.

Moving onto the outside aspects, the power switch and the pilot light can stay where they are. As for the output binding posts, you have two options. They can be removed and two input jacks installed; for instance, one normal and the other bright or high gain guitar input.

Alternatively, convert one into the speaker jack and the other into RCA output for recording purposes. Depending on the chosen option, you will need to drill one or two holes for the inputs or output(s) as required. There is plenty of empty real estate on the front plate, so you are spoiled for choice.

As already mentioned, the "BAND" switch can control the voicing or the gain of the first stage, while the "SINE-SQUARE" switch can be used for gain or tone changing purposes as a "BOOST" or "BRIGHT-DARK" switch.

The "AMPL" pot controls the output amplitude of the audio generator, so it is already wired as a master volume control in front of the output tube. 0-100% graduations are cool and easy to read.

As for the large dial on top, you can use it as a gain control or as a tone control pot. In either case, it would be wired at the output of the first 6SJ7 pentode gain stage.

TUBE PUBLIC ADDRESS (PA) AMPLIFIERS AS CONVERSION PLATFORMS

Stromberg Carlson AU57 vintage PA amplifier

Tube PA (public address) amplifiers were quite common in their day, so there are still quite a few for sale online. However, their prices have become so high that they are no longer the most viable or cost-effective platform for conversion into a guitar amp. Unless you get one cheap, that is, as in our case here.

Secondly, many are in poor cosmetic shape, dusty, rusty, battered, or all of the above. Since they are 50+ years old, all electrolytic capacitors will have to be replaced. Most likely, the film capacitors (used chiefly for interstage coupling) will need replacing too. Most suffer from significant leakage - not physical leakage of course; there is no electrolyte or insulation oil to leak out, but electrical leakage through their dielectric insulation.

Power and output transformers age too. Their paper insulation will be compromised after half a century, as will be the winding wire insulation, which wasn't very good at that stage of technological development, not nearly as good as the varnishes of today. That means it is very likely that the mains or output transformer will start smoking upon the power up, the fuse may blow, and you will need to fork out $100-$200 for a replacement unit.

Perhaps the worst scenario is buying a PA amp that had already been "converted" into a guitar amp by someone with far more confidence than knowledge or skills, an amp that has been butchered by some self-proclaimed "expert". Since these "improvements" are never documented, you'd have to spend a few hours deciphering the twisted logic behind some people's thinking, what was done, how, and why. Unless the amp is in its original, pristine condition with no mods at all, I don't think you should buy it.

The SC AU57 amp that we will use as a case study was certainly in its original condition, making its conversion much easier.

The outside looks, functions & controls

SC AU57 is a handsome-looking 4-channel PA amp. The three microphone channels are identical, two cascaded common cathode stages using 12AU7 duo-triode (one per channel). The 4th channel (marked "AUX" at the front and simply "Input" at the back) can be configured for a microphone, a magnetic cartridge (turntable), or a true auxiliary input for another source by a switch is located at the side of the amp ("MAG-MIC-AUX"). That channel uses a 12AX7 tube.

The first common stage is another half of 12AX7, followed by the common Bass and Treble tone controls and another gain stage with 12AX7. The paraphase phase inverter uses another 12AU7, and two 6L6 lovelies work in the engine room, i.e. the push-pull output stage. So, there are four 12AU7 and two 12AX7 lovelies in total.

The output transformer has many secondary impedances to choose from, so matching this head with any speaker box will be a walk in the park. Apart from the usual 4, 8 and 16Ω, there are line outputs for 12.5V (or 6.25Ω), 25V(25Ω), 33V (50Ω) and 70V or 200Ω.

Both its power and output transformers are relatively large and solidly built units.

The speaker outputs are distributed over two screw-type terminal strips. Since the 6.25Ω, 25Ω, 50Ω, and 200Ω terminals won't be used, remove the whole top strip and install a 1/4" jack and a 3-position rotary switch into its slot in the chassis.

There will be a small opening in the middle of the slot, between the new switch and the speaker out jack; you could cover it somehow if it bothers you. Leave the bottom speaker terminal strip in place.

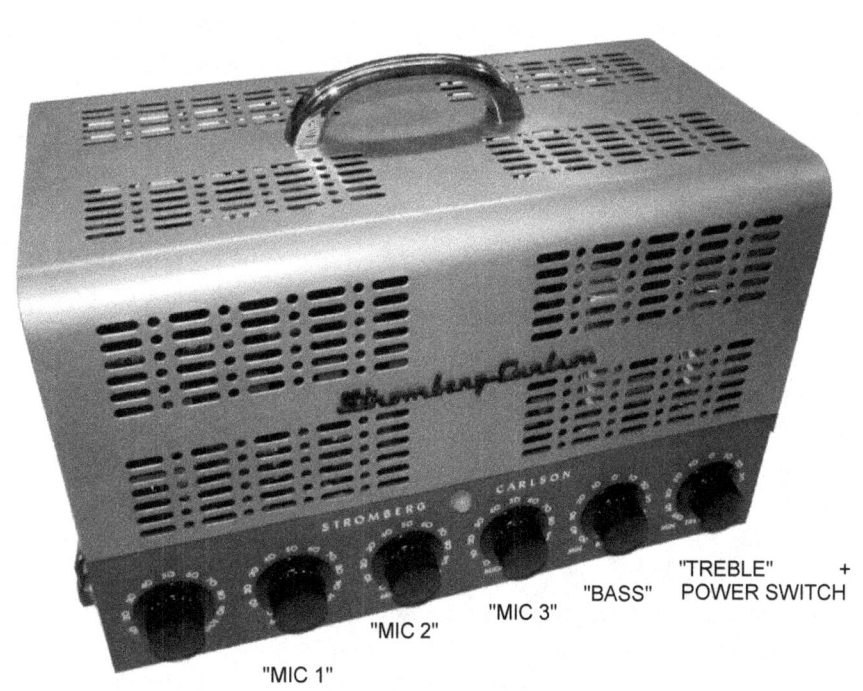

1) Mains (power) transformer
2) Output transformer
3) 5U4G rectifier tube
4) 6X4 rectifier tube
5) 12AX7 tubes (two of)
6) 12AU7 tubes (four of)
7) Step-up signal transformers for the three microphone inputs
8) Hum balance potentiometer
9) "Remote control socket" with inserted shorting plug

Unless you are planning to play more than two guitars simultaneously, which is highly unlikely, there is no need to modify all four channels. In fact, it may be useful to keep channels #3 and #4 as microphone input channels; you never know when you may need another mike or two. So, we will only modify the "Aux" and "Mic 1" inputs.

Since the selector switch isn't needed anymore, remove it and install a 1/4" phono jack in its chassis hole. Likewise, remove the large coax input connector marked "input" and install another 1/4" phono jack in its chassis hole. That way, there is no need to drill any new holes in the chassis.

Most of the coupling capacitors (IMP and Pyramid brands) were leaky (had very high dielectric losses) and had to be replaced. Interestingly, the two multi-section power supply filtering elcos were fine; their losses were in the same ballpark as the measured losses of currently made-in-China elcos, so we left them in place. This indicates that the amp was in service until very recently!

Remember, electrolytic capacitors age faster when sitting on the shelf without any power applied to them than when in service!

Remove the top speaker terminal strip and replace with one speaker 1/4" jack and a speaker impedance selector switch

Replace with two 1/4" jacks for two guitar inputs

The component layout and simple preamp mods needed

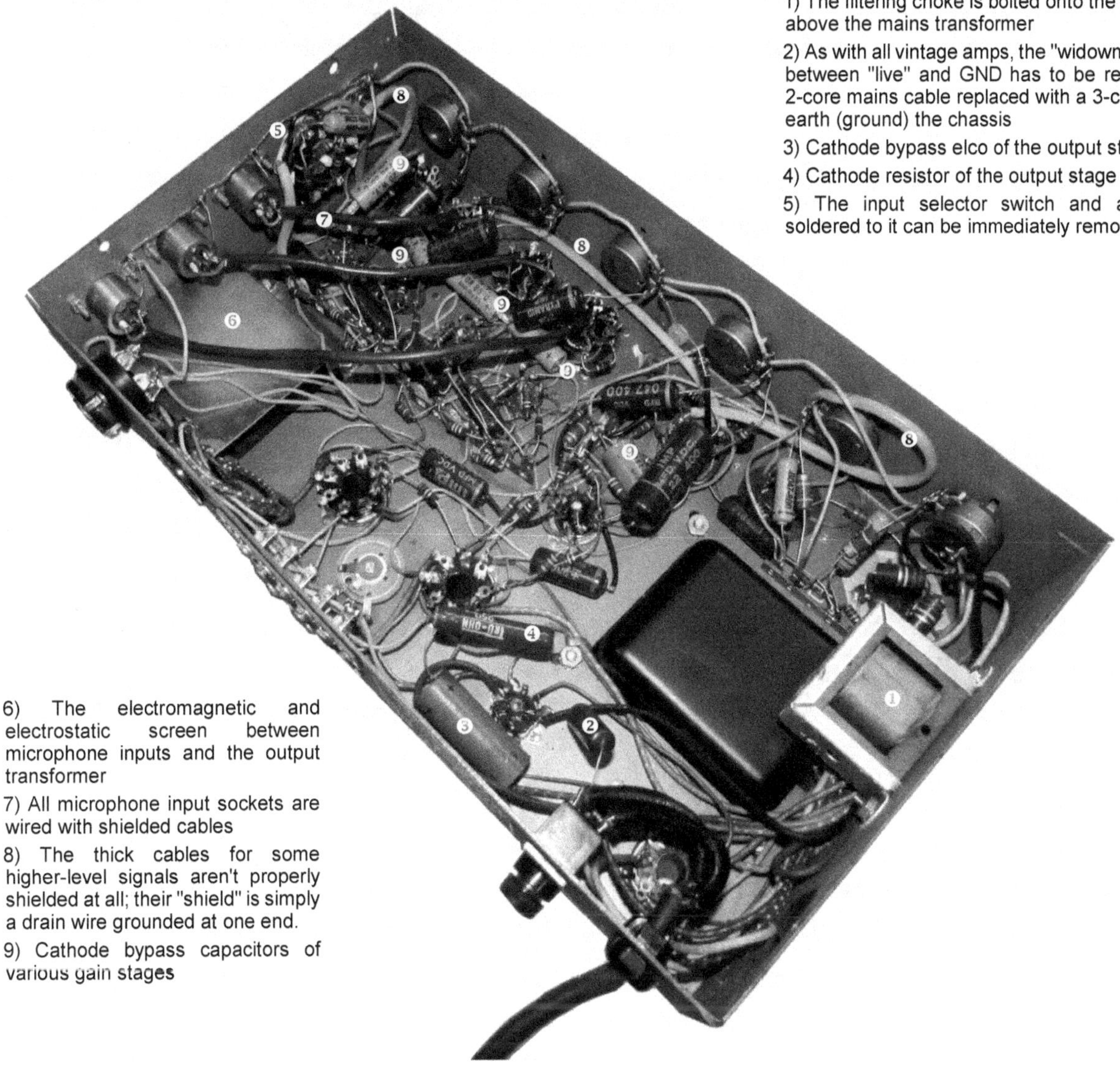

1) The filtering choke is bolted onto the amp's side wall, above the mains transformer

2) As with all vintage amps, the "widowmaker" capacitor between "live" and GND has to be removed and the 2-core mains cable replaced with a 3-core one that will earth (ground) the chassis

3) Cathode bypass elco of the output stage

4) Cathode resistor of the output stage

5) The input selector switch and all components soldered to it can be immediately removed

6) The electromagnetic and electrostatic screen between microphone inputs and the output transformer

7) All microphone input sockets are wired with shielded cables

8) The thick cables for some higher-level signals aren't properly shielded at all; their "shield" is simply a drain wire grounded at one end.

9) Cathode bypass capacitors of various gain stages

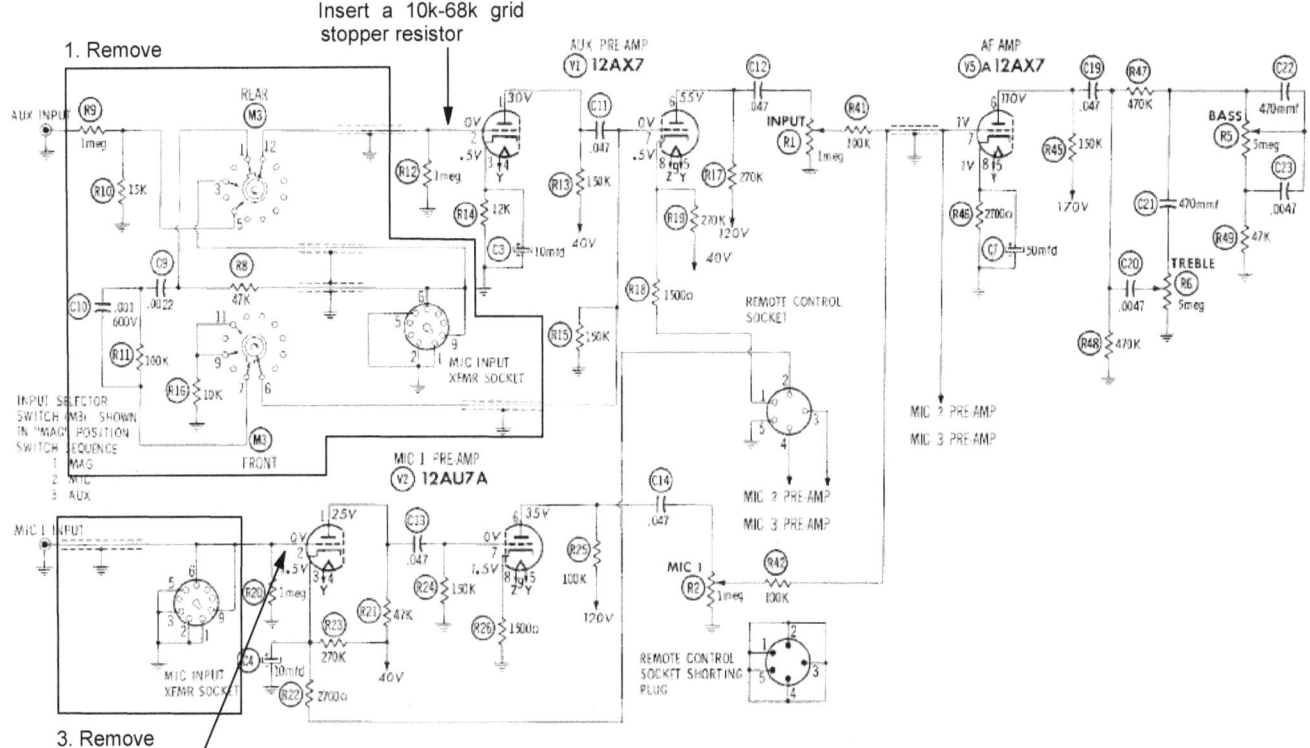

The cathode bypass capacitors of various stages were tucked in towards the bottom of the chassis and were quite large, so we replaced them as well with much smaller (physically) 10mF 16V Elna Cerafine lovelies. Although "Mic 1" input socket won't be used anymore, leave it in place but remove the shielded cable connected to it (#7 on the previous page photo).

Power supply

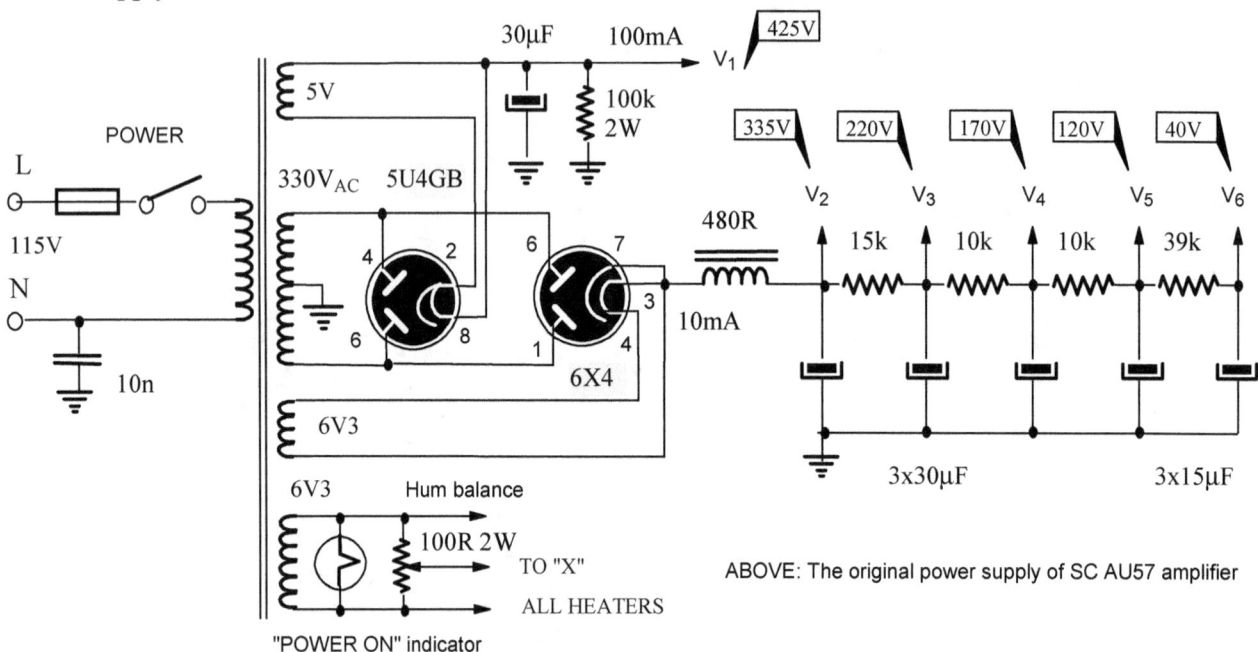

ABOVE: The original power supply of SC AU57 amplifier

The unusual power supply features two rectifier tubes. The large 5U4GB feeds a simple capacitive filter and provides $425V_{DC}$ for output tubes' anodes. Its AC ripple component is very high. Notice a current draw of only 100mA, so the use of 300mA+ 5U4GB rectifier is silly. Also, this is a directly heated tube, meaning it will heat up much faster than the indirectly heated 6L6 output tubes and supply high voltage to their anodes before the 6L6s are ready, which will shorten their life. An indirectly- heated smaller rectifier such as 5V4G should be used here. Luckily, the two tubes are pin compatible.

The 425V supply will rise by about 10V with 5V4G but that is fine. The output tubes' operating conditions are determined mostly by their screen voltage, supplied by a small 6X4 rectifier. It feeds an LC filter, providing 335V for the screens and then cascading through four additional RC filters, supplying four different voltages to various amplification stages. Notice that input stages work at a very low anode supply voltage of 40V!

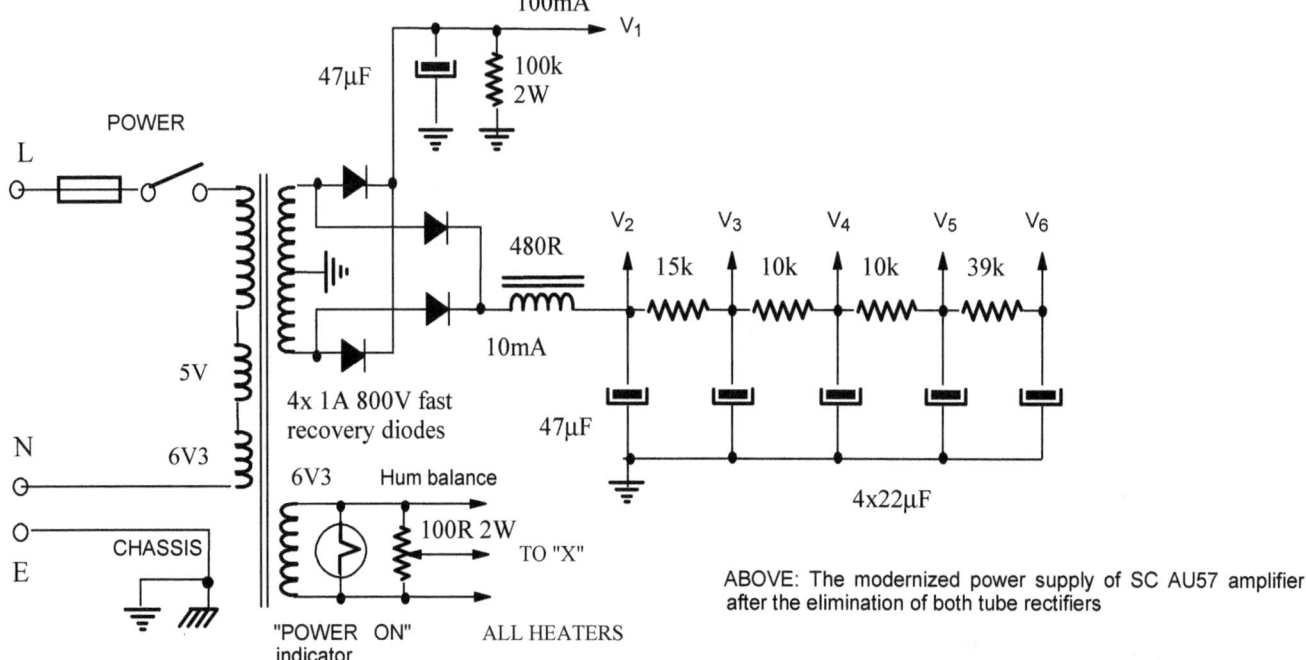

ABOVE: The modernized power supply of SC AU57 amplifier after the elimination of both tube rectifiers

Another interesting feature is the hum balance potentiometer, in particular the fact that its wiper or slider isn't grounded as per the most common practice, but is connected to the cathodes of the output tubes. Most current amp designers have no clue about the main benefit of this old trick, and that is a significant reduction of hum compared to simply grounding that wiper!

So, we have a decision to make. Leave both rectifiers as they are, replace 5U4GB with 5V4G or with silicon diodes, replace both rectifiers with silicon diodes (the power transformer will run cooler and quieter in that case), leave the two HV chains separate or combine them into one, and so on and so forth!

If you replace both tubes with SS diodes, you will have two unused heater windings, one 5V the other 6.3V. Since vintage amps were designed for 115V, USA mains voltage at the time, and modern utilities provide 120-125V or even higher voltage, connect those two windings (11.6V in total) in series with the primary. This will reduce all secondary voltages by about 8%, compensating for a rise in all DC voltages by about $15V_{DC}$ due to the elimination of the voltage drop across rectifier tubes. The $6.3V_{AC}$ heater voltage may rise a bit or may not, depending on your mains voltage levels, but even if it does, it should still be within acceptable tolerances!

AMATEUR TUBE TRANSCEIVERS

Heathkit HW30 ("The Twoer")

This cute little portable amateur VHF transceiver measures approx. 250×200×150mm and weighs only 3.6 kg, a true "lunch box" design. It is called "The Twoer" since it operates at a 2-meter wavelength. These were usually well looked after by their owners, ham operators, so even after 45-50 years, most will be in excellent cosmetic condition. They can be found on eBay and similar websites for US$50 upwards.

There are five tubes, 2x6BA8, 1x6BS8, 1x 12AX7, and 1x6AQ5. We aren't interested in 6BA8, a 9-pin triode-pentode tube, or 6BS8, a miniature medium mju twin triode designed for cascode operation.

6BA8 could be used in a guitar amp. The pentode can be configured in a high gain input stage, and the low mju triode (μ=18, similar to 12AU7) could be used as a lower impedance driver for tone stack or reverb pan or as a phase inverter/driver.

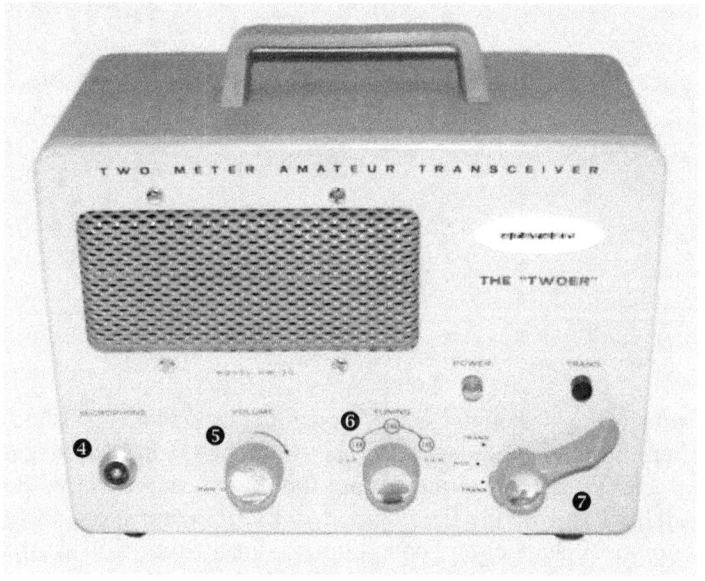

Apart from its transmitter and receiver circuitry with the three already mentioned tubes, it has a complete audio amplifier, two preamp stages with 12AX7, and the 6AQ5 (the Noval equivalent of the octal 6V6 tube) output stage, producing 5 Watts.

A conversion to a guitar amp requires very few modifications to the actual circuitry. If you look at the front and rear panels (photos on the right), practically no changes are needed to the metalwork. The mains connection is through the male octal inlet (1) and the female plug (2). These days these connectors are not approved for mains service, so we recommend that an IEC inlet or permanently wired 3-core mains cable is installed to make everything safe and legal.

The "meter jack" (3) will be used as an external speaker jack; the RCA socket next to it, marked "antenna," can become "line out." The "ground" connector can be removed, and its hole used for any purpose, a triode-pentode switch, on-off switch, etc.

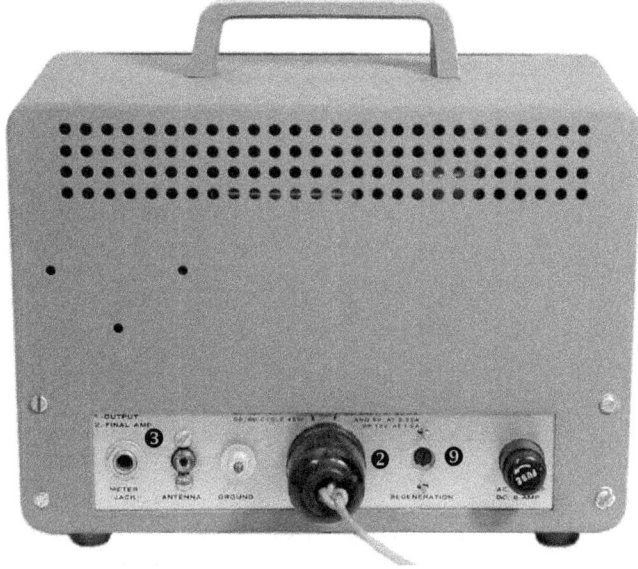

The microphone input will be replaced by a guitar input jack (4), volume control (5) stays, the tuning knob (6) becomes tone control, and the three-position Trans.-Rcv. switch (7) can be an off-standby-on switch or a bright-normal-dark voicing switch

.Notice the two audio tubes (8) right in front of the input jack (4), so even the topology is optimized, no need for shielding cables at the input or anywhere else.

The three other tubes can be removed or used for a tremolo oscillator or to drive a send-receive effects loop. In that case, the two jacks can be added at the back at (9); notice that one hole is already there. The possibilities are almost endless!

The internal speaker is a tiny 3.5 incher but adequate for practice.

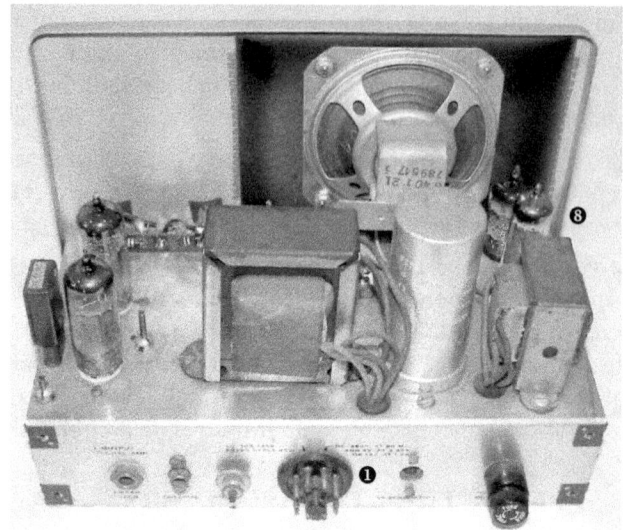

BELOW: The audio section of Heathkit HW30

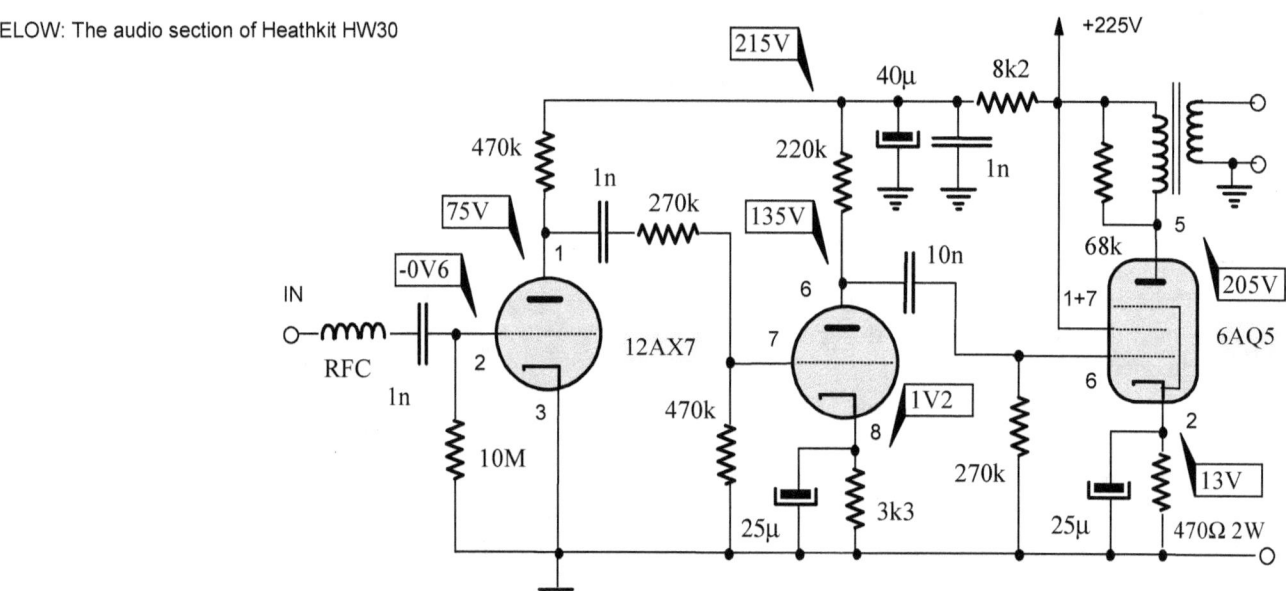

Grid leak bias through the 10M resistor is used in the 1st stage, so you should convert it to the standard cathode bias. Insert the single pot tone control between the 1st and 2nd stages, replace the 470k between pin7 and GND with a gain pot and 270k between the output tube's grid and GND with a 250k master volume pot, and you are in business. Installing and wiring the required jacks and connectors should take you an hour or so; allowing another hour for component changes and potentiometer installation, you should be able to test it with your guitar in two hours.

SONY SSA-464 AMPLIFIER SPEAKER SYSTEM

This elegant mono amplifier with an inbuilt elliptical speaker cost around 6,000 Yen in the mid-to-late 1960s (US$50.- today). A single RCA input, volume control, mains voltage selector switch, and pilot light are all its features. A single rectifier tube and 6BM8 triode-pentode (ECL82) are in the "engine room." A single triode is driving the single-ended pentode output stage.

The amp worked well with a guitar input, but there wasn't enough gain, so it stayed at a clean level; distortion could not be achieved.Replacing the obscure 5M-K9 half-wave rectifier tube with solid-state diodes would free its 7-pin miniature socket for a preamp pentode such as 6AU6, turning it into a high gain input stage.

CONVERTING VINTAGE TUBE GEAR INTO GUITAR AMPS

Alternatively, by removing the aux. mains outlet from the back panel, some room inside the tiny chassis could be freed, and a simple bipolar or JFET transistor input gain stage could be added there, right at the input jack. The pilot light can be replaced by a small neon indicator or an LED, and in its place, a tone control pot can be added.

The circuit diagram is fairly standard. The only interesting feature is the relatively complex frequency-dependent negative feedback from the speaker back to the cathode of the first stage (pin 8 of 6BM8).

1) Power transformer
2) Mains voltage selector plug
3) Output transformer
4) 5M-K9 rectifier tube
5) 6BM8 audio tube
6) Input (at rear)
7) Aux. mains outlet (can be removed)
8) Pilot light
9) On-off switch and volume control

Adding a 6AV6 triode input stage

The 7-pin miniature rectifier socket limits the choice of input tubes, most of which use Noval sockets. Due to the proximity of the power transformer and the possibility of induced hum into a sensitive 6AU6 pentode, we opted for a triode stage.

6AV6 contains two diodes that will be left unused and one high gain triode similar to 12AX7, with a mju of 100 and internal resistance of around 63kW in the common operating point 1.2mA of anode current.

The power transformer's heater winding could not cope with the additional load of 300mA for 6AV6's heater. Since the current draw of the incandescent "Power on" indicator was also 300mA, we moved it to the now unused 5V secondary and powered up 6AV6 heater instead!

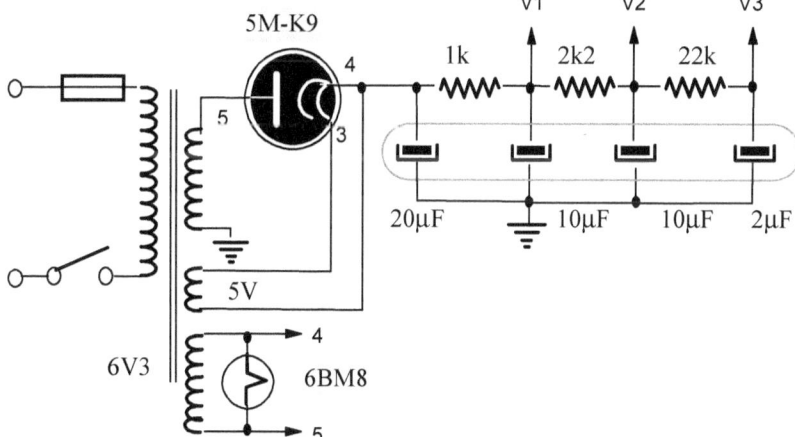

SONY SSA-464: The original circuit diagram

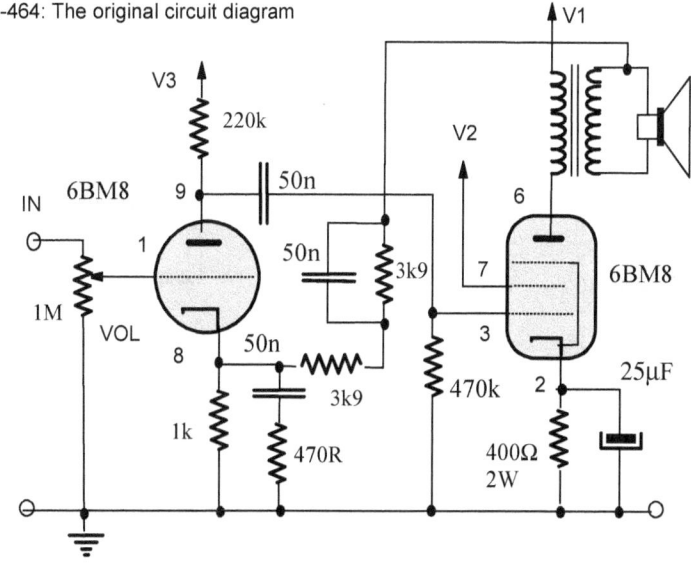

TUBE PROFILE: 6AV6

- Indirectly-heated twin diode-triode
- 7-pin mini socket, heater: 6.3V, 300 mA
- MAXIMUM VALUES TRIODE:
- V_A=300V, P_A=0.5W, V_{HK}=200V
- Typ. operation:
- V_A=250V, V_G=-2V, I_0=1.2 mA
- gm=1.6 mA/V, μ= 100, r_I = 62.5kΩ

With a bypassed cathode resistor, the added stage would amplify around 50 times, but now due to the local NFB in its cathode circuit, its voltage gain is about 35. However, even that is more than enough. The maximum output power was almost 4 Watts, way too much for the feeble internal loudspeaker, rated at only 2 Watts. If you install an external speaker" jack and drive a sensitive speaker box, this little monster can produce sound levels loud enough, not just for practice but for small gigs as well! `

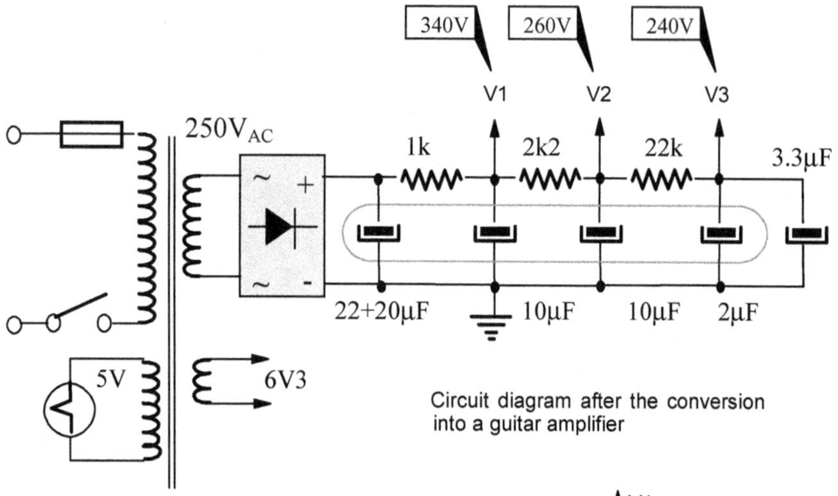

MEASURED RESULTS
- -3dB BW: 58 Hz - 25 kHz
- V_{MAX} = 5.5V
- P_{MAX} = 3.8 W (8Ω load)

Circuit diagram after the conversion into a guitar amplifier

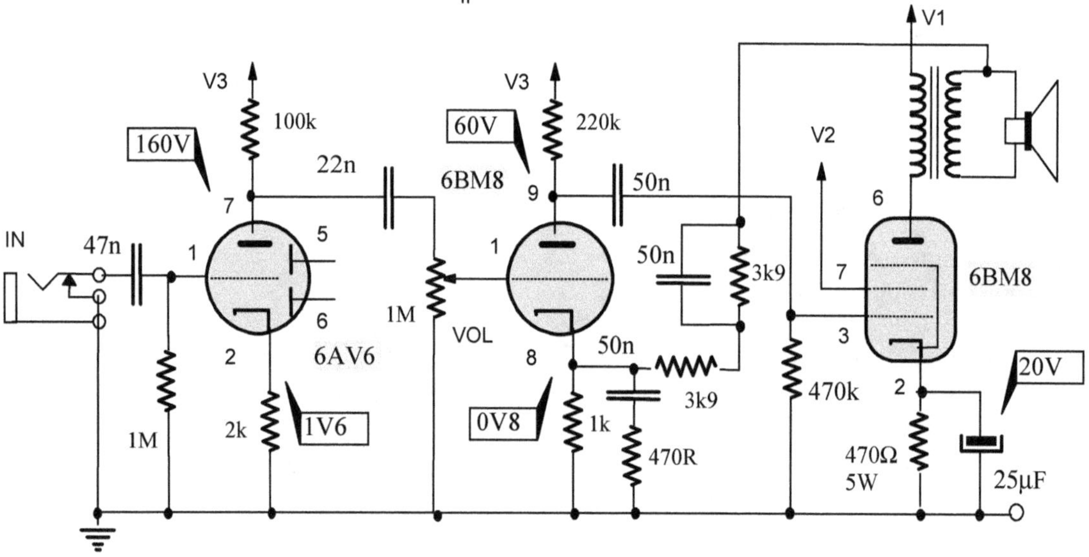

Adding a transistor input stage

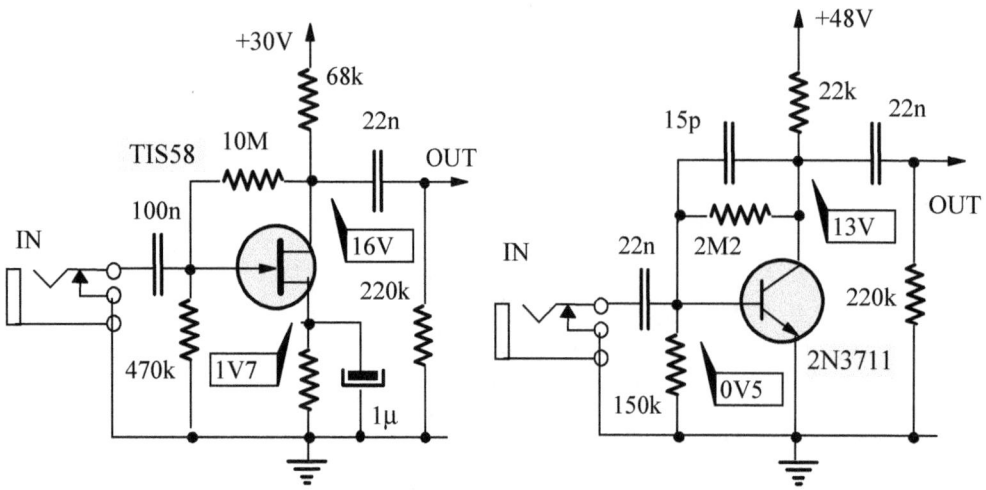

ABOVE: Input gain stage with a JFET

ABOVE: Input gain stage with an NPN bipolar transistor

OTHER TUBE AMPLIFIER BOOKS BY IGOR S. POPOVICH

Audiophile Vacuum Tube Amplifiers, Vol 1:

- BASIC ELECTRONIC CIRCUIT THEORY
- ELECTRONIC COMPONENTS
- AUDIO FREQUENCY AMPLIFIERS
- PHYSICAL FUNDAMENTALS OF VACUUM TUBE OPERATION
- VOLTAGE AMPLIFICATION WITH TRIODES - THE COMMON CATHODE STAGE
- OTHER VOLTAGE AMPLIFICATION STAGES WITH TRIODES
- TETRODES AND PENTODES AS VOLTAGE AMPLIFIERS
- FREQUENCY RESPONSE OF VACUUM TUBE AMPLIFIERS
- IMPEDANCE-COUPLED STAGES AND INTERSTAGE TRANSFORMERS
- NEGATIVE FEEDBACK
- TONE CONTROLS, ACTIVE CROSSOVERS AND OTHER CIRCUITS
- PRACTICAL LINE-LEVEL PREAMPLIFIER DESIGNS
- PHONO PREAMPLIFIERS
- SINGLE-ENDED TRIODE OUTPUT STAGE
- PRACTICAL SINGLE-ENDED TRIODE AMPLIFIER DESIGNS
- PRACTICAL SINGLE-ENDED PSEUDO-TRIODE DESIGNS
- SINGLE-ENDED PENTODE AND ULTRALINEAR OUTPUT STAGES

Audiophile Vacuum Tube Amplifiers, Vol 2:

- PRACTICAL SINGLE-ENDED PENTODE AND ULTRALINEAR DESIGNS
- PUSH-PULL OUTPUT STAGES
- PRACTICAL PUSH-PULL AMPLIFIER DESIGNS
- BALANCED, BRIDGE AND OTL (OUTPUT TRANSFORMERLESS) AMPLIFIERS
- THE DESIGN PROCESS
- FUNDAMENTALS OF MAGNETIC CIRCUITS AND TRANSFORMERS
- MAINS TRANSFORMERS AND FILTERING CHOKES
- POWER SUPPLIES FOR TUBE AMPLIFIERS
- AUDIO TRANSFORMERS
- TROUBLESHOOTING AND REPAIRING TUBE AMPLIFIERS
- UPGRADING & IMPROVING TUBE AMPLIFIERS
- SOUND CONSTRUCTION PRACTICES
- AUDIO TESTS & MEASUREMENTS
- TESTING & MATCHING VACUUM TUBES

Audiophile Vacuum Tube Amplifiers, Vol 3:

- THE FRONT-END: SUPERIOR INPUT & DRIVER STAGES
- FROM SHOCKING TO SUBLIME: LESSONS FROM COMMERCIAL LINE STAGES
- DIY LINE-LEVEL PREAMPLIFIERS: $10,000 SOUND ON $500-$1,000 BUDGET
- THE STARS OF THE AUDION ERA: ANCIENT TUBES IN MODERN AMPS
- CHEAP & CHEERFUL: PREAMP & DRIVER TUBES FOR AUDIO EXPLORERS
- SLEEPING GIANTS: OUTPUT TUBES FOR THOSE WHO WANT TO BE DIFFERENT
- THE QUEEN OF HEARTS: SINGLE-ENDED AMPLIFIERS WITH 300B TRIODES
- TRIODES, PENTODES AND BEAM TUBES: MORE SINGLE-ENDED DESIGNS
- BIG BOTTLES: SET AMPLIFIERS WITH HIGH VOLTAGE TRANSMITTING TUBES
- THE WAY IT USED TO BE: VINTAGE PUSH-PULL AMPLIFIERS
- NEW? IMPROVED? MODERN PUSH-PULL AMPLIFIER DESIGNS
- CUTE, CLEVER OR CONTROVERSIAL? INTERESTING IDEAS FROM TUBE AUDIO'S PAST AND PRESENT
- THRIFTY TIPS & TRICKS: TIME & MONEY SAVING IDEAS
- OUTPUT AND INTERSTAGE TRANSFORMERS: FROM COMMERCIAL BENCHMARKS TO YOUR OWN DESIGNS
- MEASUREMENTS VERSUS LISTENING AND OTHER AUDIO DESIGN DILEMMAS

OTHER AUDIO-RELATED BOOKS BY IGOR S. POPOVICH
Available from Amazon, Barnes & Noble, Book Depository and all other major online bookstores

Audio Tests & Measurements: How to Test Electronic Components, Audiophile & Guitar Amplifiers and Loudspeakers Using Modern and Vintage Test Instruments
ISBN: 978-0-9806223-9-3

- TEST INSTRUMENTS, ERRORS, LIMITATIONS & SAFETY ISSUES
- SIGNAL SOURCES, TRACERS, POWER SUPPLIES AND FILTERS
- MULTIMETERS - TYPES, OPERATING PRINCIPLES AND FUNCTIONS
- OSCILLOSCOPES - HOW THEY WORK & HOW TO USE THEM
- TESTING PASSIVE ELECTRONIC COMPONENTS (RESISTORS, CAPACITORS & INDUCTORS)
- TESTING AUDIO AMPLIFIERS AND PREAMPLIFIERS
- DISTORTION MEASUREMENTS
- TRANSFORMER TESTS & MEASUREMENTS
- LOUDSPEAKER TESTS & MEASUREMENTS
- TRANSISTOR TESTERS AND CURVE TRACERS
- TESTING VACUUM TUBES (VALVES)
-

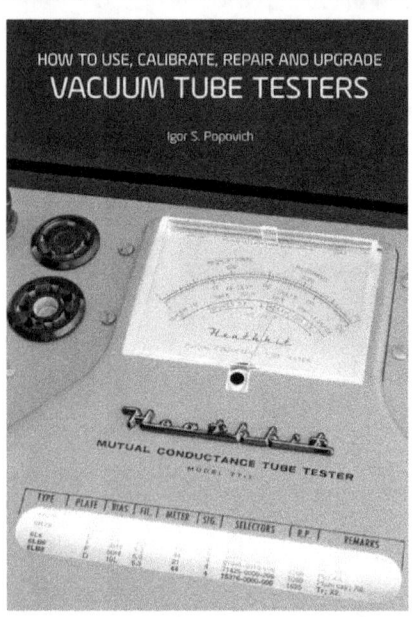

How to Use, Calibrate, Repair and Upgrade Vacuum Tube Testers
ISBN: 978-0-9806223-7-9

- HOW VACUUM TUBES WORK
- TESTING & MATCHING VACUUM TUBES
- EMISSION TESTERS
- GRID CIRCUIT TESTERS
- DYNAMIC CONDUCTANCE TESTERS
- PROPORTIONAL MUTUAL CONDUCTANCE TESTERS
- HICKOK-TYPE TESTERS
- TRUE MUTUAL CONDUCTANCE TESTERS
- REPAIRING & UPGRADING VINTAGE TUBE TESTERS
- TESTING & MATCHING TUBES WITHOUT A TUBE TESTER

Transformers For Tube Amplifiers: How to Design, Construct & Use Power, Output & Interstage Transformers and Chokes in Audiophile and Guitar Tube Amplifiers
ISBN: 978-0-9806223-8-6

- PHYSICAL FUNDAMENTALS OF MAGNETIC CIRCUITS AND TRANSFORMERS
- FILTERING CHOKES (INDUCTORS WITH DC CURRENT)
- TRANSFORMER MATERIALS, CONSTRUCTION METHODS AND ISSUES
- MAINS (POWER) TRANSFORMERS
- PHYSICAL FUNDAMENTALS OF AUDIO TRANSFORMERS
- SINGLE-ENDED OUTPUT TRANSFORMERS
- PUSH-PULL OUTPUT TRANSFORMERS
- SPECIAL MAGNETIC COMPONENTS: LOW POWER INPUT, PREAMP OUTPUT & DAC OUTPUT TRANSFORMERS, TRANSFORMER VOLUME CONTROL
- INTERSTAGE TRANSFORMERS, GRID & ANODE CHOKES
- OUTPUT AND INTERSTAGE TRANSFORMERS FOR TUBE GUITAR AMPS
- TRANSFORMER TESTS & MEASUREMENTS

BY THE SAME AUTHOR:

Sound Improvement Secrets For Audiophiles: Get Better Sound Without Spending Big

Publisher: Career Professionals
Year published: 2021
Language: English
Paperback: 328 pages
ISBN: 978-0648298205

Avoid the hit-and-miss approach and stop wasting money on overpriced high-end products in the blind hope of sonic improvement. Achieve the ultimate audio synergy and get more enjoyment from your audio system by making it as good sounding as possible.

"Sound Improvement Secrets for Audiophiles" will teach you how things work, why some circuits, designs, and technologies sound the way they do, and how to make them sound even better through simple modifications and improvements.

It is like having an audio and acoustic consultant by your side to guide you through optimizing and voicing your audio system and your listening room.

While relatively technical and in-depth, this practical manual goes way beyond "a dozen quick tips" and the simplistic advice you read elsewhere. Instead, the focus is on dozens of DIY projects, case studies, and examples of commercial audio components – turntables, preamplifiers, amplifiers, loudspeakers, power supplies, and acoustic treatments.

With over 400 photographs, diagrams, and illustrations, "Sound Improvement Secrets for Audiophiles" makes it easy for you to understand and comprehend complex technical concepts and issues.

The author does not shy away from many controversial and hotly debated topics. Tubes vs transistors, objectivists vs subjectivists, measurements vs listening, and digital vs analog: all of these are discussed in detail.

The money invested in this book would not even buy you a budget-priced pair of cables: it will prove to be one of the best financial investments you ever make. Even if you implement only a few improvements from the hundreds described within its pages – you will never look back!

BOOK CONTENTS:

1. WHY YOU SHOULD READ THIS BOOK AND HOW YOU WILL BENEFIT FROM IT
2. BEFORE YOU BUY AN AUDIO SYSTEM OR COMPONENT - THINGS TO DO & MISTAKES TO AVOID
3. WHAT DO WE LISTEN FOR AND WHAT DO WE ACTUALLY HEAR?
4. CLEANING UP THE POWER SUPPLY TO REDUCE NOISE, HUM, AND INTERFERENCE
5. CABLES, FUSES, CONTACTS, AND CONNECTIONS
6. UPGRADING & FINE-TUNING THE SOURCES: OPEN REEL RECORDERS, TURNTABLES, PHONO STAGES AND CD PLAYERS
7. AUDIO AMPLIFIERS - HOW THEY WORK AND HOW TO IMPROVE THEIR SOUND
8. HEADPHONES AND HEADPHONE AMPLIFIERS
9. LOUDSPEAKER TYPES, TESTS, AND IMPROVEMENTS
10. COMPONENT MATCHING AND AUDIO SYSTEM INTEGRATION ISSUES
11. LOUDSPEAKER POSITIONING
12. OPTIMIZING THE ACOUSTIC PERFORMANCE OF YOUR LISTENING ROOM
13. ACOUSTIC TREATMENTS
14. MINIMIZING UNWANTED VIBRATIONS & OSCILLATIONS
15. TROUBLESHOOTING YOUR AUDIO SYSTEM

INDEX

A

AC circuits, 71-75
Alnico, 24
Amplifier,
    frequency range, 60
    input impedance, 59
    output impedance, 31
    voltage levels and stage gains, 30
Audio generators as conversion platforms, 221-222
Autotransformer, 75
Attenuators,
    L-pad, 33
    Weber Mass Lite, 33
    Dr Z Air brake, 36
    Marshall power Brake, 37

B

Back-to-back power transformers, 160, 176
Bitmo™ modifications, 128-129, 187
Biyang (Wangs) VT-1H amplifer, 166
Blackheart Killer Ant amplifier, 168
Blackheart Little Giant amplifier, 106-110
Blackstar HT1 amplifier, 167
Blended output stages, 210-213
Blown fuse causes, 49-50
Blue glow inside a tube, 52
Bugera Vintage 5 amp, 38-43

C

Cathode-coupled inverter, 125, 128, 150
Cathode follower, 128
Capacitor
    leakage, 50
    motor-start, 85
    testing, 50, 59
Cathodyne phase splitter, see "Split-load inverter"
Cathode follower, 128
Choke tests and replacement, 53
Classic tone controls, 179, 202-204
Cold solder joints, 46
Causes of failures, 46
Contact bias, 88, 91
Crate Palomino V8 amplifier, 190-192
Crate Palomino V32 amplifier, 53

D

Damping factor, 31
Distortion,
    causes, 54
Dual rectifier modification, 110
Dummy load, 60-61

E

Earthing, see: Grounding
Electric shock, 70
Electrodynamic speakers, 24-28
Epiphone Electar Century amplifier, 138-140
Epiphone Electar Tube 10 amplifier, 103-105
Epiphone Valve Jnr. amplifier, 182-189

F

Fault location methods, 47-48
Fender Champ amplifier, 98-99
Fender Excelsior amplifier, 130-135
Fender Greta amplifer, 168
Ferrite rings, 196
Fixed bias, 87
Fletcher-Munson curves, 121

G

Grid stopper resistors, 51
Groove Tubes performance rating system, 64-65
Ground bus, 79
Ground-lift switch, 80
Grounding
    chassis, 72
    rules, 79-80

H

Headphone output circuit, 41, 43-44
Health hazards from solder, 76
Heater wiring, 78-79
Heathkit A7-C amplifier, 200-205
HF oscillations, 51
Hole punching, 61
Hum, 78-81
Hum balance, 226

I

Interstage transformers, 13, 190-191
Isolation transformer, 71, 89

J

Jet City JCA20H amplifier, 126-129
Joyo Sweet baby amplifier, 92-96

K

Kay K503A amplifier, 87-89

L

Long tailed pair, see: Cathode-coupled inverter
LCR meter, 56-57
LF oscillations, 52
Line matching transformers, 19-21, 151
Line out circuits, 35, 44
Lorden TL-15R amplifier, 136-137
Loudness control, 121-122
Loudspeakers,
    efficiency (sensitivity), 31-32
    plotting impedance curve, 29
    quick checks, 29
    SPL-vs-frequency curve, 32
    SPL control, 25-28
Low voltage tube guitar amplifiers, 214-217

M

Mahaffay HiLo Watt Plexi amplifer, 168
Marshall JTM-1C amplifier, 166
Matching tubes, 65-66
Mesa Simul-Class transformer, 16
Modification guidelines, 68
Mooer Little Monster amplifier, 167
Motorboating, 52
Mutual conductance tube testers, 62
Multimeters, 55-56

N

Negative feedback, 99, 101, 104, 108, 202

O

Orange Tiny Terror amplifier, 124-126
Output attenuator, 41-42, 198
Output stage biasing, 172, 174-178, 195, 207-208
Output transformer, 10-12,
Oscilloscope, 57, 69
Overvoltage protection, 84

P

PA amplifiers as conversion platforms, 223-227
Paralleled tubes, 211-212
Panama Conqueror amplifier, 97-101
Pentode input stage, 88, 91, 202-204
Percussion test, 38
Permanent magnet speakers, 25
Plates (anodes) glowing red, 52
Power supply troubleshooting, 48-50
Power transformers as output transformers, 21-22, 137
Powering up amplifiers, 49, 81-82
Presence control, 128
Push-pull stages
    composite curves, 149, 159
    DC imbalance, 54

INDEX, cont.

Q

Q-factor (transformer), 10-12, 20-22, 40, 93, 98, 126, 133, 137, 140, 156, 185, 195

R

Randall Diavlo RD1H amplifier, 169-170
RF interference, 51, 84-85

S

Safety rules and precautions, 70-71
Screen voltage,
    reduction, 94
Selenium rectifiers, 49
Self-inverting output stage, 152-157
Shielded cables, 77
Short life of tubes, 53
Signal tracers as conversion platforms, 220-221
Signal tracing, 48
Snubber, 84
Solder, 76
Solid state amplifiers,
    as cabinet donors, 142-147
Solid state diodes, 120
Spark killer (arrestor), 84
Split-load phase inverter, 132, 160, 202-204, 217
Stomberg-Carlson AU57 amplifier, 223-227
Substitution boxes, 57

T

Test adapters (jigs), 55
Thermistors, 85, 96
Tone control stack, 128, 139, 150, 153, 156, 159, 163, 186,
Topology issues, 78, 81, 209
Transfer characteristics, 63, 216
Transformer
    capacitance tests, 18
    leakage inductance, 18
    phasing check,
    power transformer mounting, 80
    primary impedance, 18,
    primary inductance tests, 18
Transistor,
    testing, 58
Tremolo
    depth, 133
Triode-pentode switching, 89, 101, 107, 135, 185, 188, 206, 210, 215, 217
Troubleshooting amplifiers, 46-54
Tube rectifier voltage drop, 94
Tube rolling, 112-120
Tube socket jigs,
Tube testers, 61-63

U

Un-bypassed cathode resistor, 114

V

Variac, 49
Voltage doubler, 215
Voltage regulation, 109,
VOX
    AC4TV amplifier, 193-198
    AC15 amplifier, 12
    Lil' Night Train amplifer, 168

W

Wiring
    color code, 77
    order, 77
    practices, 77

Z

Z.Vex Nano amplifer, 166
Zener diode, 58, 86